Metabolism of the Anthroposphere

Metabolism of the Anthroposphere

Analysis, Evaluation, Design

second edition

Peter Baccini and Paul H. Brunner

The MIT Press
Cambridge, Massachusetts
London, England

This book was set in Sabon by Toppan Best-set Premedia Limited.

Library of Congress Cataloging-in-Publication Data

Baccini, P.
Metabolism of the anthroposphere : analysis, evaluation, design / Peter Baccini and Paul H. Brunner. — 2nd ed.
    p.   cm.
Includes bibliographical references and index.
ISBN 978-0-262-01665-0 (hardcover : alk. paper)
ISBN 978-0-262-54954-7 (paperback)
1. Human ecology—Methodology. 2. Urban ecology (Biology) 3. Urbanization. 4. Urban design. 5. City planning. 6. Metabolism. 7. Environmental monitoring. 8. Environmental engineering. 9. Environmental protection—Planning. I. Brunner, Paul H., 1946– II. Title.
GF21.B23   2012
304.2—dc22
                                                                              2011016637

Contents

Preface to the Second Edition

Revisiting an old book is like wandering through an old house where one had lived for some years a long time ago. The façades show the traces of weathering. The inside exudes a faint and almost forgotten odor of a pioneering time when the house was newly built. Is it worthwhile to revise a book that has its cultural roots in the 1970s and 1980s?

When the MIT Press asked us to draft a second edition of our book *Metabolism of the Anthroposphere* (published in 1991 by Springer-Verlag), we felt honored by the invitation but reacted skeptically. The first author remembered a disappointing visit to a renowned scientific bookstore where he discovered that the new book was put in order among the monographs with the key word "Atmosphere." The surely well-educated bookseller obviously rejected the enforced coupling of two distant terms, of which the second did not even exist in his subject index. The editor in charge at Springer-Verlag had warned us against insisting on our exotic book title. Instead, he proposed terms in the title that embodied the zeitgeist; for example, "Environmental Protection" or "Waste Management." The book did not become a bestseller. The main focus in the rising environmental sciences, welling from the "Green Movement" in society, was on the environmental compartments and their reactions toward changing impacts from the anthroposphere. Re-reading our preface to the first edition, we must admit that our appeal to turn scientific curiosity toward "the cybernetics of regional economies" was more naïve than lucid.

The two greatest privileges of an academic chair in a liberal faculty are (1) not to follow the zeitgeist and (2) to attract nonetheless excellent students with master's degrees from various disciplines and curious professionals in practice from industrial and administrative branches. It took us more than 10 years to illustrate the concepts, to develop the methodological instruments, and to convince users of the potentials of our tool in problem solving. After 20 years, our book title is still not a bull's-eye. We are grateful to the MIT Press for tolerating our suboptimal "unique selling proposition." For us, the old title has become part of our "corporate identity."

Returning to the house metaphor, we must say that the old building did not just receive a façade lifting. We kept the basic structure, but we changed its main function. Consequently, we had to extend it. During the past 20 years, stimulated by fascinating experiences in transdisciplinary research, we have gradually moved the main focus from the environmental engineering topics of the anthropogenic outputs toward the design of metabolic systems within the anthroposphere. Therefore, we have chosen the sequence "Analysis, Evaluation, Design" in the subtitle. The first edition had to stop after the first evaluation steps. In the meantime, we went through a change of paradigm. We are convinced that the moving target "sustainable development" needs a commitment to design the anthroposphere *and* our environment. The houses of *Homo sapiens* are no longer special islands in a natural wilderness but rather are primarily surrounded by gardens that we inherited from our ancestors. Some of these gardens are endangered by destroyers; others embody the value added of a long cultural evolution. The old Voltaire put it in a simple imperative: "Il faut cultiver notre jardin!"

Römerswil and Vienna, December 2010
Peter Baccini
Paul H. Brunner

Preface to the First Edition

In public discussions on the quality of our environment, scientists encounter repeatedly the two following questions:

1. How much time is left to reduce efficiently existing man-made hazardous impacts on our essential resources, on water, air, and soils?
2. What should be done in first priority to prevent hazardous anthropogenic material fluxes with respect to man and the biosphere?

Both questions are very important, simple and clear. To provide good, simple and clear answers seems to be an almost unresolved problem. Natural scientists, confronted with the first question, have to admit that they do not have sufficient knowledge to predict correctly the type and rate of reactions in complex natural and anthropogenic systems. With respect to the second question, they emphasize the importance of the political boundary conditions for any actions and the need to do more research, mostly in the field of environmental systems.

Both authors of this book are natural scientists, chemists to be precise. We believe that during the past two to three decades, a good arsenal of methods and models has been elaborated to estimate essential processes in our environment, from the stratosphere to the oceans, from the Arctic to the tropical forests. We believe that there is a set of satisfactory fundamentals to justify first-quality standards for air, water, and soil to be respected by man, and further that these fundamentals need extension and permanent revision.

If we look at our own sphere of life, the anthroposphere, it appears that our knowledge of the dynamics of goods and processes that we develop and maintain is yet marginal and far from sufficient to answer the second question. We must understand better the metabolic processes of the anthroposphere. In a first step, we want to develop an instrument; that is, a method of material flux analysis. Obviously, this instrument is essential, but not the only tool to characterize man's activity. Both questions can only be answered satisfactorily if we know more about the cybernetics of regional economies. Besides the "master variables," resources, capital, and labor, we have to include the essential material fluxes entering and passing through the anthroposphere and eventually reentering the environment. The second question could be restated as follows: How can we optimize the anthroposphere in order to meet the environmental quality standards to prevent a collapse?

The analogy to medicine as a "synthetic science" is obvious. To define the healthy state of an individual is even more difficult than to ascertain a disease. Today, therapy in medical practice is more important than prophylaxis. The same is true for environmental protection. In either case, however, a sound knowledge of physiological processes is indispensable, and methods to elucidate the metabolic state of the human body are necessary. What do we have to measure, where and how should we measure? Without answers to these questions, a diagnosis is not possible, and no appropriate measures can be taken. We consider "metabolic studies of the anthroposphere" as a new branch of "synthetic sciences." Systematists might classify it as a natural science branch of anthropology. Our book is meant to be a tool for the new multidisciplinary workshop "cybernetics of the anthroposphere," an academic institution still poorly equipped.

Dübendorf, March 1991
Peter Baccini
Paul H. Brunner

Acknowledgments

We are grateful to many colleagues for supporting our efforts to write this second edition: Clint Andrews (Rutgers University), Claudia R. Binder (University of Graz), Hwong-Wen Ma (National Taiwan University), Daniel B. Muller (Norwegian University of Science and Technology), and Helmut Rechberger (Vienna University of Technology) reviewed the draft manuscript and supplied valuable comments that were instrumental in writing of the final version. Inge Hengl created all the artwork and was of precious help in managing the manuscript. Her competent and passionate efforts to improve the design of the book never ended and were greatly appreciated. Our colleagues of the Centre for Resources and Waste Management at the Vienna University of Technology (Oliver Cencic, Johann Fellner, Maria Gunesch, Ulrich Kral, David Laner, Jakob Lederer, and Stefan Skutan) contributed in various ways to the making of this book. We are grateful to the MIT Press for taking the initiative for this second edition and particularly enjoyed the kind and valuable guidance from Susan Buckley.

1

Introduction

Homo sapiens has physically transformed the surface of planet Earth. For the past 6000 years, agricultural settlements have developed into urban cores. During the past 500 years, these regional settlements became tightly connected because of development of a diverse infrastructure transporting persons, materials, and information. Urbanization formed a global network of urban systems, including colonized terrestrial and aquatic ecosystems. This network is called the *anthroposphere*. The notion *metabolism* is used to comprehend all physical flows and stocks of matter and energy within the anthroposphere.

Security is reached through the sending of colonizers to a conquered country where they take the form of a castle and a guard as it were, thereby keeping the rest in bondage. Even without such a measure, a province cannot maintain a constant population, since one part will die out on account of a lack of inhabitants and the other will be affected due to overpopulation. Since nature has no remedial solution to this specific problem, human ingenuity will have to come up with an answer. Where nature is no longer capable of helping itself, unhealthy cities become healthy if a great many people inhabit them at once and sanctify the earth through farming and purify the air through fire. Venice, which is located in a marshy, unhealthy region, constitutes a good example of how a city can be restored to health by the great number of inhabitants which flock to the town simultaneously.
—Niccolò Machiavelli (1469–1527)

Mankind's sphere of life, a complex technical system of energy, material, and information flows, is called the anthroposphere. It is part of planet Earth's biosphere. We think of the anthroposphere as a living system that evolves with its own history. In analogy to the physiologic processes in plants, animals, and ecosystems, the metabolism of the anthroposphere includes the uptake, transport, and storage of all substances, the total chemical transformations within the sphere, and the quantity and quality of all refuse. Anthropospheres have evolved over thousands of years and show different properties depending on space and time. At the beginning of the twenty-first century, four concepts dominate discussions on the future of planet Earth; namely, urbanization, resource scarcity, climate change, and sustainable development.

Urbanization of Planet Earth

Human beings have developed into urban creatures. From a metabolic point of view, the contemporary idiosyncrasies of urban culture are high population density, high stocks and exchange rates of information and goods, and a vital dependency on sources and sinks for energy and matter far beyond the settlement borders. Furthermore, human societies not only managed the needed physical resources but also developed a capital of technological know-how, of governing rules, and of organizing institutions. The evolution of all human societies has passed through similar forms of urbanity, independent of the boundary conditions given. Both the Occidental and the Oriental worlds have had and have their cities of different sizes and characters, all with the essential role to dominate regional development. Cities became creators and managers of markets and centers of political, economic, and cultural power.

When Niccolò Machiavelli, at the beginning of the modern era, wrote down his reflections on the making of provinces with urban centers, he chose Venice as an example to illustrate the potential of humans to overcome nature by technological ingenuity to achieve a "healthy and secure" settlement. Five hundred years later, we are impressed not only by the sober scientific analysis of Machiavelli but also by his lucidity. A philosopher of our time (Sloterdijk, 2004) describes the current society as a "therapy and insurance society." The *Homo sapiens* in modern times is a Homo habitans, who "thinks and resides." His or her life is based on a technical infrastructure that ensures safe supplies of the essential goods and a healthy environment. To pray is good, to insure is better. Science is one important basis for the evolution of urban systems. A majority of the current *polis* in a so-called developed state has withdrawn from an exclusive religious interpretation of the fate of life. Even societies with a theocratic political system adapt the emerging urban technologies. Within the timescale of about six human generations, from the middle of the nineteenth century to the end of the twentieth century, the technological innovations have led to a global urban network, essentially independent of the ideological background of the individual nodes connected to each other.

In the "greenhouse climate of large cities" (Braudel, 1979), the seeds of new cultural ideas and even revolutionary movements start to sprout. Over thousands of years, ideas and concepts to change the world may have originated in small religious groups or intellectual circles, be it in monasteries, within feudal staffs, or on university campuses, many of them situated in a rural environment. However, the application of ideas and concepts was made feasible in the fostering soil of cities.

Mandarin, the lingua franca of the Chinese, started in the cities of the imperial officials; Christianity needed Rome, colonization needed Amsterdam, London, Lisbon, and many others; Henry Ford's Model T needed Detroit, and microchips needed the Bay Area of San Francisco. Cities became stone-built symbols, materialized ideas of humans. An inscription above the entrance of the Aq Saray summer palace of the great Timur in Shakhrisabz, Uzbekistan, reads: "If you challenge our power—look at our buildings!" Cities flourish and decline with the goods they produce and market. Cities as a whole are the most complex anthropogenic systems.

In the twentieth century, the global influence of cities has grown dramatically. Around 1900, there was only one "City," London, with more than 5 million inhabitants. In 1980, 34 cities, most of them in developing countries, formed a group of continually expanding centers with each center possessing between 5 million and 16 million people (Brown & Jacobson, 1987). In 1950, only about 14% of the world's population lived in an urban environment; at the beginning of the twenty-first century, almost half of the people, 3 billion, live in cities (United Nations, 2004). The current picture in population growth worldwide shows a higher rate in urban areas (1.8% annual rate) than in rural areas (0.1%), mainly because of migration from rural to urban environments (UN-Habitat, 2005). A rough estimate predicts that approximately 60% of the next generation will live in cities; that is to say, about 5 billion people in 2030. Today, it appears that the urbanization of planet Earth is an irreversible process. The expression "let's go to town," known in most languages in analogous terms, has manifold meanings.

A closer look at the definitions and corresponding statistical data on which the distinction urban/rural living is based shows a high variety (Salvatore et al., 2005) and raises the question of the suitability of some well-established criteria, such as size and density of population settlement areas. The rural part of a country comprises then, simply defined, all the areas that are not urban. With new methods using satellite photos, combining them with statistical data, and mapping the results with the support of Geographic Information System (GIS) technique, the geographic definition of urbanity has been enlarged. It includes indicators such as land-cover types, drainage, transportation structure (railroads, roads, nautical and aeronautic utilities), cultural landmarks, and so forth. One of the most striking maps is the visualization of the night-lights of our planet (figure 1.1) giving probably the best first overview of the urban regions on a global scale. It raises also the question whether the definition of urbanity needs a principal revision.

The polis as a social concept and as a dense morphological manifestation within the landscape goes back to prehistoric times. Urban life started, according to the hypothesis of archeologists, in different continents and different cultures on this globe, most likely independent from each other, 6000–8000 years before the present. From a physiologic point of view, the rate-determining step of the growth of cities

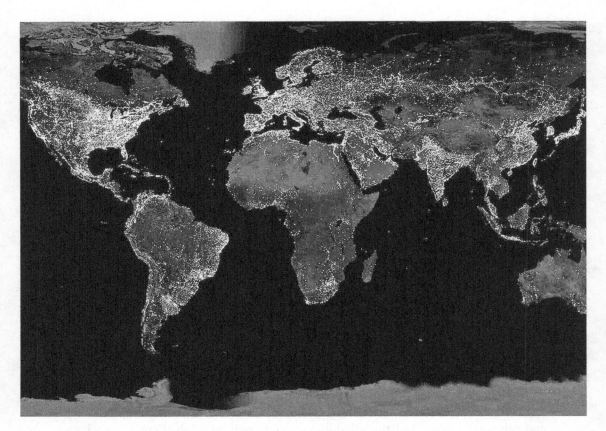

Figure 1.1
Earth by night (from Google Earth; reprinted with permission from NASA).

(meaning primarily the growth in population) was their acquisition of the necessary agrarian *hinterland*. Rome, for example, in its climax as a world power of Western antiquity, had to develop a logistic master plan to feed daily approximately 1 million inhabitants based mainly on relatively small vehicles and an average transport velocity of less than 5 km/h. In the first half of the nineteenth century, cities were still perceived in the old paradigm of the feudal hierarchy. An impressing illustration for this is Thünen's model of the "City State" (after Thünen, 1826). It is mainly an economic model of the city that is "driven by solar energy." The consequent spatial arrangement is the radial concentric pattern of the primary and secondary producers of goods with the city in its center (Thünen's circles). The urban people lived within the walls, the rural people outside of them. In contemporary terms, the agrarian culture was a "solar system" and, in principle, a sustainable one.

In the twentieth century, a fundamental and dramatic change occurred in urban development. The large-scale exploitation of fossil energy and the technical inventions and innovations in the transport infrastructure (railways, individual road vehicles with combustion engines, large vehicles in air transport, electronic telecommunication) allowed on one hand a rapid liberation from the limits of renewable biomass and on the other hand a high exchange rate of people and goods over large distances. This is a dramatic change in the evolution of human settlement within three to four generations. The twentieth century urbanization led to a "dilution" of urban settlement from a dense center into a network, with a high variety of nodes and connections (Baccini & Oswald, 1998). The distinct separation of rural and urban segments within a cultural landscape disappeared. The once concentric and regional hinterland diffused into a global set of hinterlands. The notion *hinterland*, borrowed from German, means literally "the land behind" and was used by geographers to describe the rural or colonial land controlled by dense settlements. In the context of metabolic studies, *hinterland* is used as a terminus technicus indicating all entities outside of a given system and connected with it through flows of matter and energy.

This new form of urbanity started within the culture of Europe and North America. In North America the notion *urban sprawl* and in Germany the term *zwischenstadt* (Sieverts, 1997) were introduced to describe the emerging morphological phenomenon. Today, the great majority of the population lives an urban life in settlements outside of the classical centers of the nineteenth century. Sociologists discover a "siteless" lifestyle (Siebel, 2002). The "new urbanity" is defined by access to all relevant goods and services a city can offer within one-half hour travel time (be it on foot or by private or public transport vehicle). Therefore, the term *urban system* is introduced. A name of the model to qualify and quantify an urban system is *netzstadt* (Oswald & Baccini, 2003). The netzstadt concept will be applied in chapter 5 when the case study on mobility is introduced. The definition is as follows:

An urban system is composed of open geogenic and anthropogenic networks that are connected with each other. The nodes of these networks are places of high densities of people, physical goods (geogenic included) and information. These nodes are connected by flows of people, goods, energy, and information. The system's boundary is given by political conventions in the case of anthropogenic subsystems, by climatic properties for geogenic subsystems.

In urban planning, as a part of territorial planning, the old paradigm of centers and their agglomerations is still en vogue. This has to do with the fact that in developing countries, this concept is still applicable. More and more people migrate from rural areas to the outer bounds of large cities, where they settle in poverty in dwellings called slums, bidonville, favela. It is uncontrolled urban growth without the essential resources. These settlements, in the statistics as part of the urban area, do not have access to the "relevant goods" urbanity can offer, as stated above. In these cities, the disparity of the winners and losers of globalization is manifested. The netzstadt method is one tool to design the urban by reconstructing existing settlements to achieve the qualities strived for by its inhabitants.

The economics of the new urban systems has been driving globalization, especially through the lowering of communication and transport costs. Within the twentieth century, the nation-state's power in controlling economics has been squeezed by globalization (Stiglitz, 2006). In the current race between economic growth (the great promise of globalization) and population growth, the latter is about to become the winner in large regions of the developing world. Instead of basing economic growth solely on monetary units (GDP/capita), indicators such as "health" (infant mortality, life expectancy), "education" (literacy, technological know-how), "living standard" (comprising the accessibility of "essential resources"), and "happiness" should become the units of measure for growth. A balanced growth, or a "sustainable growth," is to be supported by a "sustainable design" of the urban infrastructures. In other words: The anthroposphere is about to start another transformation process at the beginning of the twenty-first century. For a better understanding of the situation at the start of this transformation, a short review of the evolution of the anthroposphere is presented.

Evolution of the Anthroposphere

The scientific definition of *evolution* with regard to life on Earth comprises all processes that are involved to bring forward new forms of life (Fischer & Wiegandt, 2010). For biologists, evolution is the grammar of life. During billions of years, an unknown number of species appeared and died out, mainly due to changing environmental conditions. With growing complexity of living organisms, more and more species became emancipated from the local environment. The current model of evolution is not deterministic (i.e., the outcome of future forms of life cannot be predicted). This is not yet accepted by everybody as the mechanics of our solar system already are. *Homo sapiens*, a product of evolution, has a brain that allows decoupling the instructions for an action from its realization. The brain is the "biological premise" to start a "cultural evolution," a further step of emancipation from environmental conditions. The human brain, the key resource for the design of the anthroposphere, is able to model its environment, make tools (prostheses) to adapt the environment to human needs, test thereby the quality of the model, and improve it by trial and error. The knowledge gained was given to the next generation by oral and, much later, by written communication (the invention of written language). It is the emancipation from the purely biological way of storing new qualities in the genetic code. Therefore, we speak of the cultural evolution that allows a higher velocity of adaptation to a changing environment and an active role in designing the environment.

The formation of anthropospheres that are large enough to control large-sized areas took thousands of years after the end of the last ice age. Small family groups and tribes, comprising hundreds or even a few thousand individuals, mostly hunter-gatherers, some of them starting with agriculture and becoming "settled," were not yet able to dominate the processes of their regional biosphere. The breakthrough came with the formation of kingdom states in large river valleys (Euphrates/Tigris; Nile; Indus; Yellow River) approximately 4000 to 5000 years ago and in Central America with the Olmecs 500 to 1000 years later. Here, the inventions of domesticating plants and animals for a controlled production of food and the engineering of water flows for irrigation and fertilizing the soils went through a process of large-scale innovation (Diamond, 1997). Astronomic and meteorologic knowledge was applied to optimize agricultural production. The population grew faster and the population density increased. The first urban-like sites were formed to give the top of the pyramidal hierarchy not only its residence but also its stone-built icons, the symbols of identity of a political system (kingdoms and kleptocracies). In the new "anthropospherical climate," a man-made landscape, a human society emerged that

- had a high degree of diversity with regard to its tasks and tools;
- planned its resource management over several years to satisfy the needs of the whole population;
- guaranteed a certain degree of security with regard to protection from aggressors and to daily supplies of essential goods;
- based its economic power primarily on agricultural production (civil engineering);
- secured its political power with professional armies (military engineering);
- showed cooperation far beyond its kin;
- had a permanent platform to "invent and innovate," documenting the findings and improving the techniques;
- developed trade of various goods (import and export) over large distances far beyond its state borders; and
- had religious institutions to give the population a transcendental vision of life.

The relative weight and the different combinations of these characteristics led to a variety of types of anthropospheres in time and space. However, for thousands of years, until the dawn of "globalization" (starting in the late Middle Age, roughly 800 years before the present), the economies of the anthropospheres of the "first wave" (Toffler, 1980) were all based primarily on agricultural production (first pillar). The strength of an anthroposphere was directly proportional to the size of its agricultural territory and the technical standard to achieve high productivity (yields of food per labor and capital invested). Water-flow management for irrigation and fertilizing became one of the first large-scale engineering skills of mankind. The second pillar of economic activities was the development of trade over larger distances (Bernstein, 2008), especially between sites of precious resources, such as obsidian (a volcanic glass, the black gold of the Stone Age), silk, and metals (e.g., copper ores). The rising demand for such commodities stimulated the transport technology, mainly on waterways, and made the formation of remote specialized cities possible. Already in the Bronze Age "copper cities" were built (e.g., Dilmun on the Persian Gulf, run by the anthroposphere of Mesopotamia). The miners and metallurgists in Magan shipped the metal to the "copper trade City" Dilmun. Its inhabitants received aliment by maritime flows of food from their "motherland" and filled the returning ships with the red metal. A third pillar of the rising anthroposphere became the construction of residential homes and buildings for craftsmanship, for public, administrative, and religious institutions, all arranged spatially in a rational way to support the "inner flow" of goods and information, to secure the essential supplies, and, in very dense settlements, to organize the waste management.

The three economic and technological pillars (agriculture, trade, urban construction) can be comprised by the notion *civil engineering*. A fourth pillar consists of all skills to conquer new territories or to defend the ones cultivated over generations. Worded in a simple trilogy: Trade, raid, or protect! "War is the father of all things," a hypothesis first stated by the Greek Heraclitus (living in Asia Minor around 2500 BP), gives military engineering a higher ranking than today's historical analysis can support. The evolution of the anthroposphere is surely not a function of just one main variable. This same pre-Socratic philosopher came to a conclusion that is much closer to an insight that is still in agreement with the current view: "We both step and do not step in the same rivers. We are and are not." Simplicius, almost 1000 years later, gave an interpretation of Heraclitus' sentence by stating: "Panta rhei," meaning "everything flows." Confucius, the sage of ancient China, a contemporary of Heraclitus, reflected the characteristics of a feudal system of his time as follows (Kelen, 1983):

"A state needs three things: sufficient food, sufficient military equipment, and the confidence of the people in their government." When he was asked which one he would eliminate first, he answered: "Military equipment." And next? "Food, because everyone must die, and life is not worth much unless people have confidence in their government."

Confucius' comments reflect a crucial and constant debate on the type of governance emerging from the evolution of the anthroposphere. Roughly 500 years later, Marcus Tullius Cicero (2106–2043 BP), a Roman politician, lawyer, and writer, published *De re publica* ("Treatise on the commonwealth") in which he described, dramatized in a Socratic dialogue (an antique talk show), three types of governing societies: (1) the kingdom, (2) the oligarchy, (3) the state of the people. He pleads for a balanced constitution bringing the best of each into political power.

All four personalities witness the first highly developed anthropospheres of "the first wave." The cultural heritage, be it political, economic, or technical, never got completely lost but was transformed within new societies choosing other goals, mainly based on religious beliefs, such as Buddhism, Hinduism, Judaism, Christianity, and Islam. The first "World City" is Venice according to Fernand Braudel (Braudel, 1979). Machiavelli's statement at the beginning of this chapter emphasizes the great respect for the achievement of man forming an urban system out of the wilderness. Braudel, 500 years later, reminds us that Venice was, at its zenith between the thirteenth and fifteenth centuries (700 BP and 500 BP), the strongest economic power of the "world." The "global scale," grasped at that time from a European point of view, had a span from the shores of the Atlantic to eastern China. Why did Venice, the master of "world trade" during 300 years, lose its hegemony during the sixteenth century and be "replaced" by Amsterdam and eventually by London?

The answers are found in the changing systems of trade. The first long-range trading system was developed, roughly 4000 years before the present, between the anthropospheres of the great river valleys Euphrates/Tigris (Mesopotamia), Nile (Egypt), Indus (Harappan civilization), and Yellow River and Yangtze River (China). A trade road was established to transport Chinese silk and Indian pepper from East to West and move craftsmen's goods the opposite way. It was a pluralistic pattern of trade, over manifold stages, and poorly integrated; meaning that the first sender never saw the final receiver, and the consumer had no idea about the life of the producer and its environment (Bernstein, 2008). However, it was the start of a development that would reach, after roughly 2000 years, a first culmination peak in the Pax Romana (figure 1.2a). The Roman Empire, although having its center in the Mediterranean, was able to establish a trade axis from Rome to the Han Empire in China. The exchange of goods was supported by a silver currency (silver became the first "internationally" accepted money) and by a lingua franca (Latin or Greek).

The decline of the Roman Empire gave new options for the trade system. From the eighth to the fourteenth centuries AC (1200–600 BP), the rise of the Arabian traders, the followers of Prophet Mohammed, led to a unified commerce between Cordoba (southern Spain) and Canton (southern China). It was unified because of one culture, one religion, and one law. The centers were in Bagdad and Damascus. The Islam period tied together Arabia, Persia, India, and China in one commercial network (figure 1.2b). Now, African ivory and gold could be exchanged with Scandinavian furs, Baltic amber, Chinese silk, Persian metal craft, Indian pepper, and slaves from everywhere. The Arabic Muslims re-created the Roman Empire to a larger extent. The three East–West routes (Silk Road, Red Sea, and Persian Gulf)

a)

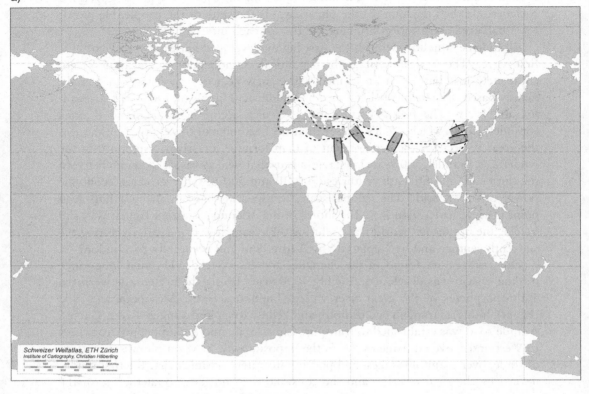

Schweizer Weltatlas, ETH Zürich
Institute of Cartography, Christian Häberling

Figure 1.2
Three steps in the evolution of the anthroposphere. (a) The rise of the first anthropospheres with urban elements in large river valleys beginning in the Bronze Age (4000 BP), with manifold trade connections, culminating in the Rome–Han Axis (2000 BP). (b) The first large-scale "unified world trade system" during the Islam period (1200–500 BP), integrating several East–West routes and forming a trade arch (shaded area) from the eastern Mediterranean to China. (c) The global network of the anthroposphere (from 500 BP to the present), a globalized transport and communication system, integrating both East and West Hemispheres and all continents ("Modern Times") (maps courtesy of Spiess, 2008).

were not competitors but rather parts of an integrated logistic system (Bernstein, 2008). Arabic was the lingua franca.

The Western European societies, grouped in the Christian "Holy Roman Empire of the German Nations," in France, England, and the northern Iberian Peninsula, all former provinces of the Roman Empire, had lost their dominance in the West–East trade and tried to regain free access to the big trading routes. Their first attempts, eventually not very successful, were connected to the Crusades, mainly in the thirteenth century (800 to 700 BP), opening partly the frontiers in the

b)

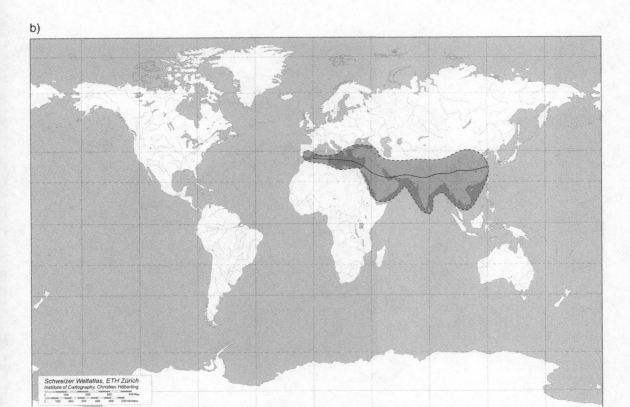

Figure 1.2
(continued)

Mediterranean Sea and the gates to the Near East. From the end of the fifteenth century (500 BP), the technological progress in seafaring and navigation brought them to America and around Africa into the Indian Ocean. Before Columbus there were hardly any important exchanges between the civilizations of America and those of Europe and Asia. It was the start of what became the "globalization of trade," a network of commerce embracing the whole globe (figure 1.2c). The lead in this process (eventually to be defined by historians as "a colonization within imperialistic regimes") was taken by European countries, such as Spain, Portugal, France, the Netherlands, and England. In the nineteenth century, Great Britain was at the top of the world's economic powers. The sun would never set on Queen Victoria's empire. The French writer Jules Verne (1828–1905) created the first global figure, the English gentleman Phileas Fogg, who traveled around the world in 80 days. Fogg was using very cleverly the transport logistics, originally not constructed for rich

c)

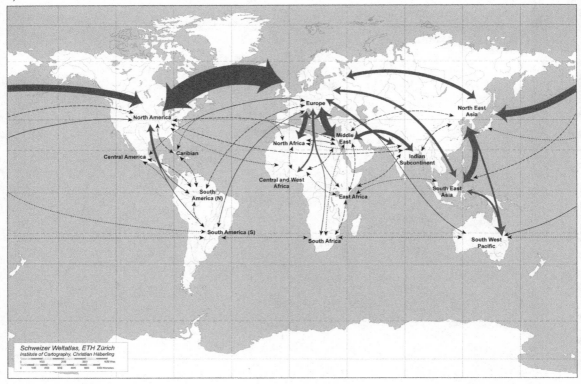

Figure 1.2
(continued)

and betting tourists, but for trading goods. Verne's narrative showed the world the actually existing connections between the anthropospheres on a global scale. It was not an Odysseus-like voyage, although spiced with many adventures, but a carefully planned and well-controlled enterprise in a precisely measured world. Fogg came back to London, just in time.

This world is no longer a patchwork of oasis-like settlements with a terrestrial and aquatic wilderness around them. Because of the manifold physical connections between the settlements, the various anthropospheres grow together. Industrialization, the "second wave" (Toffler, 1980), beginning slowly in the eighteenth century and growing exponentially in the nineteenth and twentieth centuries, brought a giant thrust in production, transportation, and communication technology. Following the geologic nomenclature for time-scaling, the term *anthropocene* was suggested

(Crutzen & Stoermer, 2000). In North America and Europe, the political institutions changed from feudal to democratic systems. Historians use the term *modern age* to title the result of transformation in society and characterize it by the following properties (Huntington, 1996):

• Industrialization
• Urbanization
• Low illiteracy
• Education
• Welfare
• Social mobility
• Complex and diversified professional structures.

The increase of economic efficiency, within the rules of capitalism and socialism, is based on mass production. Transport logistics now enable transport of mass goods over large distances. Fossil fuels, apparently unlimited in supply, decouple the physical growth of the anthroposphere from the limits of traditional solar energy systems. The human being has become the "Unbound Prometheus" (Landes, 1969). After less than 200 years, the innovations shorten the travel time and lengthen the distances between the producers and consumers by orders of magnitude. Population growth, due to better nourishment and health care, is increasing dramatically in the context of human development during the past 10,000 years (figure 1.3).

In the second half of the twentieth century, it becomes clear that the global network of anthropospheres has become a decisive physical player in the evolution of the planet's metabolism. When well-developed countries started to enter into a "post-industrial society," that is, to withdraw step by step from mass production of physical goods, turning more and more toward "information services" (the third wave), a politically controversial publication with the title *The Limits to Growth* (Meadows et al., 1972) questioned, on the basis of scientific resource models (Forrester, 1971), the globalization of the Western lifestyle. The scholars of the Club of Rome concluded that a collapse of the anthroposphere will take place within two or three generations due to lack of resources (energy, food, water, and minerals) and/or environmental pollution. For many historians, this publication was just a revisit of the Malthusian hypothesis stated 170 years earlier.

The prognostics of resource limits were rejected by a majority. However, the facts and biological effects of man-made environmental pollution had to be accepted—somewhat reluctantly at the beginning—because of evidence on the local scale (e.g., water pollution) and scientific measurements on a regional scale, showing even the history of man's industrial activity (figure 1.4).

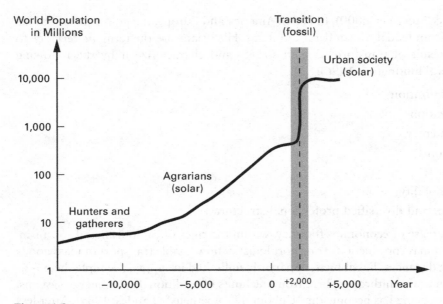

Figure 1.3
Population growth of *Homo sapiens* from the end of the last ice age. It is assumed that global population can be stabilized at a high level (between 8 billion and 10 billion at the end of the twenty-first century) living in a new "solar system" (reprinted with permission from Baccini, 2002).

Environmental protection laws were created. The ecological movement turned, at the beginning of the last decade of the twentieth century, toward a concept called "sustainable development." "Climate change," a scientifically well-founded hypothesis, was not readily accepted as a global phenomenon with substantial man-made contributions. Under the auspices of the United Nations, political, economic, and scientific representatives met in 1991 in Rio de Janeiro to launch a long-term project to adapt the metabolism of the anthroposphere to the resource limits and the carrying capacity of planet Earth. It became clear that the state of development of each country required tailor-made measures to move toward the "moving target" of sustainable development. It is a survival strategy. Common to all cultural periods is the strategy to minimize the risks, especially with regard to fatal shortages of the basic resources. However, it seemed paradoxical (after three to four generations of exponential physical growth of the anthroposphere) that human beings knew much more about the stocks and flows of their money, essentially an immaterial good, than about the quantity and qualities of the physical stocks and flows of materials they use in their sphere. In the story of cultural evolution, technological inventions and innovations play a key role. Cultural evolution is always interwoven with technological evolution.

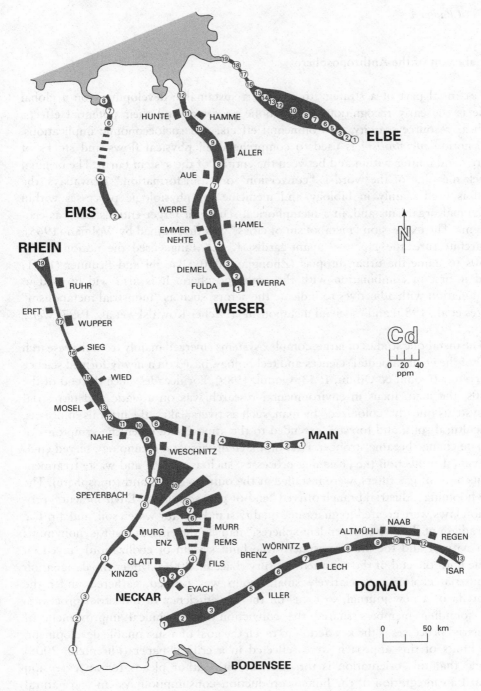

Figure 1.4
Cadmium in the clay fractions of river sediments in the Federal Republic of Germany (reprinted with permission from Förstner & Müller, 1974).

Metabolism of the Anthroposphere

An essential part of a strategy to achieve a sustainable development on a global scale is the early recognition of metabolic processes and their potential effects, such as resource scarcity, environmental effects, and socioeconomic implications. The notion *metabolism* is used to comprehend all physical flows and stocks of energy and matter within and between the entities of the system Earth. The original Greek meaning of the word is "conversion" or "transformation." Nowadays, the term is used mainly in biology and medicine for physiologic processes within individual organisms and, in a metaphorical sense, for larger entities, such as eco-systems. The expression "metabolism of cities" was first coined by Wolman (1965). In architecture, the Japanese avant-gardist Kenzo Tange used the notion in the 1960s to name the urban utopias (Zhongjie, 2010). Baccini and Brunner (1991) used it first in combination with the anthroposphere. It is also widely used in combination with adjectives to indicate the actors, such as "industrial metabolism" (Ayres et al., 1989) and "societal metabolism" (Fischer-Kowalsky et al., 1997; Weisz et al., 1997).

The metabolic studies of large, complex systems emerged mainly from the research field of the environmental sciences and technology, based on a newly formed science of ecology (Odum & Odum, 1953; Odum, 1989). For decades since the end of the 1960s, the main focus in environmental research was on aquatic and terrestrial ecosystems (mostly "colonized" by man, such as rivers, lakes, the littorals of oceans, agricultural soils, and forests), extended to the atmosphere when the symptoms of climate change became stronger. First of all, environmental technology served environmental protection (i.e., cleaning processes, such as sewage and waste treatment plants and off-gas filters, were installed at the outputs of the anthroposphere). The new booming industrial branch offered "end-of-pipe solutions." Environmental protection laws were created to guarantee quality standards for water, soil, and air. The "paradigm of the two independent spheres," namely the environment (synonymous with nature) and the anthroposphere (the built system of civilization) stayed en vogue until the end of the twentieth century (Baccini, 2006). Under the designation "industrial ecology," a relatively small group was formed, gathered under the umbrella of a new journal, to focus on the eco-efficiency of industrial processes. The founding members shared the conviction that technical improvement of industrial processes is the key action to reach the goal of a sustainable development. The limits of this approach were reflected in a critical paper (Ehrenfeld, 2008) stating that this orientation is too narrow. The author pleads for an extension toward an integration of the human production–consumption system and natural systems into one ecosystem.

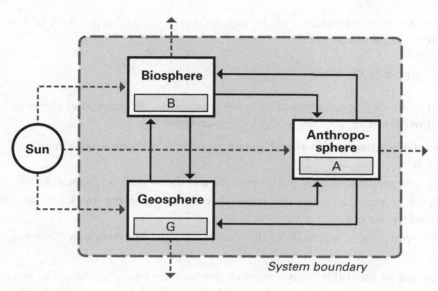

Figure 1.5
Scheme of main metabolic processes of planet Earth. The solid lines with arrowheads indicate the material flows between the processes, the dashed lines with arrowheads indicate the energy flows, and the letters A, B, and G indicate the stocks of materials in each process (MFA method, see chapter 3). In a first approximation, material inputs (e.g., meteorites) and material outputs (e.g., spacecraft residues) are neglected. Therefore, the system is considered as a physically closed system. To separate process A from B and G is a first sketch and has to be partly corrected (see chapters 2 and 4) because A contains large areas of colonized elements from B and G.

Metabolic phenomena have to be grasped on different scales. On the global scale, planet Earth is seen, in a first approximation, as a closed system (in the thermodynamic definition) consisting of three processes; namely, the geosphere, the biosphere, and the anthroposphere (figure 1.5).

In this scheme, the method of material flow analysis (MFA) is applied. The vocabulary of the "language of metabolic studies" stems from science—from physics, chemistry, biology, and geology. The basic grammar is MFA (see chapter 3). The same language is used for the smaller scales, from regional to local scales and eventually to individual technical processes. In this monograph, the focus will be on the regional scale (chapter 4). Contrary to the global system, the regional system is an open system. In consequence of the Earth's evolution and the history of the anthroposphere, the metabolic diversity of regional systems has a large spectrum at present. A region is a scale that comprehends all essential parts of an anthroposphere; namely, critical sizes of settlements and of economic enterprises, terrestrial and aquatic ecosystems, and political and cultural institutions (see chapters 4 and 5).

In a first overview, the phenomena of the metabolism of the anthroposphere in its facets are introduced in chapter 2.

Scope and Purpose of the Book

The second edition of *Metabolism of the Anthroposphere*, 20 years after the first edition, follows the same strategy as the first; namely, to

• show the characteristics of material stocks and flows of human settlements in space and time (chapter 2);

• summarize the methodological instruments for metabolic studies (chapter 3);

• illustrate, on a regional scale, the material systems generated by four basic activities (chapter 4); and

• exemplify, with four case studies, the design of optimal metabolic systems (chapter 5).

Since the end of the 1980s, some relevant developments have taken place and have to be considered in the revision. The worldwide political debate on sustainable development started the same year when the first edition was published. The scientific hypotheses on "climate change" are more strongly supported now. The topic will stay on the political agenda during the coming decades. The declining stocks of fossil energy and of freshwater are more widely accepted as future "real limits" worldwide. The topic of urbanization became more differentiated in the sense that a broader concept of urban life is on the academic and political agendas, not restricted to the large city (or megalopolis) in the old paradigm. In academic education, universities have installed a manifold set of environmental engineering and science curricula at the undergraduate and graduate levels. With regard to metabolic studies, new research groups have stepped in, produced new data, and enlarged the set of methodological instruments. Therefore, the second edition

• focuses more strongly on the design of the anthroposphere and less on the protection of the environment;

• concentrates more on resource utilization, as a function of activities and of technological innovations, and less on resource scarcity;

• includes results from transdisciplinary projects based on MFA; and

• updates figures and tables from the first edition still suitable for the second edition.

The book is written by two scientists, chemists to be precise, who have worked for more than three decades in the rising faculty of environmental science and technology. Most research projects were realized by interdisciplinary teams (i.e., physicists, chemists, biologists, geologists, civil engineers, mechanical engineers, economists, and architects took part). However, after 20 years, the outcome of tools extending MFA with additional parameters is still small in quantity. Sets of other tools are widely applied, such as TMR (total material requirement), LCA (life cycle analysis), and ecological footprint (see chapter 3). However, the material presented to illustrate metabolic properties is dominated by the works of the authors who focused on the MFA method. Therefore, the book is not a balanced encyclopedic presentation of the results achieved in metabolic research within the broad spectrum of methods applied today. Some of the data presented seems to be quite old. It is chosen in order to show, especially in chapter 2, the process of perceiving metabolic phenomena in time. In some cases, it underlines the observation that since the 1980s, no updating and revision has been undertaken.

The first edition addressed "natural scientists, engineers, and economists working in the fields of regional development, environmental protection, and material management, whether in private industries, counseling, or administration." The second edition strives, in addition, to reach undergraduate and graduate students taking courses in the fields of resource management, be it under the label of "industrial ecology," "environmental engineering," "resources management," or others with similar contents. It is not a textbook in the strict sense (i.e., it does not offer questions, exercises, and reading lists). Furthermore, the chapter on methodological instruments does not comprehend the current diversity because there are already various books covering these aspects in detail (see references in chapter 3). Under the same label there are also socioeconomic approaches, actor-oriented investigations, and policy studies. All these achievements are not covered. Literature on the topic "sustainable development" has grown exponentially within the past 20 years. A linkage between sustainability and industrial ecology is already on the market (Graedel & van der Voet, 2010).

The main purpose of this book is to give the reader an introduction to the general topic "Metabolism of the Anthroposphere," seen from a cultural historical point of view, supported with an instrumentation of the natural sciences and of engineering, and oriented toward designing metabolic systems. The second edition gives, in a revised and more sophisticated way, answers to the questions: "How can we grasp metabolic phenomena, and how can we understand their essence?" In addition to the first edition, restricted to analysis and evaluation, the second edition gives answers to the question: "What do metabolic designers do?"

2

Metabolic Phenomena in the Anthroposphere

Exploration of the world of metabolic phenomena is done in a narrative way. This chapter tells a story about the questions posed and the answers found in their times. It starts with a medical doctor and the human body and follows chemists, physicists, biologists, and geologists investigating physiologic properties of societies living in entities of growing scales (private households, cities, regions). It chooses a few regional examples to illustrate behavioral patterns and to show initial attempts to design metabolic processes on larger scales. This chapter provides answers to the following questions:

• How did and do scientists perceive metabolic phenomena?
• How did societies experience metabolic changes and how did they react?

Once as a child, I experienced that material in itself had an incredible power of expression, which can be of great importance to the world, that's how I experienced it. Or that the whole world depends on the constellation of a few pieces of material; on the constellation of where a thing stands, of its geographic location and of how the things stand towards each other, quite simply.
—Joseph Beuys (Zweite, 1986)

The Metabolism of the Individual Human Being

Dottore Santorio Determines his Input–Output Balance

About 400 years ago, the medical doctor Santorio Santorio (1561–1636), professor at the University of Padua, published his experimental findings (*De Medicina Statica Aphorismi*, 1614) and became "the father of the science of human metabolism." He recognized the necessity of measuring physical processes in the body. Therefore, he developed methods and instruments that permit the physician to quantify properties of the body. His most famous experiment lasted over a period of 30 years. He weighed himself, all the food he ingested, and all the excreta that he passed (figure 2.1). His input–output balance showed convincingly that about half of the input

Figure 2.1
Santorio's weighing apparatus: The experimental setup of Santorio Santorio (1561–1636) to analyze the material metabolism of a person. (a) The person is sitting on a chair attached to a scale. The weight of the food and the person are measured. (b) Despite the fact that all human excreta were collected and weighed, input A and output B do not balance. What is missing? (c) Santorio's measurements could not confirm his hypothesis that an unknown fluidum leaves the body during the night, but they proved that more than half of the input mass leaves the body by an unknown pathway.

b)

Figure 2.1
(continued)

was not found, neither in the increase of the body stock nor in the output via excreta (urine and feces). He concluded that the missing material was leaving the body by an "insensible perspiration." At Santorio's time, air in general as well as air of respiration was not considered to have a mass. Therefore, no instruments were developed to measure this metabolic path.

Monsieur Lavoisier Discovers the Chemical Principles of Combustion and States the Law of Mass Conservation

Antoine Laurent Lavoisier (1743–1794), almost 200 years after Santorio, made experiments in the laboratory in which he weighed carefully all educts and products of controlled chemical reactions. His findings (and those of others like Black, Cavendish, Lomonosov, Priestly, and Rey) led to the falsification of the "phlogiston theory." Phlogiston was hypothesized to be a component in each combustible substance to explain the "loss or gain of mass in a combustion process." Lavoisier

c)

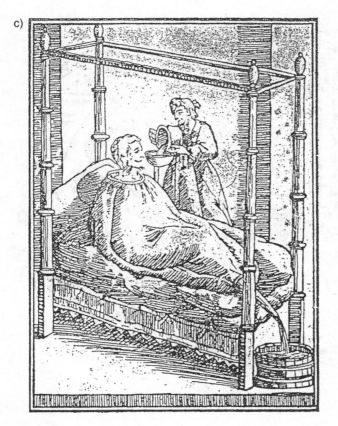

Figure 2.1
(continued)

identified experimentally the element oxygen, a chemical component of the atmosphere that takes part as a reactant in every combustion process. In his *Traité Elementaire de Chimie* (Lavoisier, 1789), he presented the first principles of stoichiometry to guide future chemical experiments. The balance was his crucial instrument. He finally unraveled Santorio's mystery of the missing link in the human metabolism by stating that respiration is a combustion process, in principle not different from a burning candle. The mass and the chemical compositions of the respiratory gases can be measured and identified. They are carbon dioxide and water.

It is worth noting that Lavoisier's chemical theories were not accepted by a great majority of his contemporary scientific colleagues. It was the next generation that became convinced of his concepts. The rise of chemistry as a serious science within academia led some leading scholars to name Lavoisier as the father of modern chemistry. In Geneva, at the end of the nineteenth century (i.e., 1892), 34 chemists continued the work on chemical nomenclature that Lavoisier, together with Claude-Louis Berthollet, had started in his paper "Méthode de nomenclature chimique" of 1787 (figure 2.2). Today, the International Union of Pure and Applied Chemistry (IUPAC) in Research Triangle Park, North Carolina, is responsible for this task in the worldwide community of chemists.

Designing the Human Metabolism in Space Habitats

Whereas during Santorio's time period, material flows through human beings were known on a very crude mass level only, inputs and outputs of modern men have been extensively studied by medical professionals and are well known today at a very detailed level of organic substances and even trace substances (e.g., Lentner, 1981). This development culminated to date by material balance studies performed for space travel in the second half of the twentieth century. The metabolisms of cosmonaut Gagarin and astronaut Glenn did not differ from those of Santorio and Lavoisier. The first human space missions were of short duration. The space engineers worked with an "open life support system"; that is, essential consumables (food, water, oxygen) were provided at launch (Wydeven & Golub, 1991). Wastes generated have been returned to Earth. Human space missions of long duration, on space stations and missions to Mars, ask for a fully regenerative physical life support system. Controlled ecological life support systems (MacElroy et al., 1989) would eventually allow an integration of biological subsystems to support air, water, and waste processing or recycling and to provide for food. For very long space trips, even small human outputs such as tears, sweat, saliva, or scurf have to be accounted for because of harmful accumulation in the spacecraft.

Mankind was beginning to design extraterrestrial closed environments and took the terrestrial closed environment, planet Earth (see figure 1.5), as a reference. Space travel engineers emphasize that the main differences of the two closed systems exist in the gravitational forces and the scales of time and biodiversity. It is expected that space missions can rely on adapted metabolic technologies of the reference system. However, it is hoped that innovations in space engineering for extraterrestrial anthropospheres leads to useful spin-offs for the terrestrial base station. In table 2.1, the design parameters of a life support system for the human metabolism in a space habitat are given.

The set of parameters and its flows show that

• Water is the dominating material in the overall mass flow (>90%).

• The mass flow of the "cleaning process" to keep up the necessary hygiene standard of the human body is about fivefold that of the "nourishing process."

• The input of the necessary carbohydrates, proteins, fats, minerals, vitamins (etc.), summarized as "dry food," is only 2% of the total material flow.

	Noms nouveaux.	*Noms anciens correspondans.*
Subſtances ſimples que appartiennent aux trois règnes & qu'on peut regarder comme les éléments des corps.	Lumière......................	Lumière.
	Calorique	Chaleur.
		Principe de la chaleure.
		Fluide igné.
		Feu.
		Matirè du feu & de la chaleur.
	Oxygène......................	Air déphlogiſtiqué.
		Air empyréal.
		Air vital.
		Baſe de l'air vital.
	Azote...........................	Gaz phlogiſtiqué.
		Mofete.
		Baſe de la mefete.
	Hydrogène	Gaz inflamable.
		Baſe du gaz inflamable.
Subſtances ſimples non métalliques oxidables & acidifiables.	Soufre	Soufre.
	Posphore	Phosphore.
	Carbone.......................	Carbon pur.
	Radical muriatique........	Inconnu.
	Radical fluorique	Inconnu.
	Radical boracique	Inconnu.
Subſtances ſimples métalliques oxidables & acidifiables.	Antimoine	Antimoine.
	Argent.........................	Argent.
	Arſenic	Arſenic.
	Biſmut.........................	Biſmut.
	Cobolt	Cobolt.
	Coivre.........................	Coivre.
	Étain...........................	Étain.
	Fer..............................	Fer.
	Manganèſe	Manganèſe.
	Mercure.......................	Mercure.
	Molybdène...................	Molybdène.
	Nickel	Nickel.
	Or...............................	Or.
	Platine	Platine.
	Plomb..........................	Plomb.
	Tungſténe	Tungſténe.
	Zinc............................	Zinc.
Subſtances ſimples ſalifiables terreuses.	Claux...........................	Terre calcaire, chaux.
	Magnéſie	Magnéſie, baſe du ſel d'Epſem.
	Baryle	Barote, terre de paſante.
	Alumine	Argile, terre de l'alun, baſe de Falun.
	Silice	Terre ſiliceuſe, terre viſiſtable.

Figure 2.2
Table of chemical elements from Lavoisier's paper of 1787 on "Méthode de nomenclature chimique."

Table 2.1
Metabolic parameters for the design of space habitats

Input per Person		Output per Person	
Material or Substance	Flow (kg/d)	Material or Substance	Flow (kg/d)
Oxygen	0.83	Carbon dioxide	1.00
Food, dry	0.62	Respiration and perspiration water	2.28
Water in food	1.15	Urine	1.50
Food preparation water	0.79	Water in feces	0.09
Drinking water	1.61	Sweat solids	0.02
		Urine solids	0.06
		Feces solids	0.03
Total for nourishing (% of total) [of which water in %]	5.0 (16) [71]	Total of human body outputs (% of total)	5.0 (16)
Oral hygiene water	0.36	Hygiene water	7.18
Hand/face wash water	1.81	Latent hygiene water	0.44
Shower water	5.44	Clothes wash water	11.87
Clothes wash water	12.47	Latent clothes wash water	0.60
Dish wash water	5.44	Latent food preparation water	0.04
Flush water	0.49	Dish water	5.41
		Latent dish wash water	0.03
		Flush water	0.49
Total for cleaning (% of total)	26.0 (84)	Total for cleaning (% of total)	26.0 (84)
Overall total (of which water in %)	31 (95)	Overall total (of which water in %)	31 (96)

After Wydeven and Golub (1991).

• A technology to recycle wastewater to get again clean water is a primordial task (otherwise a huge reservoir of water would have to be carried into space for long-term missions).

• Santorio's and Lavoisier's findings are applied in the design of space habitats, although in a more sophisticated way.

Evidently, the overall metabolism of the space habitat is not shown in the first set of parameters (shown in table 2.1); namely,

• the material and energy used by astronauts/cosmonauts to *reside and work* in space;

• the material and energy to *transport* humans into space, return from there, and to *communicate* with their terrestrial base; and

• the infrastructure needed in the "base camp" Earth, the *hinterland* of a space mission in a closed habitat. The *system border* for the metabolism is not identical with the walls of the closed space habitat.

A large-sized experiment to test ecologically designed closed systems was launched in Oracle, Arizona, at the beginning of the 1990s. A "small Earth," called Biosphere 2, was set up for "Biospherians" to live in about 1 year on the basis of their "closed solar system." It was meant to be a learning kit to "colonize Mars." The published intermediate results provoked much criticism from scientific groups who missed a transparent and reviewed description of the whole project. Therefore, they are not used here with data to illustrate a more complex closed anthroposphere. However, such types of small "closed ecosystems" will further serve as experimental entities to test metabolic setups supporting human life in space.

The Metabolism of Human Societies

Santorio's metabolic experiments comprise only those materials that are necessary from a physiologic point of view to sustain human life. However, modern man—in contrast to his ancestors several hundred thousand years ago—relies on many more materials for his activities. This so-called anthropogenic metabolism of a person is at least 10 times larger than the physiologic metabolism and comprises all goods needed to reside, to work, to communicate, to eat, to breathe, to clean, and so forth. It is interesting to note that information about the physiologic metabolism of men is exceptionally abundant, whereas information about the also highly important anthropogenic metabolism is still rather scarce.

Humans are societal beings. Aristotle coined the metaphor of "the political animal" that builds its *polis*. In analogy to the characterization of social insects, "One ant is a disappointment; it is really not an ant at all. The colony is the organism" (Hölldobler & Wilson, 2008), one can postulate that the physiologic metabolism of a human individual is not the key process of the anthroposphere. It is the anthropogenic metabolism of a human society—made up of millions of inhabitants—that gives the full picture. During the cultural evolution, different metabolic systems developed, although the physiologic metabolism of the human individual did not change significantly. How do we measure the entire metabolism of large-sized anthropogenic systems such as cities, nations, continents, and the global network? The essentials of regional metabolic studies are presented to give a first answer (Brunner & Baccini, 1992; Brunner, Daxbeck & Baccini, 1994; Baccini, 1996; Daxbeck et al., 1996; Baccini, 1997).

In the following text, a selective picture of the anthropogenic metabolism is exemplified for average European conditions by focusing first on the level of private households, next on the city level, and finally on the regional level. On all three levels there is a hinterland where products for the household, city, or region are produced, respectively where off-products are disposed of. This hinterland is generally not included in this section of chapter 2. It will be discussed in chapter 4 when main human activities are introduced.

Metabolism of the Private Household

The kind and amount of goods used to feed the anthropogenic metabolism are mainly determined by the availability of resources, by the technologies of exploiting resources and of producing and marketing goods, by economic conditions, and by the consumer demand. Thus, to focus first on private households is motivated by the fact that it is the consumer who pulls on the product chain.

To describe the metabolism of private households in a systematic way, the main activities of human beings are identified and the corresponding material flows and stocks quantified. The goal is to give an overview of the total material stock and turnover of an "average" individual in his household in order to use this "elementary cell" to synthesize larger metabolic units such as cities and regions.

In an average private household of a present-day region (table 2.2), the per capita turnover of materials amounts to approximately 80 to 90 Mg per capita and year, with water responsible for two thirds of this flow. The stock of materials is even larger comprising between 200 and 300 Mg per capita. As investigated by Santorio, the first and most important activity from a physiologic point of view is to eat food and drink beverages. In addition, the uptake of air for breathing is essential to supply the necessary oxygen to oxidize biomass carbon to CO_2 and hence to supply the human body with energy. The combined uptake of food, drink, and air for breathing is subsumed by the activity TO NOURISH. The entire activity (see chapters 3 and 4) comprises the whole food chain from production of crop and produce in agriculture, the processing of food in industry, the distribution, and the consumption of final food products by consumers. In table 2.2 only those material flows and stocks are included that are associated with private households. Hence, TO NOURISH within the system "private household" includes only purchasing, storage, preparation, and consumption of food by individuals living in this household. It ends with digestion of food by humans. Included in mass flows and stocks are not just the streams of food but also the necessary appliances such as refrigerator, kitchen ware, dishes, storage space, and so forth.

The activity TO CLEAN derives as a necessity from the activity TO NOURISH. Human digestion results in the following products: feces, urine, and off-gas mostly from breathing. These products have to be collected and discarded from households.

Table 2.2
Material turnover of a private household as a function of four activities (see also chapter 4)

Activity	Input (Mg/c.y)	Output (Mg/c.y)			Stock (Mg/c)
		Sewage	Off-gas	Solid Residues	
To nourish	5.7	0.9	4.7	0.1	<0.1
To clean	60	60	0	0.02	0.1
To reside and work	10	0	7.6	1	100 + 1
To transport and communicate	10	0	6	1.6	160 + 2
Total	86	61	19	2.7	260 + 3

Note: Figures are average values representative for Western Europe around the year 2000.

TO CLEAN comprises other processes as well; for example, personal hygiene (bath and shower), washing of textiles, dishes, room surfaces, appliances, and so forth. Because in today`s household, the main "conveyor belt" to remove dirt is water, the activity TO CLEAN is responsible for a huge flow of water, which is the single most important material flow through a modern private household. The activity TO RESIDE comprises all material flows and stocks needed to live in an apartment or a house. It includes the building itself as well as the constituents (windows, floor liners, carpets, wallpaper, furniture, appliances for heating, air conditioning, illumination, etc.). The system boundary in space is the building or apartment that hosts the household. Hence, the activity TO RESIDE&WORK in private households comprises the flows and stocks of all construction materials that are included in the building, but not the construction materials discarded or wasted during the construction process. In other words, the activity TO WORK is excluded here. These material flows are allocated to the "hinterland" of the private household; they will be included in chapter 4 within the activity TO RESIDE&WORK. The activity TO TRANSPORT&COMMUNICATE stands for the transport of persons, goods, energy, and information. Because it includes information, too, the entire communication sector is included in this activity. The reason for this is further discussed in chapters 3 and 4.

The Metabolism of a City
Cities, especially megacities with populations exceeding 10 million inhabitants, are the most complex systems that mankind has created. To understand the functioning of cities, many disciplines ranging from natural sciences to engineering and social sciences are required. A necessary but by no means sufficient base for the understanding is the knowledge of the urban metabolism as it manifests itself in flows and stocks of materials and energy. The focus on the material dimension is based

on one hand on the fact that these are crucial issues for the sustainable supply, management, and disposal of cities. On the other hand, because both reliable metrics for assessing urban material flows and stocks as well as corresponding data are available today, they form a sound database for investigations into the functioning of a city.

Although some cities are shrinking, most cities are growing, some of them dramatically (UN-Habitat, 2004). They are growing in numbers, in population, in the utilization of material and energy, and in the accumulation of material stocks. The material turnover of a modern citizen is about one order of magnitude larger than that in an ancient city of the same size; in contrast to old cities, material flows consist mainly of water and air for the human activities TO CLEAN, TO RESIDE, and TO TRANSPORT. With a growing fraction of global population living in cities, and due to high population densities and high per capita spending power, cities have become the hot spots of material flow and stock density.

Regarding metabolism, cities are facing two huge challenges: (1) the supply of plenty of goods, energy, and information to enable urban activities; and (2) cleaning of off-products of these activities (wastewater, off-gas, and wastes) and finding adequate sinks for the remaining emissions (e.g., $CO_2 \rightarrow$ atmosphere). On the input side, the provision of goods is highly controlled by market forces; the supply is potentially constrained by scarcity and availability of resources. On the output side, the disposal of off-products is governed by incomplete and sectorial sets of regulations. Because in cities—except for water and air—man-made urban material flows surpass geogenic flows by far, the urban environment is heavily affected by the mixture of substances that are mined, synthesized, and used by men. Even if sophisticated pollution abatement techniques are applied, diffusive sources like corrosion and weathering of surfaces or wear of materials by transport systems (e.g., erosion of brake linings and overhead lines) will slowly increase pollutants in water, soils, and sediments. The question arises if the capacity of sinks within the city and its hinterland suffices to accommodate all off-products without impairing the biosphere. Thus, the knowledge of urban material flows and stocks is a necessity for successful future management of cities.

The metabolism of cities, comprising private households as the "primary cells," private businesses for trade, commerce, and industry, public institutions, as well as the built infrastructure (roads, networks, supply of energy, water, information, etc.), has previously been investigated by several researchers (Duvigneaud & Denayer de Smet, 1977; Boyden et al., 1981; Daxbeck et al., 1996; Warren-Rhodes & König, 2001; Kennedy, Cuddihy & Engel Yan, 2007). Whereas most of these studies focus on the level of goods, some include also balances of substances (e.g., Daxbeck et al., 1996; Bergbäck, Johansson & Mohlander, 2001). In the following, the project METAPOLIS (Baccini et al., 1993) is used to exemplify an investigation into urban

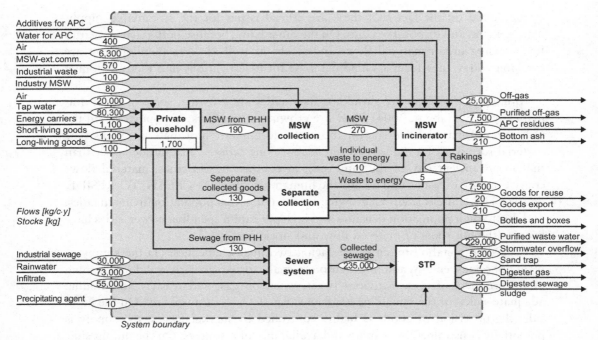

Figure 2.3
Material flow of a dense, medium-sized city (St. Gall with 70,000 inhabitants) with special emphasis on the private households (Baccini et al., 1993). The private households are responsible for roughly 80% of the total wastewater and 70% of the total municipal solid waste. APC, air pollution control; MSW, municipal solid waste; PHH, private households; STP, sewage treatment plant.

metabolism. In this project, the following hypothesis was tested: In affluent societies, the urban economy tends toward higher shares of working places in the tertiary branch, and the size of private households (number of persons per household) is getting smaller. This leads to a dominant role of the private households (PHHs) within the whole urban metabolism. This hypothesis was investigated in St. Gall (Switzerland), a small city of 70,000 inhabitants with 35,000 PHHs. The input data for the overall supply and consumption were given by market research panels and official statistics for the supply of water, energy, and other materials; the flows of wastes were measured and calculated by city authorities and the research team.

The material flows are illustrated in figure 2.3. PHHs consume approximately 100 Mg per capita and year (100 Mg/c.y) of short-living goods (residence time in households between minutes to several years) of which 80% is water and nearly 20% is air (this quantity corresponds with the one given in table 2.2 of 86 Mg/c.y). The rest, namely food and nonfood products (energy carriers included), is only 1%. The stock of mobiles within the households amounts to 1 ton per capita with an

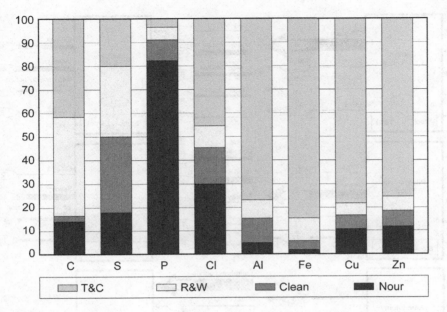

Figure 2.4
Relative contributions of the four activities TO NOURISH (Nour), TO CLEAN, TO RESIDE&WORK (R&W), and TO TRANSPORT&COMMUNICATE (T&C) to eight element flows (carbon, sulfur, phosphorus, chlorine, aluminum, iron, copper, zinc) in the city of St. Gall (after Baccini et al., 1993).

annual growth rate of 2%. The automobile is responsible for about 50% of this stock. The long-living stock of buildings and infrastructure is discussed in chapter 4. In comparison with the city's industrial, commercial, and trade activities, PHH dominates the overall metabolic processes by a two-thirds share.

A more detailed flow analysis of eight indicator elements (figures 2.4 and 2.5) shows the characteristics of an urban metabolic pattern. Practically all carbon (as organic carbon) is transformed into carbon dioxide (combustion of fossil fuels and food) reaching the atmosphere. Phosphorus is a good indicator for the activity TO NOURISH, which is responsible for more than 80% of the total urban P flow. Chlorine flows are dominated by polyvinylchloride (PVC), an important polymer for the building and transportation sector and—at the time of the study—for packaging systems; in the meantime, PVC has been banned as a material for packaging in many countries. The flow of four metals (aluminum, iron, copper, zinc) investigated is dominated (>80%) by the contribution of transport vehicles (automobiles, motorcycles, bikes).

The results of the project METAPOLIS, illustrated here by selected examples, allow us to postulate the following conclusions (Baccini et al., 1993):

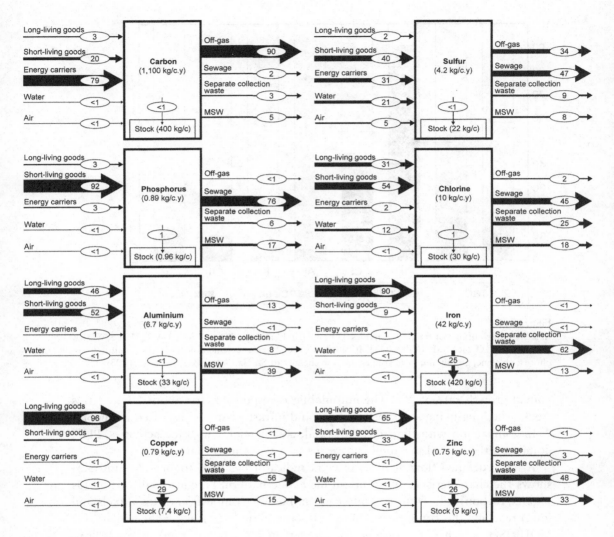

Figure 2.5
Element balances of the private households in a city (after Baccini et al., 1993).

Dense urban settlements of the St. Gall type, small in size with a regional-center role in political administration and a service economy, show no significant recycling processes. They are pure "flow through reactors" in which the PHHs play the major role. This type of city is typical for the metabolic pattern in industrialized countries on a global scale and exists in numbers of about 10^4 units (i.e., of about 20% of the total population). They depend on various hinterlands for the supply of goods, for the export of their products in trade, and for complete waste management. Because of high dependencies on hinterlands, these types of urban settlements are not suitable units to study metabolic processes for resource management strategies. The system border has to include larger-sized urban systems (see chapter 4). From an ecological point of view, the most important inputs into the overall metabolism are energy carriers (80% fossil fuels), supporting mainly the activities TO RESIDE&WORK and TO TRANSPORT&COMMUNICATE (for definitions see chapter 3). Therefore, the main effort in improving urban resource management must be laid on long-living goods that determine energy consumption (transport infrastructure and buildings) and type of energy carriers. On the output side, metals and plastics are already handled quite efficiently via separate collections. This is established by motivating and training the urban population with incentives (e.g., applying the polluter–payer principle), the most efficient procedure to prevent diluting and mixing concentrated substances in various materials. It is important to strengthen the recycling processes (technically and economically) on larger scales (regional, national, global) to improve the overall resource management for nonrenewable substances.

The urban sewerage system, including the sewage treatment plants, is based on the use of large amounts of water for the transport of wastes. It seems to be a relatively successful concentration step for phosphorus. However, as we shall see in the RESUB study below, the main P turnover is due to agricultural practice and does not happen in the PHH. For most other substances, sewage is not an important pathway. Therefore, this process can only be supported by (1) hygienic reasons, for which it was originally invented since Babylonian times; and (2) water protection measures as introduced in the second half of the twentieth century. However, it is not relevant from a point of view of an efficient urban resource management. Thus, taking in mind the high infrastructure costs for the sewerage system, the cleaning paradigm of urban systems has to be questioned if resource efficiency, in the context of a sustainable development strategy, has to become an equivalent parameter besides hygiene and environmental protection.

Based on regional and local metabolic studies (Duvigneaud & Denayer de Smet, 1977; Boyden et al., 1981; Baccini et al., 1993; Brunner, Daxbeck & Baccini, 1994; Lohm, Anderberg & Bergbäck, 1994; Daxbeck et al., 1996; Bergbäck, Johansson & Mohlander, 2001), it became clear that the perception of physical flows and stocks of urban systems was not part of the common knowledge either of the entrepreneurs or the administrators, not to speak of the designers of new urban settlements. Although cities were seen as complex "supply and waste treatment machines," the relevant indicator was and still is the flow and stock of money and not the differentiated set of physical resource indicators. After recent experiences with environmental pollution, climate change, and financial breakdowns on a global scale, it seems evident that human societies suffer from "the late-recognition syndrome" with regard to metabolic phenomena. The metabolic problems encountered with technically sophisticated closed anthropospheres (space habitats) remind us of the tragedy of open anthropogenic systems experiencing a collapse due to an erroneous concept of their metabolism.

The Metabolism of a Region

A *region* as introduced in this chapter can be defined as a system comprising cities and/or rural areas as well as derelict land, deserts, and the like. Its borders are usually denominated by administrative criteria and are often based on natural geomorphologic or hydrologic conditions. The regional metabolism differs from the urban metabolism insofar as (1) a different set of processes is usually used on the regional level (urban as well as agricultural or industrial activities) and (2) flow and stock densities are often lower. The region is presented here first to present an example of a regional material flow analysis showing its feasibility and implications. The second reason is to discuss the so-called hinterland, which comprises the many regions needed (1) to supply all the goods used in a particular region and (2) also to dispose of and dissipate the off-products (wastewater, off-gas and particulates, solid wastes) of the region.

A material flow analysis (MFA) was performed for a Swiss Lowland region having an area of 66 km² and a population of 28,000 inhabitants (figure 2.6). Measurements and data collection were performed at the end of the 1980s. For each of the four main processes, an in-depth, differentiated MFA (see chapter 3) was run over at least 1 year. Four substances, namely nitrogen, phosphorus, chlorine, and lead, were chosen as indicators to get an initial picture of the whole system. For the four main processes, additional substances were included. The landscape of this region is a valley with area shares of 56% agricultural land, 30% forests, 13% settlements, and 1% running waters. The population density of approximately 400 capita/km² corresponds with the average value found in most large-sized urban systems in developed countries. In such regions, between 70% and 80% of a national population is settled. Therefore, by purpose, a classical "dense city" was not included here, because on a national scale, only 20% to 30% of the people live in cities having densities >1000 cap/km².

The anthroposphere shows a material flow of the order of magnitude 200 Mg/c.y (figure 2.7). The water share in the input amounts roughly to 70% and reminds us

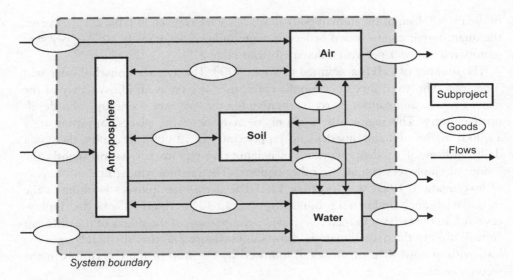

Figure 2.6
System analysis for the RESUB region. The spatial system boundary is identical to the geographically defined border of the region (Brunner et al., 1990).

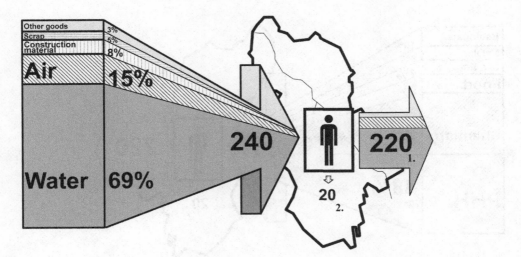

Figure 2.7
Material flow through the anthroposphere of the RESUB region (in Mg/c.y) (Brunner et al., 1990). The output consists mainly of water (166 Mg/c.y) and off-gas (36 Mg/c.y). The rest (18 Mg/c.y) consists of a large variety of solid products, such as intermediate and consumer products and wastes treated in central plants outside the region. Twenty Mg/c.y stay in the region, mainly as construction material. This quantity indicates the material growth per year of the regional anthroposphere.

of the metabolism of the human body in space travel as given in table 2.1. However, the quantity per capita is two orders of magnitude higher, namely 10^2 Mg per year (compared with 1 t per year calculated from table 2.1).

The number of PHHs amounted to about 9000. The average household size was 3 persons. The given mix of economic enterprises is seen as an idiosyncrasy of the chosen region and cannot be representative for the "average economy" of a developed country. The region offers about as many working places (approximately 11,000) as their inhabitants demand (approximately 12,000). In reality, there is a daily outflow of residents working in neighboring regions surpassing slightly the inflow of employees residing in other regions. (The unemployment rate in the year of investigation, 1989, was less than 1%.) The share of employees working in the secondary branch (industrial production) is about 40%. About 50% of the employees work in the tertiary branch (services). A comparison of the shares of the different contributors to the overall material flow shows (figure 2.8) that the PHHs have the lead with almost 40% of the total, followed by the sum of the enterprises in the secondary branch.

Two substance flows are chosen to illustrate the metabolic characteristics on the level of chemical elements. The phosphorus flow through the region is shown in figure 2.9. The MFA leads to the following conclusions:

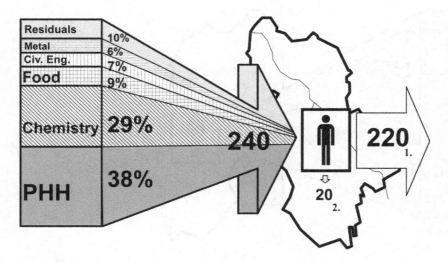

Figure 2.8
Shares of economy branches and private households in the overall material flow through the anthroposphere of the RESUB region (in Mg/c.y) (Brunner et al., 1990). 1. The output, 220 units, is mainly sewage (166 units) and off-gas (36 units). 2. The annual stock input (20 units) is mainly construction waste.

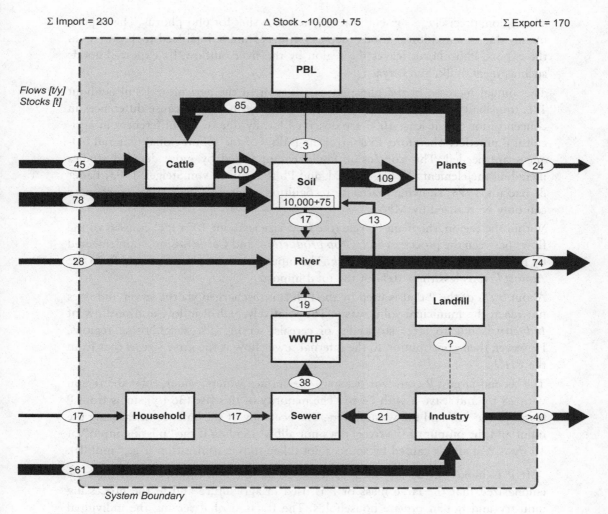

Figure 2.9
Phosphorus flow of the RESUB region (flows in t P/year; stocks in t) (reprinted with permission from Brunner & Baccini, 1992). PBL, planetary boundary layer.

The region, precisely its agricultural soil, acts as a sink for phosphorus. The import, mainly due to agricultural goods such as fertilizers and animal feedstock, surpasses the export. Phosphorus leaves the region by the river outflow, by exported goods such as meat, milk, and cereals.

The annual increase of the phosphorus reservoir in the agricultural soil is about 1%, too small to be measured directly on a short-term basis because differences in concentrations of at least 10% are observed locally due to the differences in agricultural practice. Therefore, to control the P flow of the soil, regional material balances are needed. This conclusion could be supported by more detailed studies including the elements N, Cu, Zn, Cd, and Pb (Baccini & von Steiger, 1993; Bader & Baccini, 1993) showing that early recognition of element concentration changes can only be realized by MFA.

Within the region, the main P cycle (the flow size is about 10^2 t P/y) consists of the flows between the processes *Soil, Crop production,* and *Cattle raising,* supplemented by imported *fertilizers* and *foodstock.* The input from *sewage sludge* and from the *Atmosphere* is less than 10% of the total import.

About 90% of the P that is used by the PHHs is discharged via the sewer and does not reach the municipal solid waste. The relatively high P inflow and outflow of *Industry* is due to large stockpiles of cereals, serving also neighboring regions. However, their contribution to the internal sewer flow is the same size as that from the *PHHs*.

The second largest P conveyor belts are the surface waters, which enter the region with 28 t/y and leave it with 74 t/y. The majority of this rise (36 t/y) stems from P losses of the *Soil* and the wastewaters. The accuracy of the balance (measuring the input and the output of the river) does not allow deciding if the "missing input" of 10 t/y is real and is caused by sources not taken into consideration (e.g., landfills).

If P is looked at as an essential resource for the anthroposphere, it must be emphasized that the large mass of P is used in agriculture and in the processing industry and not in private households. The fraction of P serving the individual human metabolism reaches only about 10% to 20% of the total P flow in food production and processing (see also chapters 4 and 5).

It must be added that 20 years after this study was published, the Swiss agricultural policy has, based on these findings, established an ordinance demanding annual P balances of all farming enterprises to control the P flows in agriculture. State subsidies are only given if the enterprise has a balanced P (and N) household.

The regional lead household is given in figure 2.10. The following observations are noted:

The chief import consists of 330 t of lead contained in used cars and scrap metal. Within the region, it is processed by a car shredder whose residues, roughly 20%

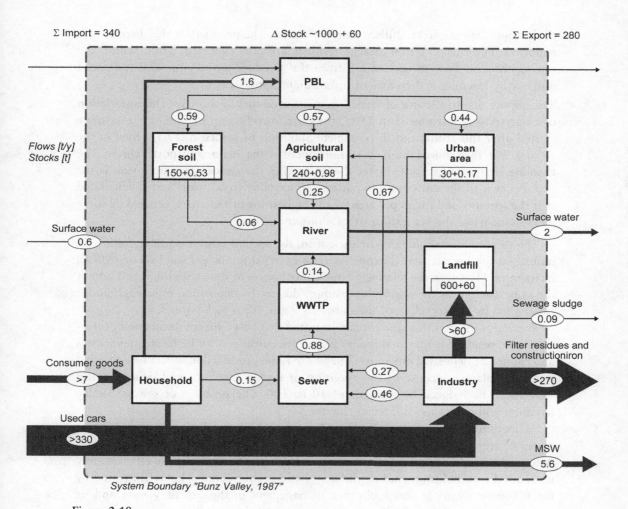

Figure 2.10
Flows and stocks of lead in the Bunzvalley (RESUB project) (after Brunner et al., 1990). The main lead flow is due to end-of-life vehicles that are recycled and turned into steel for construction. Filter dust from steel making and shredder residues are landfilled. Further sources of lead are consumer goods and leaded gasoline, which was phased out in the 1980s. PBL, planetary boundary layer; WWTP, waste water treatment plant.

of the input, are put in landfills within the region. The products of the shredder are exported. The smelter's iron and its residues are also exported. Thus, within the region, the lead flow to the landfill from the shredder is the largest. Its potential outflow by leaching is therefore of primordial importance.

The import due to the use of leaded gasoline amounts to 1.5 t/y. (The installation of catalysts was introduced in 1987, thus the introduction of nonleaded gasoline started after this study, and the complete abolition of leaded fuel happened in the 1990s). For this contribution, the atmosphere is the main transport vehicle. The gasoline lead was distributed over the soil within the region. The main sink is the *Soil*. Because of the consumption patterns of gasoline (road traffic), it is calculated that the specific load (mass per area) is the highest for urban areas, verified by some measurements of the vegetation strips along the roads.

The private households export their lead, the second largest outflow, mainly in municipal solid waste that is exported to an external municipal solid waste (MSW) incinerator. However, the relatively large contribution to the agricultural soil (about the same amount as the atmospheric input) due to the deposition of sewage sludge (motivated by the recycling of the nutrients P and N) is to be noted.

Twenty years after this study, it can be noted that based on environmental protection laws, residues from car shredder plants are considered to be hazardous wastes and cannot be deposited any more. They have to be processed in specialized thermal treatment plants. Because of the abolition of leaded gasoline, lead deposition on soils could be reduced by a factor of 10 to 100. The problem of sewage sludge handling will be discussed later in chapter 5.

As mentioned before, regions depend on their hinterland; thus, metabolic processes of regions cannot be compared without taking the hinterland into account. Also, measures to optimize (e.g., the sustainable use of materials) are efficient only if the total metabolism including the hinterland is considered. This is illustrated by the following example about nitrogen management in the city of Vienna and its hinterland (figure 2.11): To determine the impact of the city of Vienna on the water quality of the River Danube, an MFA reveals that—mostly due to the activity "to nourish"—about 11,000 Mg of nitrogen is released by the wastewater plant to the River Danube. The nitrogen flow of 26,000 Mg in the hinterland, induced by the agricultural production of food for the activity "to nourish," is much larger. Thus, in order to design effective water pollution protection measures, it is necessary to expand the system's boundary from the city of Vienna to the entire hinterland.

A similar example is given in figure 2.12 for the nutrient phosphorus. The main P flow in cities is again due to the activity TO NOURISH: Each person consumes about 1 g of P per day, equaling about 0.4 kg per year. Based on MFAs performed in the 1980s, the P losses in the food industry are a little above 50%, and roughly 5 kg/c.y of P is applied in *Agriculture* to produce the dietary P needed for human

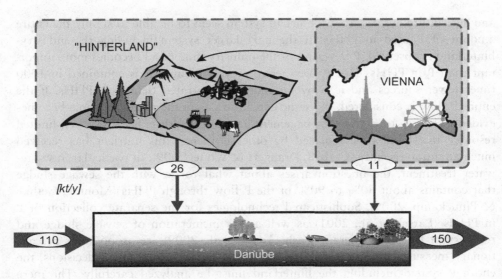

Figure 2.11
Nitrogen flows in the system "Vienna with Hinterland" (Obernosterer et al., 1998).

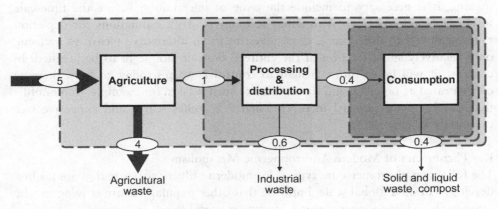

Figure 2.12
Regional phosphorus flow (in kg/c·y) (after Brunner & Rechberger, 2004).

Consumption. The low ratio of product to waste of P in *Agriculture* stems from the fact that (1) the production of meat results in large amounts of P-containing wastes that cannot be recycled but are rather stored in the soil as well as released to surface waters, and that (2) the meat fraction in the urban diet has increased greatly reducing the P-efficiency of *Agriculture* (see also chapter 4).

For decisions regarding P management, system boundaries are important: If just the urban system (mainly PHHs) is regarded without the hinterland of industry and agriculture, wastes from the urban system seem to be able to supply the entire amount of P used in PHHs. If the next larger system including the industry-hinterland is observed, P in wastes of the industry-hinterland becomes more important than P in PHHs. This P is easier to recycle because it is contained in 1000 times fewer sources and it is by far more concentrated than P in PHHs. If the entire P flow is considered, the same conclusion as in the above example becomes evident: The key to nutrient management lies in agriculture. Because P is a limited resource that cannot be replaced by other elements, this nutrient has received much attention recently (Cordell, Drangert & White, 2009). In particular in waste-water treatment, the question arises about what to do with the sewage sludge that contains about 80% to 90% of the P flow through PHHs (Montag, Gethke & Pinnekamp, 2009). Sophisticated technologies for the separate collection of P in PHHs (Larsen et al., 2001) as well as for incineration of sewage sludge and recovery of P from incineration ash (Adam et al., 2009) are proposed as precautionary measures against P scarcity. For cost-effective management decisions, the entire P system including the hinterland must be analyzed carefully: The mere focus on the two smaller systems "Private Households" and "Industry and Trade plus Private Households" leads to inefficient solutions. Also, for rational decision making, it is necessary to include the issue of uncertainty: Before the timescale for P depletion is known with some accuracy (today's estimations for depletion range from 80 to 400 years), large investments in alternative measures focusing on a relatively small fraction of the entire P flow do not seem to be justified. In contrast, it will be of future benefit to control P wastes today, to keep them as concentrated as possible, and to accumulate such wastes, for example, in monofills that can be easily recycled in times when P becomes scarce and expensive (see case study in chapter 5).

Key Phenomena of Modern Anthropogenic Metabolism

The following phenomena are typical for modern affluent societies that are leading development on a global scale, implying that other populations are striving for the same goals yielding a phenomenon of same material flows and stocks:

Increasing urbanization   The percentage of persons living in cities is constantly increasing and surpasses 50% of the global population (see figure 2.13). With regard to the anthropogenic metabolism, this has consequences: On one hand, the per capita demand for materials decreases because of more efficient networks for the transportation of people, material (products, water, sewage, waste), energy, and information. Also, because of higher densities in cities, the demand of area per capita is lower. On the other hand, the high density of people and their metabolic turnover exerts a high pressure on the regional environment because, for example, (1) all nutrients required by the inhabitants must be disposed of in water, soil, or air, and

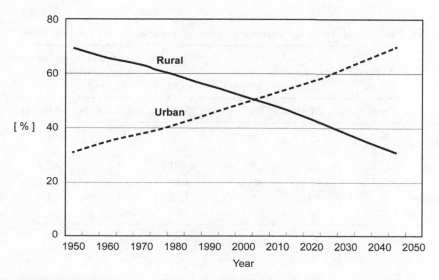

Figure 2.13
Global urban population from 1950 to 2050 (UN, 2009).

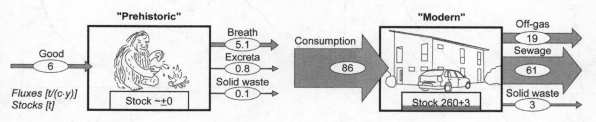

Figure 2.14
Comparison of material flows of a hunter-gatherer and a modern human being (Brunner, Daxbeck & Baccini, 1994).

(2) CO_2 resulting from combustion and biochemical processes must be disposed of in the atmosphere.

Dynamics of material flows: high growth rates and shrinking   In many parts of the world, material flows are increasing fast: As presented in figure 2.14 and table 2.3, the household consumption of goods has increased form prehistoric to modern times by more than an order of magnitude. Growth is even larger if all materials used in the hinterland for agriculture, forestry, mining, manufacturing, and distribution are considered. This "rucksack" of "gray" materials and energy characterizes the "invisible" energy, materials, and wastes associated with the total production of a good (Schmidt-Bleek, 1997).

Table 2.3
Per capita material flows and stocks for selected activities in prehistoric and modern times

Activity	Material Turnover (Mg/c.y)		Material Stock (Mg/c)		Annual Change in Material Stock (Mg/c.y)	
	Prehistoric	Modern	Prehistoric	Modern	Prehistoric	Modern
To breathe	4	4	0	0	0	0
To nourish	1.6	1.7	0	<0.1	0	<0.1
To clean	<0.1	60	0	0.1	0	<0.1
To reside	<0.1	10	<0.1	100	0	3
To communicate	0	10	0	160	0	3
Total	6	86	<0.1	260	~0	+6

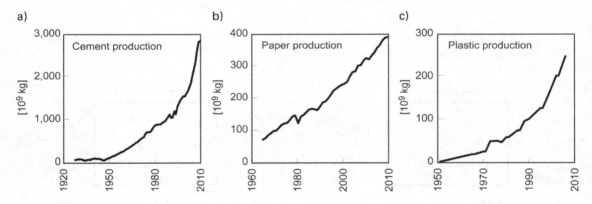

Figure 2.15
Global growth of mass good production. (a) Global production of cement, 1900–2000. (b) Global production of paper, 1900–2000. (c) Global production of plastics. 1900–2000. (a) from U.S. Geological Survey, 2009; (b) from FAO, 2010b; (c) from PlasticsEurope, 2008.

The examples given in figure 2.15 show the tremendous increase in the production of goods on a global level. This unprecedented growth was made possible through sophisticated technologies, institutions, regulations, and investments that have been developed for large-scale mining, primary production, manufacturing, and distribution of goods.

Compared with the flows of *goods*, the increase in the flows of many *substances* is even more dramatic: It has been estimated that the mining of lead rose during the past 7000 years from about 1 ton per year to more than 3 million Mg in 1990 (figure 2.16) (Settle & Patterson, 1980). The situation is similar for other metals

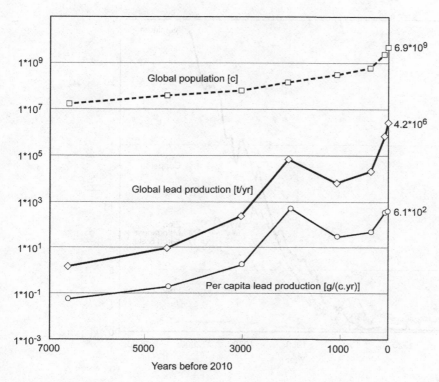

Figure 2.16
Global lead production and use (after Settle and Patterson, 1980; figure adjusted for 2010).

and for natural and synthetic organic substances (iron, copper, ammonia, and poly-ethylene) (figure 2.17). Because of technological progress, there are continuous shifts in the use of different materials, as illustrated by the changes in materials used for construction in the United States (figure 2.18). Wood has been replaced by rein-forced concrete as the key construction material, and plastic materials as well as metals are becoming increasingly more important.

Whereas growth is a characteristic of most industrialized countries, many less developed countries still consume at a much lower speed, with sometimes stagnating material flows. Based on the current viewpoint of most of these countries, it is expected that in the future, they will strive to reach the same material turnover rates as those of affluent societies.

It is noteworthy that the turnover of materials in several cities, regions, and nations is actually not growing but shrinking. The number of cities with a shrinking population during the past 50 years is as follows: United States, 59; Great Britain, 27; Germany, 26; Italy, 23; South Africa, 17; and Japan, 12 (Oswalt & Rieniets,

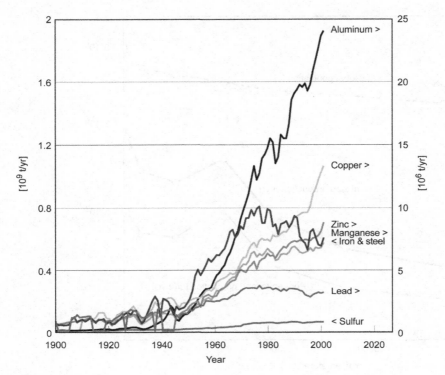

Figure 2.17
Increase of global production of selected materials.

2006). Since 1990, several states of the former Soviet Union, such as Russia, Ukraine, Uzbekistan, and Kazakhstan, experienced a smaller material turnover than that before the fall of the Iron Curtain. In some provinces of the former East Germany, population density decreased and, due to stagnation or even decrease of regional GDP (gross domestic product), material flows also decreased. For design, planning, and management of urban metabolism, it is necessary to take both potential developments, growth as well as shrinking, into account. There are also opportunities in shrinking: Whereas in a growing economy, recycling can only cover a certain (often small) fraction of the demand, in shrinking regions, recycling is able to supply, for example, most of the resources used for construction.

Imports Exceed Exports Resulting in Stock Formation
As illustrated by figure 2.10 (regional flows and stocks of lead in the Bunzvalley) and figure 2.19 (example Vienna), total material imports into most regions are exceeding exports, hence increasing the stock of materials. In Vienna, the material

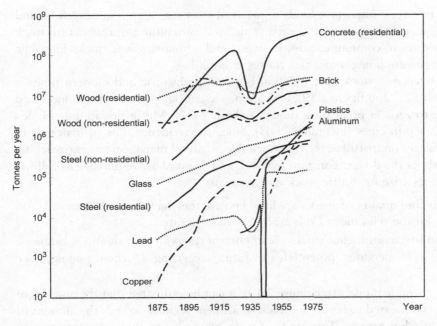

Figure 2.18
Construction materials used in the United States from 1875 to 1975 (from Wilson, 1990; with permission from Elsevier).

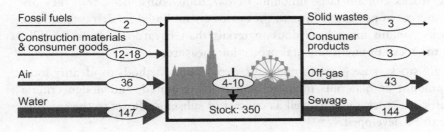

Figure 2.19
Urban material flows and stocks (Vienna example) (in Mg/c.y) (Daxbeck et al., 1996).

stocks of private households, the public and private sectors, and the infrastructure doubled in about 50 years. In the Bunzvalley, imports of the substance lead exceed exports, too, resulting in a large regional stock that doubles up every 15 years. On a global scale, material stocks grow faster than ever due to the new stocks that are currently being accumulated in China and India. Exemptions comprise shrinking regions mentioned earlier and regions that produce on a large scale primary products such as minerals (metals, construction materials) or fossil fuels for export.

There are several important kinds of material stocks: mining residues left behind as tailings; material stocks in industry, trade, and agriculture; infrastructure stock for transportation, communication, business, and administration; stocks in public works and private households; and wastes in landfills.

The difference in stock accumulation between prehistoric and modern times is striking (table 2.3 and figure 2.19). Per capita, present-day material stocks increased from close to zero in prehistoric times to about 2–300 Mg in private households and 3–400 Mg in cities (including PHH). Since the residence time of materials in the stock ranges up to 100 years, this stock needs careful management: Far-reaching decisions about the design, constant renewal, and disposal are required. The following challenges arise for future stock management:

1. Quantity and quality of stocks are little known, they have to be determined, and awareness for the relevance of this stock has to be created.

2. Based on first assessments, stocks are important reservoirs of valuable substances; they hold a tremendous potential for future recycling ("urban mining," see chapter 5).

3. Based on simple models (see figure 2.19), it can be estimated that the amount of materials accumulated today in the stock of affluent regions exceeds the amount of waste produced at present. Thus, in the future, when this stock accumulation turns into waste, amounts of waste are likely to increase, even if waste prevention measures are implemented.

4. Because stocks contain large amounts of hazardous substances, too, they are long-term sources of pollutant flows to the environment. Based on first estimations, urban stocks contain more hazardous materials than hazardous waste landfills, which are the focus of environmental protection measures.

5. For future stock reuse ("urban mining"), exploration methods to identify location and amount of materials in urban stocks are required, and design criteria allowing efficient stock reuse as well as control of substance flows to the environment have to be developed.

6. The economic question arises how to maintain high growth rates, building up even larger stocks, and providing sufficient resources to maintain this growing stock properly over long periods of time.

Linear Material Flows

Today's man-made material flows are mainly linear (throughput economy). Thus, regions are heavily dependent on their hinterland as a source for raw materials, as a partner for trade of manufactured goods, and as a sink for dissipation. The hinterland of a region is not a specific geographical area, but it comprises several scales, depending on the materials discussed. In the case of Vienna (figure 2.19), the flow of water, which comprises the largest single flow of a material for all current cities, comes from regional mountain springs, whereas most construction materials are drawn from local resources. For metals and nutrients, Vienna depends on a global hinterland. Even if some of the essential materials could still be found locally, it is often more economical to import these materials from a globalized world market.

The linear flow means that the hinterland has to cope with the residues of anthropogenic activities, too. The hinterland is also a disposal system, a function that is becoming increasingly important: The growth in material flows implies more wastes as well. Because of the buildup of long-term stocks, there is a time lag between material input and output. Future amounts of waste materials will be larger than nowadays, even if consumption were to remain at current rates. Hence, the city's metabolism and corresponding stock determines tomorrow's emissions and state of the environment in the hinterland.

The interdependence between an urban region and its hinterland regarding linear material flows can be illustrated by the example of nitrogen management: For nutrients such as nitrogen, the receiving waters of the hinterland are crucial because of potential eutrophication problems. Whereas anthropogenic nitrogen emissions such as NO_x from industrial sources or automobiles can be minimized by technical means (e.g., DENOX-processes, electric vehicles), the direct nitrogen flows from humans cannot be influenced significantly; they can only be decreased by expensive wastewater collection and treatment. In addition, the lifestyle of the urban population is important: If nitrogen is consumed as animal protein, the hinterland emissions for the production of this protein are high. If the dietary nitrogen stems from plant sources, the hinterland emissions will be 50% smaller. Thus, to protect the hydrosphere from linear nutrient flows, nutrient management programs on a large scale are necessary, facilitating nutrient cycles in agriculture as the primary source and integrating food industry emissions, traffic emissions, and other urban nutrient flows.

Whereas linear nutrient flows are of concern on a (regional to continental) watershed level, greenhouse gas emissions are of global significance. Besides the hydrosphere, the most important disposal hinterland of every city is the global atmosphere. Without this large conveyor belt and intermediate sink for carbon dioxide, the energy metabolism of cities could not survive a single day!

In addition to the large global and regional conveyor belts for emissions, there are local sinks for urban wastes, such as landfills. Today, the larger part of minerals imported into cities is finally disposed of in sanitary landfills. Although these "anthropogenic sediments" contain at present not more than about 10% of the stock of materials within a city, they will grow and may become an important source of resources in the future. In particular, if specific wastes are concentrated and individually accumulated for future reuse and are not diluted with other wastes or additives such as cement, future amounts may become large enough to make reuse economically feasible.

Consumption Emissions Surpass Production Emissions

A fourth phenomenon of the metabolism of advanced societies is that production emissions are decreasing, whereas consumer emissions are increasing. Data from Bergbäck et al. (1994) indicate that for heavy metals in Sweden, during the 1970s end user–related non-point emission sources became more important than industrial point sources. On one hand, this is due to advanced legislation and technology in the field of industrial environmental protection. On the other hand, the high and still growing rate of consumption has led to large stocks of materials that have to be fueled, operated, and maintained. These stocks are subject to weathering, corrosion, and deterioration, discharging materials that are hardly noticed on the individual level but that are significant for the quality of water, air, and soil if looked at as a total quantity from all sources. Hot spots of such non-point sources are cities. Examples of consumer-related emissions are carbon dioxide and other greenhouse gases due to space heating and transportation, the weathering of surfaces of buildings (zinc, copper, iron, etc.), wear, corrosion, and erosion of vehicles and infrastructure for transportation (chassis, tires, brakes, catenary wire), and the growing problem of nitrogen overload mentioned earlier.

In figure 2.20, an example is given to illustrate the significance of consumer emissions versus production emissions: Modern electroplating plants produce only small amounts of production residues; most of the metals used to protect surfaces leave the factory incorporated in the product. In general, the solid and liquid metal residues are recycled to a large extent, resulting in low losses of metals to the environment. Thus, from a company point of view, the goal of pollution prevention has been achieved, and most of the zinc or chromium leaves the factory as coatings on products. Nevertheless, during the lifetime of the product, processes such as corrosion and weathering will remove the zinc from the surface resulting in losses of heavy metals to the environment. Because the corrosion processes are slow and the residence time of metals on surfaces can be rather long, the metal flows to the environment last for years to decades. The legacies of today's protected surfaces are tomorrow's environmental loadings.

To prevent consumer-related emissions is more difficult than to stop industrial pollution: (1) In general, emissions of industrial sources can be reduced with a much higher efficiency than that for small-scale household emissions. (2) The number of (consumer) sources is orders of magnitude larger than that of production facilities. (3) The individual consumer emission may be very small, and only the multiplication by the large number of sources may cause a risk for the environment. (4) It is a bigger challenge for a company to change or discontinue a product than to add a pollution control device to its production facility.

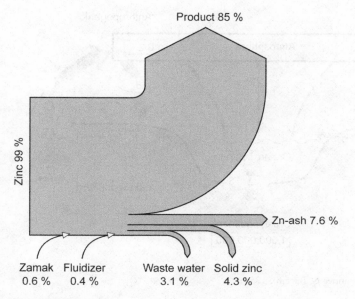

Figure 2.20
Zinc balance of a modern electroplating factory (W. Enöckl, personal communication, 1994). Except for small amounts in wastewater, all of the residues of this modern electroplating factory are recycled. The product, zinc-coated iron sheets that will be exposed to wear and corrosion during use, thus becomes the major future emission source.

Anthropogenic Material Flows Exceed Geogenic Flows

Because of the large growth rate of the exploitation of essential minerals, man-made material flows are approaching and even surpassing natural flows of several substances. As a consequence, the flows, stocks, and concentrations of some substances such as heavy metals and nutrients are rising in the environment, causing the biosphere to change more rapidly and in different directions than it would without anthropogenic material flows. In the following, cadmium is used as a case study to exemplify the importance of today's man-made flows. For this metal, the gap between actual and toxic concentrations in soils is smaller than that for other metals.

In figure 2.21, the global natural and anthropogenic flows of cadmium during the 1980s are summarized. The total man-made flows of about 17 Gg per year are more than twice the size of the natural flows of about 6 to 7 Gg per year due to erosion, volcanic eruptions, and sea spray. Because the input into the anthroposphere is about 6 Gg larger than the sum of all outputs, about 30% of the cadmium annually mined from the earth's crust is actually accumulated in goods with long residence times (previously: paints, plastic additives, surface coatings; today: NiCd batteries). In the 1980s, the most important pathway for cadmium from the anthroposphere to the environment was the atmosphere. This is due to the properties of

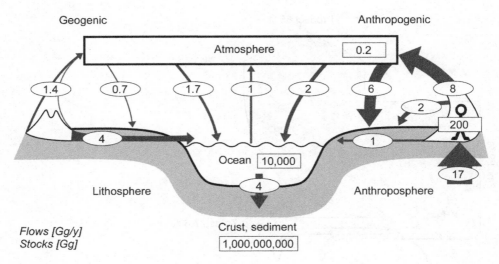

Geogenic

Anthropogenic

Figure 2.21
Global cadmium cycle (after Brunner & Baccini, 1981).

cadmium and its compounds having a comparatively high vapor pressure and due to the lack of efficient flue gas cleaning technology three decades ago. Thus, the man-induced global flows of cadmium into the atmosphere were an order of magnitude larger than the natural flows. Today, the emissions into the atmosphere are substantially lower because of advanced air pollution control technologies in metallurgical and waste treatment processes.

The main sink for cadmium is the soil, followed by ocean sediments. The global input into the soil by atmospheric deposition (6 Gg/y) and land filling (2 Gg/y) is much larger than the output by erosion and river systems (1 Gg/y). Comparing the natural side (ratio of deposition to leaching of 0.17:1) with the anthropogenic side (ratio of deposition to leaching of 8:1) in figure 2.21, the large accumulation of anthropogenic Cd in the soil becomes evident. This is confirmed by regional cadmium balances, too, and applies to other metals as well (Bergbäck et al., 1994; Azar, Holmberg & Lindgren, 1996).

Increasing Complexity

Since around 2009, about 1 million new substances are identified each year, and about 30,000 chemical substances are marketed in amounts above 1 Mg, with about 10,000 in amounts over 10 Mg. In parallel to the growth of flows and stocks of goods, the composition of goods has become more complex, too. Modern goods are composed of numerous matrix and trace substances. Most of these substances have been specially designed to be incorporated in goods and have not been introduced as impurities or by negligence. Consumer goods are usually composed of many substances; only few applications require the need for a single "pure" chemical. To give a few examples: The number of elements used for information technology products increased from 11 elements in 1980 to 15 elements in 1990 and to 60 elements in 2000 (Mc Manus, 2006). The number of goods contained in building plaster has increased during the past 50 years from 4 to 13 (Rebernig, Müller & Brunner, 2006).

Complexity of product composition has implications regarding resources and the environment. First, modern products are dependent on the availability of a whole array of substances: If a single substance such as rhenium is missing, a modern jet engine cannot run at the designed high speed because the materials could not stand the extreme temperatures such devices experience at full load. Hence, more goods in general become more vulnerable toward resource scarcity and availability. Second, recycling complex products is a challenging and difficult task: Often, valuable substances in waste products are of small concentration, and hazardous constituents jeopardize the potential income generated by the recycling of the valuables. Because modern products are often mixtures that cannot be taken apart by mechanical means (e.g., plastics as mixtures of polymers, plasticizers, softeners, fire retardants, stabilizers, pigments, etc.), more sophisticated chemical or thermal methods are needed to recycle substances from these goods. Also, waste management processes have to be designed considering the many substances that potentially can be present in waste goods. Incineration of municipal solid wastes as well as of construction and treated wood wastes must take into account trace metals as well as persistent organic substances contained in the complex waste mixtures. A strategy of prevention to reduce emissions from waste treatment will become effective only if products with complex compositions can be removed from the waste stream.

Learning from Metabolic Accidents and Collapses in History

As exemplified in the previous sections of this chapter, the characteristics of anthropogenic metabolisms are changing over time. Most often, these changes, induced actively by the ingenuity and capability of progressive civilizations, improve quality of life and benefit societies. However, looking at long-term historical developments, the induced anthropogenic changes can result in disaster, too.

Easter Island
Easter Island (Rapa Nui), located in the middle of the Pacific Ocean in a subtropical marine climate, is one of the most isolated inhabited spots on Earth. When in the eighteenth century the first Europeans visited the small island (about 160 km^2), they were somewhat puzzled by a large number of giant stone monuments (Moai). It

seemed to them that the low technical standard of the approximately 1000 natives could hardly suffice to build such monuments. Today, based on a series of archeological, anthropological, and ecological research projects, a well-supported hypothesis states that the ancestors of Polynesian origin lived for several hundreds of years on this island and experienced, during the seventeenth and eighteenth centuries, an ecological collapse due to their self-made deforestation. Their terrestrial hinterland, namely forests and forest-based agricultural soils, was reduced drastically and led eventually to famine, sickness, and the collapse of the island society (Diamond 2005). There was no easy escape from this small ecosystem and no easy compensation by imports of essential goods from neighbors. When the first Europeans arrived, this process had already taken place. Why did the natives destroy their ecological basis of life? Did they not see the correlation between their limited resource stock and their consumption rate? Although the natives had developed their own script and had high skills in craftsmanship on a Neolithic level, the answers to these questions will stay speculative. However, the myth of the ecological wisdom of Stone Age societies cannot be supported in this case. It is more probable that human societies, in every period of their evolution, are prone to make deadly mistakes.

Greenland

A second example is the case study on the Viking settlements on the shores of Greenland in a climate that is surely less hospitable than that of their nearest neighbors in Iceland and Norway, not to speak of the subtropical island in the Pacific. Most likely (we do not know from sound historical documents), immigrants from South and East of Greenland took the warm period between the tenth and thirteenth centuries AC to settle on juvenile soil in an area without indigenous farming. The European farmers, raised in a Christian culture, met there only Inuits, a hunting culture specialized in getting their food from the fauna in pack ice. Furthermore, the new settlers became, institutionally defined, a colony of Northern European states and based their technological know-how and their trade on their motherlands. They reached a population of several thousand inhabitants until the fifteenth century. When the Little Ice Age started in the sixteenth century and was about to last for 300 years, the vegetation periods for farmers became shorter and shorter and the crop yields smaller and smaller. The longer winters asked for more wood to warm the houses and led to deforestation and increased erosion of soils. The European settlers, in their Christian faith, reacted by building more churches to convince God to correct the climate change to their favor. Eventually, the Norse settlers had to give up and clear the place. Although they had their Inuit neighbors who showed them how to survive in this climate (an impressive evolution over thousands of years), they were not ready to change their basic supply system from farming to hunting (Diamond, 2005). The cultural paradigm was stronger than the ecological reasoning to adapt the metabolic lifestyle to the new environmental conditions.

Angkor

In Southeast Asia between the ninth and fifteenth centuries AC (1100–500 BP), the people of Khmer developed a mighty kingdom whose center became Angkor (etymologically rooted in the Sanskrit name *Nagara*, meaning "city"). In its maximum size, the city area covered roughly 1000 km² and had about 1 million inhabitants. The city engineers built two large water reservoirs with a total capacity of approximately 70 million m³ to irrigate the agricultural fields in the vicinity of the urban settlements. With this artificial irrigation, three rice harvests per year were possible with a total yield of about 2.5 Mg per hectare. It follows that a total area of about 2000 km² of rice fields could secure the basic food supply of Angkor's population. Today, the temples of Angkor are one of the most famous witnesses of a once flourishing and now practically deserted city. According to the state of the art in archeological and historical research, there are several factors that led to the decline of this huge urban system (such as wars, moving political and economic power to other centers). There is also a plausible hypothesis that a city of this size must eventually collapse when the water management cannot be maintained properly. Water is the key resource in such a system because all food production depends on it. There must not necessarily be a long-term drought (as historians postulate as one of the main reasons that cultures in the American continent collapsed): It suffices that the built infrastructure (water collection and water distribution) is falling apart and no means are available anymore to repair or to rebuild it.

Rome

In the Western Hemisphere, history offers several examples of shrinking cities. The city of Rome, the center of the Mediterranean empire, had approximately 1 million inhabitants 2000 BP. It needed a logistic master plan to feed daily the urban population, based mainly on relatively small vehicles and an average transport velocity of less than 5 km/h. When the Roman Empire started to fall apart, its center started to shrink. Historians estimate that within 200 years, Rome had reduced its population to approximately 40,000 inhabitants (1300 BP) and started to grow again during the Middle Ages as the geographical center of Christianity (the Holy Roman Empire of the German nation, created around 1000 BP), named itself "the eternal city," became eventually, in the nineteenth century, the capital of the new nation Italy, and has, at the beginning of the twenty-first century, a population of nearly 3 million. As with Angkor, some stony witnesses of the ancient urban system remain as ruins (e.g., the Colosseum, the Forum Romanum), but contrary to the South Asian example, they are embedded as protected monuments within a pulsing megalopolis of the present. Rome is an example of a system that could only maintain its size (in antiquity) on the basis of a political system that guaranteed the supply flows. The drastic shrinking was not due to an ecological collapse but to an institutional breakdown. The metabolism of such large systems is not robust because it cannot maintain itself without a huge colonized hinterland. It has to reduce its population to a size that is in balance with its economically and ecologically defined hinterland.

What can be learned from these examples and many more (see also Diamond, 2005) is the insight that metabolic mismanagement (overuse of resources, late recognition of environmental changes, lack of agility to adapt in time) can lead to a collapse or at least to severe damage of human societies independent of geographical site, climate zone, cultural idiosyncrasies, or level of development. Until the past decades, history writing has underestimated the importance of metabolic properties of the anthropospheres. Sustainable development has to do with survival strategies that have to include metabolic strategies. The scientific analysis of the metabolic history in the evolution of anthropospheres helps to develop a strategy of risk minimization. Anthropological studies have shown (e.g., Groh, 1992) that in each step of the evolution of human societies, the strategy of risk minimization has been applied. The fact that there have always been societies that were successful (otherwise the species *Homo sapiens* would not exist anymore) leads to the questions about their skills in adapting themselves to the ecological limits and the environmental changes. These questions can be summarized in a set of three categories:

• How does a society recognize the problem to be relevant for its future ("early recognition")?
• How does a society tackle the problem when it becomes acute ("political awareness and solution solving")?
• How does a society implement its decisions and manage the technical and economic processes ("learning by doing")?

All three categories must be based on good governance. The first ability, "early recognition," needs a "political insight" that a critical mass of scientific activity is indispensable to maintain and sharpen this ability. The second and third abilities ("political awareness and problem solving" and "learning by doing") depend strongly on the institutional system established in a society and on its size. A feudal or feudal-like organization proceeds differently than a democratic one. Smaller-sized sovereign societies seem to have a lower moment of inertia but are less effective in provoking feedbacks from their hinterlands. Therefore, the quality of metabolic processes, with regard to a sustainable development, has to be seen in the context of these categories. However, the experiences, gathered so far in the actual globalization process, are still fragmentary and do not give a solid picture of "best solutions." The following section illustrates a few case studies to show some characteristics of "first moves in the right directions."

Learning by Adapting the Anthroposphere

Forestry and Timber Consumption, a Prototype of Sustainability

The notion *sustainability* with regard to resource management was applied first in forestry. The Saxon engineer von Carlowitz demands in his book *Sylvicultura oeconomica*, published in 1713, that a "continuierliche, beständige und nachhaltende Nutzung" (a continuous, permanent and sustainable use) of forests is indispensable for the survival of a country (Müller, 1998). The feudal system established strict consumption limits for its underlings to protect its timber stocks. It took more than 100 years to establish a timber management that was based on measured stock quantities of timber in forests, its annual growth, and the measured flow of the annual timber output. Until the beginning of the twentieth century, many industrialized countries fought with the still growing reduction of the "timber capital" in their forests, mainly due to the exponential increase of population and the growing timber demand per capita. One of the measures taken was to replace slower-growing trees (such as oak and beech) by faster-growing species, namely spruce. However, the decisive step to release the demand pressure on forests was the substitution of timber as energy source by fossil fuels (coal and oil) and as construction material by concrete and steel. Not before then could countries enforce laws to secure a "sustainable forest management." The simplest principle to follow was to keep the forest area constant. In figure 2.22, the different strategies of forest use over centuries are visualized in a scheme. Today, in some developed countries, the forest area is growing again because timber consumption is decreasing due to economic reasons. In some developing countries, the opposite move is the case.

In the context of the growing demand of renewable resources, the question arises what roles timber could play in the future, in "the second solar civilization," as a construction material, as a raw material for paper and paperboard production, and as a fuel. If a future overuse has to be avoided, this question needs to be answered with the help of a dynamic material and energy flow analysis (Müller, Bader & Baccini, 2004). The analysis comes to the conclusion that a standard scenario ("business as usual") for the twenty-first century can hardly offer an essential contribution (ca. 1%) to the overall energy demand if the forests should be managed in a sustainable way. However, in an adapted anthroposphere (scenario "restructured"), timber can contribute up to 20% of the total heating demand (1) due to a threefold decrease of heating energy for buildings, from 400 MJ/m^2.y (year 2000) to 130 MJ/m^2.y (year 2100) and (2) by reducing the carpentry wood share in the second half of the century

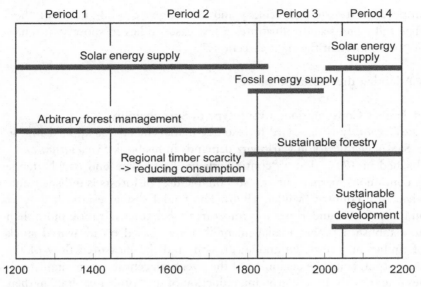

Figure 2.22
Scheme of the four phases in timber management (after Müller, 1998).

(figure 2.23). The setup of such scenarios will be shown and discussed in chapter 3, 4, and 5.

The example forestry and timber management, seen in the historical perspective and in the potential contributions for a sustainable development, gives the following clues:

The timescale for a metabolic analysis has to be adapted to the "lifetime of the stocks"; that is, in the production (forestry) and in the consumption (the goods with the longest residence time, namely buildings). Therefore, an observation and adaption period of at least 200 years is appropriate for a timber management strategy.

The concept of sustainability, a reasonable first approach in the eighteenth century for a colonized ecosystem such as a forest, cannot be restricted to the forest (or an agricultural area, or an aquatic system). An ecosystem is more or less intensively connected to urban systems with continually changing resource demands and, thereby induced, to changing metabolic patterns. When no substitutes are found or technologically invented, the stocks of the ecosystems will eventually be raided. In other words, a subsystem of the anthroposphere cannot, in the long term, be run separately in a sustainable way. Only a larger entity, such as a region, with included colonized terrestrial and aquatic systems and with a defined hinterland on the global scale, can tackle the problem of a sustainable development.

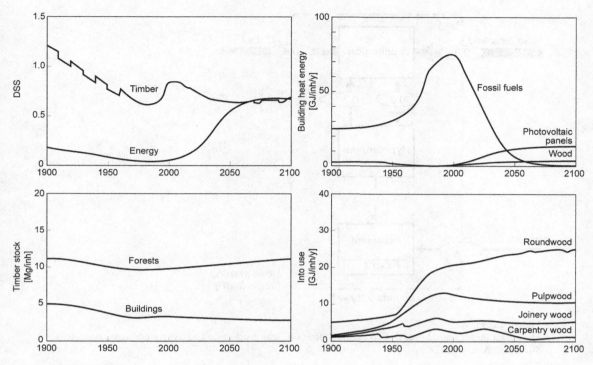

Figure 2.23
Modeled timber management within a restructured region attaining a sustainable metabolism (after Müller, Bader & Baccini, 2004). DSS, degree of self-sufficiency; inh., inhabitant.

Managing Oligotrophic Ecosystems within a Eutrophic Anthroposphere

Freshwaters, lakes and rivers, are sensitive indicators of the metabolic state of their environments. Lakes, having drainage areas that are used intensively by agricultural production and/or are receptors of wastewaters from private households and industry, react relatively fast with changes in their biocenose. At the beginning of the 1970s, the first scientific models for the primary production in lakes could support the hypothesis that phosphorus is the growth-limiting nutrient in lakes. If phosphorus flows in the input increase, the algae production grows proportionately. This process is called eutrophication. Most lakes are originally in an oligotrophic state (i.e., the natural phosphorus input does not allow a high algae production). The higher the primary production in the epilimnion (figure 2.24), the higher the oxygen consumption in the hypolimnion, due to the decomposition processes of sedimenting organic material by microorganisms. If the oxygen flow from the atmosphere to the hypolimnion is too low due to physical reasons (stratification of two water layers

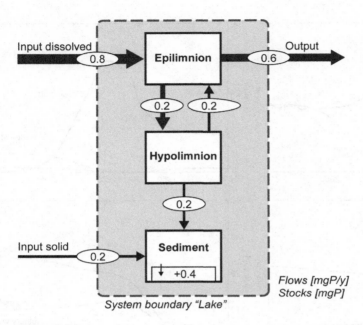

Figure 2.24
Flow diagram for phosphorus in a lake and its effect on algae production and oxygen consumption (after Stumm & Baccini, 1978).

by density difference), the deeper lake waters become anaerobic. Eventually, the lake changes drastically its biological composition.

The experimentally founded lake model gives the scientific basis for qualifying and quantifying the metabolic properties of an aquatic ecosystem. For each individual lake, the critical phosphorus load (i.e., the load pushing the lake out of the oligotrophic state) can be determined (figure 2.25). If a society sets freshwater quality goals, it has to control, among other substances, the phosphorus input of lakes. In many industrialized countries, the phenomenon "water pollution by sewage" led the politically responsible to act. This recognition of the problem took place in the 1950s. The technological measures, namely installing sewage treatment plants, started and took at least two decades until the great majority of polluters were connected to these plants. It took at least two decades to set the quality standards in scientific units and numbers (fixed in compulsory legal ordinances). The first effects could be observed 30 years after the political recognition of the problem. In the case of some lakes, a restoration to the oligotrophic state could be achieved showing again the original biocenose.

From the regional phosphorus study (figure 2.9), we have learned that the largest anthropogenic P flows are created within the subsystem agriculture. It follows that

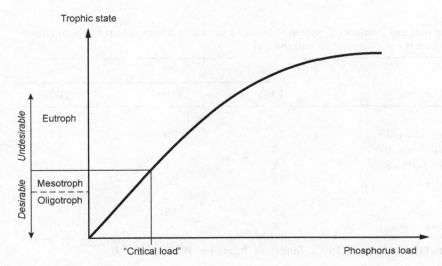

Figure 2.25
Schematic illustration of the trophic state of a lake as a function of its phosphorus load (after Stumm & Baccini, 1978).

measures to limit the P flows for water protection have to be taken in various ways, namely (1) in agriculture to prevent nutrient losses in run-offs and luxury fixation in soils, (2) in limiting or banning additional P coming from other sources such as additives in detergents, and (3) in increasing the P storage in sewage sludge. The figures in table 2.4 illustrate the following effects:

Agricultural P control could reduce its input by approximately 25%. However, the soil (landfilling included) is still the main P sink, increasing its P concentration, because the export by products varies between 50% and 80% of the input (i.e., the P flow of the agricultural soil is not yet near a steady state).

A phosphorus ban in detergents (to clean the textiles) reduced its flow by 80%.

The output into the hydrosphere could be reduced by 50%.

  First experiences with the control of the essential nutrient phosphorus show that

• The goal to keep aquatic systems in an oligotrophic state within a eutrophic anthroposphere is difficult to attain.

• Metabolic models for ecosystems are needed to evaluate in a qualitative and quantitative way the role of a nutrient. Scientific models are based on scientific research and need also decades to pass a sound reviewing process.

• The time period between political recognition of a metabolic problem, the implementation of various solutions (technological equipment, laws, bans, incentives, etc.), and first effects of the measures comprises several decades.

Table 2.4
Intermediate national P balance (Switzerland) after a period of additional measures to reduce the P flows into the surface waters and the soil

Input and Output	Flow (Gg/y)		
	1983	1994	2006
Input			
Fertilizers and fodder	27	20	14
Detergents/chemicals	5	1	1
Atmospheric deposition	1	1	1
Total	33	22	16
Output			
Soil	25	18	14
Hydrosphere	8	4	2

After Baccini (1985); Siegrist (1997); Binder, de Baan, and Wittmer (2009).

Because of this large time lag, quite often political decisions for substance flow control have to be taken before all open scientific questions are answered satisfactorily. It is less costly to stop a measure after a decade, due to a corrected scientific analysis, than to wait until the model is highly sophisticated but the system is in a state of irreversible negative effects or even near collapse.

Metabolic Problems of a Subsistence Economy
The oldest type of economy, since the Neolithic period, is the subsistence economy. People meet all or most of their daily needs from their local natural environment. Food is grown, hunted, and gathered. Goods for running the household (energy carriers, construction) and for agricultural and hunting tools are mostly self-made. Trade with neighboring communities is only possible if a surplus of food or material is produced. At present, this type of economy still comprises a large part of the human population (approximately 60%). Their regions are named "developing countries (DECs)" and are based mainly on an agrarian economy. The complementary economy type is a market economy that comprises the "developed countries (DCs)." In a historical perspective, these countries started from a subsistence economy and moved, in most cases supported by an intensive industrialization process within their own borders, to a complex combination of diversified products that could be traded over large distances. Because of the globalization process, the countries with subsistence economies became interwoven with the market economy. From a socio-economic point of view (i.e., applying the indicators of the market economy), DEC means poverty, and DC means welfare.

When Hardin published his article with the title "Tragedy of the Commons" (Hardin, 1968), a scientific debate arose on the best governance to protect common goods from overuse. Although much criticism was given to Hardin's interpretation of the case study with which he supported his hypothesis, it was eventually accepted that he is correct in claiming that humanity needs to cede the unlimited freedom to reproduce.

A metabolic study of a DEC region is shown to illustrate the problem (Pfister & Baccini, 2005). In a Nicaraguan agricultural region, the focus is laid on current farm resource management. The indicators are nitrogen (the growth-limiting nutrient of soils) and the degree of self-sufficiency (DSS). The latter is essential to evaluate the performance of a subsistence economy. Farmers have different sizes of business (landless, small, medium, and big). The population distribution among these categories is as follows: 50% landless, 30% small, 15% medium, and 5% big. (This distribution type can be found in historical records of European regions in the seventeenth and eighteenth centuries). The two basic staples are maize and beans. The cash crop is coffee. The main energy source is wood from local forests. Three different management scenarios were compared; namely, the "status quo," high population growth ("scenario 1"), and population reduction ("scenario 2").

The land-use pattern for each scenario is given in figure 2.26. The results can be summarized as follows:

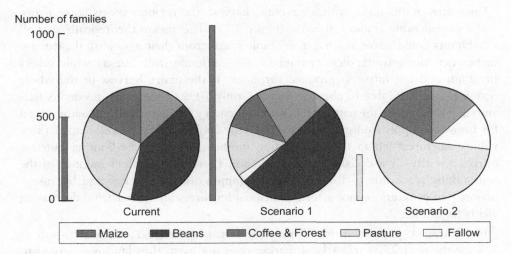

Figure 2.26
Land-use patterns in an agricultural region of Nicaragua (reprinted with permission from Pfister & Baccini, 2005).

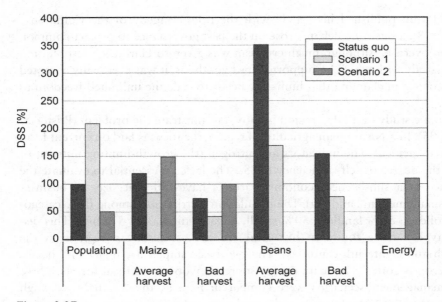

Figure 2.27
Regional degree of self-sufficiency (DSS) in a subsistence economy (agricultural region of Nicaragua) as a function of three scenarios (reprinted with permission from Pfister & Baccini, 2005).

• Currently (status quo), with an average harvest, the region exports some maize and a considerable amount of beans (figure 2.27). This means theoretically that all inhabitants could cover at least their staple needs from their agricultural earnings. In practice, there are landless peasant families suffering from hunger while others (medium and big farmers) produce surpluses. If the maize harvest in the whole region is bad (e.g., due to climatic impact), only 70% of the maize needed is harvested. Furthermore, the region faces a serious energy problem. Without a hinterland for firewood supply (indispensable for the daily cooking of beans and maize), there will be no forest left in the region after another 30 years. The limiting nutrient nitrogen is already a decisive economic factor. There is a "nutrient mining" of the soils taking place due to food production (approximately 30 kg/ha.y) because a majority of peasants cannot afford additional fertilizers and suffer from decreasing yields (figure 2.28).

• In Nicaragua, population is still growing fast. In scenario 1, the growth rate is 3% for the next 25 years. A land market does not exist, thus land is inherited to the progeny and divided equally among it. Within a generation, the population doubles. The land use changes to smaller holdings per family. The most striking effect is the food autoconsumption, which is cut in half compared with the status quo, which indicates a severe shortage of food.

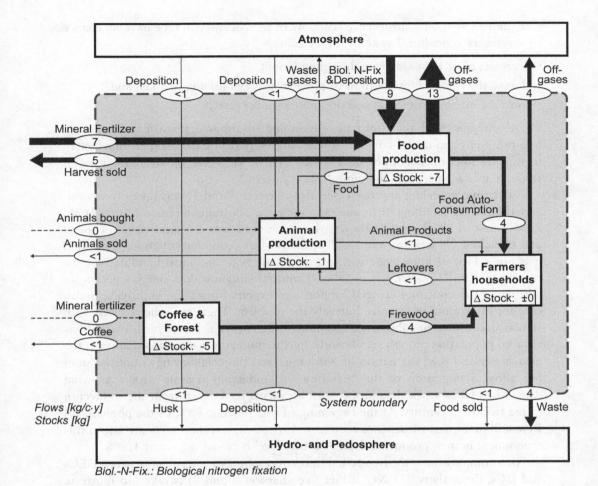

Figure 2.28
Regional nitrogen flows (in kg/c.y) in an agricultural region of Nicaragua (reprinted with permission from Pfister & Baccini, 2005).

• In scenario 2, populations sink to one half of the current population. The production type moves to pastures with animal husbandry. The great majority of the farmers move to the category "medium size." Food security is given priority. Nitrogen mining can be stopped. The energy supply is covered with firewood from each family's own coffee plot. Thus, the DSS for a sustainable system lies around 110%. It is evident that this scenario, with regard to the agricultural technique, has its experimental proofs in the green revolution of European farming in the nineteenth century.

Regions based on a subsistence economy of the Nicaraguan type have no chance to survive as agricultural systems if they cannot

- decrease, or at least stabilize, their population;
- change their production pattern to prevent nitrogen mining of soils; and
- solve the energy problem based on limited timber stocks.

Other regions with the same type of economy are exposed, because of the climate and other environmental conditions (e.g., drafts, parasitic diseases, erosions, etc.), to various other difficulties to sustain their system. However, the overall pattern of resource management has the same "tragic pattern."

A dynamic modeling approach (see also chapters 3 and 5) was taken to investigate the current handling of resources in the same Nicaraguan region (Pfister et al., 2005). First of all, the model shows why the current farmers make certain choices and that these choices are well set to optimize their economic situation. Therefore, it is not a lack of knowledge with regard to the best "technical handling" of the current system. However, the social and cultural situation does not yet permit an adequate birth control. Financial support and experts from DCs, improving health care for humanitarian reasons, intensify the problem. There are other regions with similar dominance of subsistence economies (e.g., in Asia) where birth control has reduced population growth significantly. Furthermore, the economic situation of the near hinterland (i.e., the nation of Nicaragua and the neighboring countries) does not allow a migration of the "overflow" to industrial regions with a growing demand of labor work. In Europe, such a process took place from the eighteenth to the twentieth century. At the beginning of this process, 80% of the populations were still in a quasi subsistence economy. At the end, less than 10% are left in this economical branch producing food at a DSS level between 80% and 120%.

For a comparison of the two actual states of resource management between DECs and DCs, the activity TO NOURISH (see chapters 3 and 4) is taken to illustrate some crucial differences in the metabolic and economic pattern (table 2.5).

The total energy consumption of U is approximately 10-fold that of A. For the activity TO NOURISH (including agricultural production, upgrading and distribution, consumption), the system A needs roughly 90% of its total energy demand, whereas the urban system U can manage with only 20% of its total. A's energy source is from local forests (80% self-sufficiency). U's energy carriers are mainly fossil fuels and are imported. U needs 80% of its energy demand for the activities TO RESIDE&WORK and TO TRANSPORT&COMMUNICATE (see chapter 4). A's food is mainly produced within the region (90% self-sufficiency). U needs, on the bottom line (taking import/export flows into account), a "global hinterland" for approximately 40% of its food demand. In A, the peasant household has to spend roughly 90% of its income for food, whereas the urban household can manage this

Table 2.5
Comparison of a peasant society A in Nicaragua (Pfister, 2005) and an urban society U in Western Europe with regard to their energy demand in the activity TO NOURISH and to their economic effort (Faist, 2000)

Parameter	Agrarian, A		Urban, U	
	Energy Flow (GJ/c.y)	Self-sufficiency (%)	Energy Flow (GJ/c.y)	Self-sufficiency (%)
Energy total	19	80	180	10
Energy to nourish	17		30	
Regional supply		>90		60
Ratio of total income	>80%		10%	

activity with only 10% (average values) of its total income. In the comparison of the two cases, the following additional insights must be stated:

1. The peasants in A have a reasonable strategy in their agricultural production. Because of their increasing population, their system is neither economically nor ecologically sustainable. They need a "hinterland" where they can sell their labor force. However, the "hinterland" cannot offer enough labor. A second source of income is the cash crop coffee, a product to be sold on the global market that is mostly out of reach for poor farmers. Their main problems are thus population growth and the lack of a strong complementary region offering labor and/or good prices for agricultural products.

2. The people in U are economically successful in a global market, mostly due to their products in the tertiary sector. However, because of their strong dependence on nonrenewable energy sources, their system is not sustainable on the long term (see also section "The Great Risk of the Developed Countries" below). Without reconstruction of their physical infrastructure from a fossil system to a solar system within the next two to three generations, U will collapse.

The starting positions of an A and a U society to enter a process toward sustainability are completely different, seen from an ecological and an economical point of view. Here, the various differences in political and social culture are not yet considered. Although the developed countries with their new urbanity consume, at the beginning of the twenty-first century, about 70% of the total resource demand worldwide, a strategy for a sustainable resource management must comprise two interwoven concepts, one for the type A and one for U. While the systems of type A are in a daily struggle to get the minimum for survival, the systems of type U cruise on well-furnished vehicles toward dangerous cliffs. These two processes, more or less in a parallel arrangement, have a relatively young history. The metabolic facets of it first became politically relevant in the context of environmental pollution, roughly a generation before the target of a sustainable resource management was put on the agenda. With the rise of media technology that could spread news at light speed around the globe, preferably the bad news ("good news is no news"), scandalous waste stories made the headlines for some time.

Worldwide Waste

At the end of the eighteenth century, James Watt, the Scottish engineer, together with his colleague and entrepreneur Matthew Boulton, improved the steam engine to higher fuel efficiency. This technical improvement and its innovation in manufacturing is often used by historians as a metaphor to indicate the start of the "industrial revolution." The steam engine replaced step by step the water wheel and increased exponentially the demand for fuel to produce steam. Coal, a fossil residue in the earth's crust, became the most important energy carrier for running the new engines, long before oil could partially replace it. A hundred years later, the combustion engine, a technical alternative to the steam engine with an even longer invention history, made its big step into innovation when the German engineer Karl Benz started the production of the first automobiles (1886). It is the gate opener to the fossil fuel era. After another 100 years, under the auspices of the United Nations, a protocol on climate change was signed in Kyoto in which 37 industrialized countries committed themselves to reduce their emissions of greenhouse gases (1997). At the beginning of the twenty-first century, when the Kyoto Protocol should be revisited in Copenhagen (2009) with a perspective for 2020 and 2050, the substance CO_2 has become the most famous waste gas in human history. Since Lavoisier's first understanding of the chemical qualities of combustion processes to the quantitative relevance of engines applying their exothermic nature, only six human generations have participated in this development. At the beginning of the scientific debate was a data set of atmospheric CO_2 measurements from Hawaii and from the South Pole taken between the end of the 1950s and the end of the 1970s. The increase from roughly 310 ppmv to almost 340 ppmv within two decades was an alarming signal. At first, the climate change debate was a quarrel about the significance of the anthropogenic contribution (for about 20 years), and then, after a broader acceptance of the reports of the scientific community (IPCC, the Intergovernmental Panel on Climate Change), it has become a haggling over CO_2 reduction. In the (partly) hidden agenda, the topic is controlled by national economical weighing of self-interests. It is not yet a profound discussion about new metabolic designs of the anthroposphere. It is essentially a waste reduction debate in the well-established culture of environmental protection.

Another impressive illustration for this attitude is the worldwide cadmium distribution.

In the 1940s, Japanese farmers started to suffer from an unknown rheumatism-like illness which they named "Itai-Itai." For some patients, the skeletal deformities were mortal. They all worked in the rice fields of the lower reaches of the Jintsu River. After two decades, the cause was found. It was the result of a chronic cadmium poisoning. The cadmium source was a zinc mine situated some 50 km upstream. Cadmium is a geochemical partner of zinc ores. The mining plant effluents were discharged directly into the Jintsu River. Eventually, the cadmium-containing sludge was deposited in the rice fields, which were irrigated by the river waters. The harvested rice contained more than 10 times the average trace metal concentration. This finding triggered a worldwide investigation of heavy metal pollution (see also figure 1.4) and stimulated a series of sketches of metal balances and biogeochemical cycles for different scales (see figure 2.21 for the global cadmium cycle).

Sweden was the first country to take concrete measures to ban cadmium as a component of any good in the 1970s and started a practical search for all the Cd-containing products. It was found in dyes, plastic stabilizers, NiCd batteries, corrosion protection, and many other products. More countries followed Sweden during the 1980s. It took another decade to make a first evaluation of the cadmium story (figure 2.29). For Sweden, a regional balance gave an overview of the stocks and flows for the period between 1940 and 1990 (figure 2.30).

The following lesson can be learned from the "cadmium story":

• Within 50 years of exploration, a by-product of zinc mining, originally not foreseen as a useful substance, finds a broad spectrum of technical applications. No cadmium flow control was installed. When, 40 years after its technical installation, the first toxic effects were proved, it was a very costly task to identify the sources, the intermediate stocks, and the sinks.

• On a global scale, the anthropogenic flow size of cadmium surpasses the geogenic one (figure 2.21). Because of the atmophilic property of cadmium and its handling in waste management (incineration), it is distributed around the globe via the atmosphere. Therefore, in some urban systems, the cadmium deposition was or is still 10-fold higher than that in remote areas.

• Because of cadmium bans for certain applications (e.g., dyes, stabilizers), the emissions could be reduced (figure 2.29). It must be underlined that there were substitutes available that could offer the same functions. Environmental laws and stricter control of waste outputs could prevent more poisonous and even deadly discharges. However, the cadmium stocks in the anthroposphere are still growing and have to be managed in a secure way to prevent further impacts in the environmental

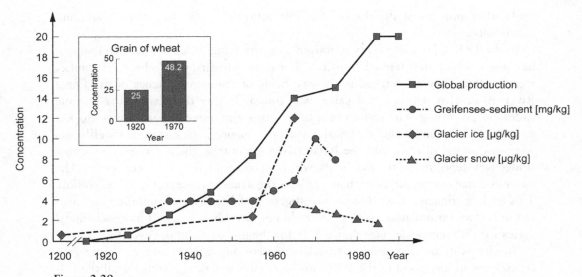

Figure 2.29
Man-made cadmium production and cadmium concentrations in the environment as a function of time (reprinted with permission from Fiedler & Rösler, 1993).

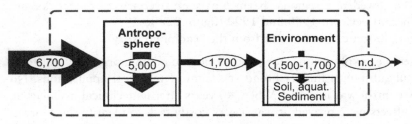

Figure 2.30
Cadmium balance of Sweden between 1940 and 1990 (reprinted with permission from Bergbäck, Anderberg & Lohm, 1994).

compartments. The Swedish balance for the period 1940–1990 shows (figure 2.30) that about 75% of the cadmium input went into the anthropogenic stocks. To manage this stock, one has to know in which goods and where the cadmium is stored and when it is to leave the anthroposphere ("residence time of the good").

The case studies with CO_2 and Cd, at first sight two completely different stories, have the following aspects in common:

• When new technologies (e.g., engines driven with fossil energy) or new substance applications are introduced (cadmium compounds) and eventually launched in mass production worldwide, the resulting metabolic processes were and still are not accompanied by scientific investigations. Phenomena such as an increase of the CO_2 content in the atmosphere or the adsorption of cadmium by rice plants from contaminated irrigation water were not taken into account as a possible metabolic scenario, although the scientific knowledge at that time was already available to do so.

• In both cases (and in many more), only the worldwide scandalized human wastes (pollution of "our climate" or of "our food") could trigger societies to adapt their science and economic policy. However, political and economic institutions are inclined to challenge scientific communities first to prove that the wastes could have possibly negative effects.

• Although the principle of precaution is part of every environmental law, the polluter–payer principle has had so far a negative rather than a positive feedback. It is still cheaper to waste first and pay, if necessary, later. Later means, in both cases shown above, the next generations. The concept of an urban infrastructure run by fossil energy was converted into practice in the first half of the twentieth century. The same is true for the application of cadmium. The great-grandchildren of the innovators have to handle the wastes and clean up the globe. The logical conclusion of this interpretation of the polluter–payer principle would be that the descendants of the developed countries—if they are still wealthy enough—with the highest resource consumption and the highest material stocks per capita should pay the bill for cleaning planet Earth.

The Great Risk of the Developed Countries

Small islands are helpful models to illustrate the essentials of a societal metabolic problem and its way of failing or of solving it. Easter Island was taken as a historical example of a subsistence economy that failed. At the beginning of the twenty-first century, the island of Samsø is about to minimize the great risk of the developed countries. This small Danish island (114 km^2) with 4300 inhabitants (2009) lies in the Kattegat Sea between Sweden to the east and the Danish mainland to the west. In 1997, Samsø won a government competition to become a renewable-energy community. At the time, the island was based entirely on oil and coal, all imported from the mainland. Ten years later, the community's electricity stems 100% from its own offshore and inland wind turbines (a total of 21 units), and 75% of its heat is produced within a central straw incinerator plant and distributed into the homes and workplaces. The straw is produced in its own agriculture. Some vehicles are already running with biofuel, also produced on the island. Contrary to Easter Island, Samsø is an open system. The technological transformation from a fossil fuel–driven system, depending on a global energy market, to a regional or even local renewable energy market was only possible by importing know-how and technological equipment from a national and global market. However, it needed the political will of the community to make this step with own efforts and investments. The main motivation was obviously the conviction to adapt first its energy household to minimize in time the risk of its own developed society. The community created neither new insights nor new technologies. It did not reduce its standard of life. All it did and still does, step by step, is pursue a new design of the community metabolism.

The Resource Management Perspectives on a Global Scale

After the Club of Rome (1972) (see chapter 1) and U.S. President Carter's Global 2000 (Global 2000 & Barney, 1980) at the end of the 1970s, the main resource reservoirs of the earth, the perspectives of population growth, and the resource consumption rates were known within the accuracy of the correct order of magnitude. Since then, a series of calculations has been made to quantify the timescales for the availability of various resources. In particular, the USGS (United States Geological Survey) periodically issues information about the reserve base of many important minerals. To reach a "physiological status of sustainability," the system has to measure its resource demand on the "global scale of scarcity." There are several studies available to illustrate the "ecological footprint" of large urban systems (see, e.g., *Action Plan Sustainable Netherlands* by Buitenkamp, Venner & Wams [1993] and *The Ecological Footprint of Vienna* by Daxbeck, Kisliakova & Obernosterer [2001]). An illustration is given in table 2.6. As with the territories, a globalization of the urban system would not reduce the available productive area for agriculture and forestry by more than a few percent. If the forestry management is limited to the current timber demand of urban systems, the global reservoir could satisfy the annual need by its annual increment of 1% to 2% of the stock. Even if there are still uncertainties about the size of fossil energy reservoirs, the oil example illustrates that a globalization of the urban system type would not be possible for 8 billion people, assuming that all of them should reach the consumption rate of the present DC population. The copper example illustrates the following phenomenon: The theoretical copper ore reservoir in the earth's crust per capita for 8 billion people is equal to the per capita stock of copper metal already accumulated in the developed urban system. It is obvious that urban systems are about to become secondary mining sites for copper. In conclusion, the following three hypotheses can be stated:

1. There is enough territory "on reserve" to continue the urban growth (1 m^2 per capita and year for the next 50 years). It is not a question of size but of design. The actual design of the "cultural landscape" decreases biodiversity (The United Nations Conference on Environment and Development in Rio de Janeiro 1992 with its declaration to save biodiversity, one important indicator for sustainable development). The opposite was the case with earlier agricultural landscapes that increased

Table 2.6
A selection of estimated resource reservoirs per capita for a world population of 8 billion compared with corresponding stocks and actual consumption rates in developed urban systems

Territories	Global Reservoirs for 8 Billion People	Stocks in Developed Urban Systems	Consumption of Developed Urban Systems
	ha/capita	ha/capita	ha/capita and year
Agriculture	0.5		
Forestry	0.3		
Settlement		0.3	0.0001
	m^3/capita	m^3/capita	m^3/capita and year
Timber	50	10	0.4
	GJ/capita	GJ/capita	GJ/capita and year
Oil	800	40	100
	kg/capita	kg/capita	kg/capita and year
Copper	300	300	10

Note: The data, presented as orders of magnitude, are based on published resource estimations by Global 2000 (Global 2000 & Barney, 1980), Buitenkamp, Venner, and Wams (1993), and Zeltner et al. (1999).

biodiversity. Urbanization does not necessarily mean that biodiversity is decreasing (Sukopp, 1990).

2. The dependence on nonrenewable energy (fossil fuels) is the most critical physiologic aspect concerning the limited reservoirs and the effects on climate change, followed by the freshwater shortage for crop production in some regions.

3. The rate-determining step is the reconstruction of the "built anthroposphere" (not a specific technology innovation) with a drastic improvement of its ecological quality.

The Essential Mass Resources in the Development of Urban Regions

To meet the ecological criteria of a sustainable development, the metabolism of urban systems has to be evaluated with regard to the regional and global stocks. In the following, the metabolism of a typical urban system is exemplified with the Swiss Lowland region (Baccini, 1997). In this region, a transportation network was built within 40 years (1950–1990) allowing practically every inhabitant, independent of his place of residence, to reach any urban activity within half an hour. The region has become a "compact" urban system (5 million inhabitants, population density 500 capita per km^2). The forest area stays constant, due to a forest conservation law. The agricultural land is constantly reduced but increases steadily its productivity. Neither territorial planning nor environmental protection measurements (since 1970) have influenced significantly the growth of the urban settlement area and its metabolic rate (consumption of joule and kilogram per capita or km^2 and year). There is enough territory "on reserve" to continue this regional growth (3 m^2 per capita and year for the next 20–30 years).

From a physiologic point of view, the survival of urban regions increasingly depends on a continental or even global hinterland. Whenever an urban region expands, a corresponding resource area must be found "on the global market."

Three working hypotheses are chosen for evaluating sustainability of urban systems (Baccini, 1997). The metabolism of an urban system is "sustainable" if the following conditions hold:

1. The demand for essential "mass goods" such as water, biomass, construction materials (e.g., stones), and energy carriers can be satisfied autochthonously by more than 80% in the long term. The degree of self-sufficiency, arbitrarily chosen with respect to a set of essential goods, determines the ecologically defined border of the urban system.

2. The remaining needs can be covered from the "external market" in such a way that the global resource stock is not reduced significantly. The mean flows of available goods per capita from the external market are determined by the annual growth rate of the "global hinterland" divided by the world population.

3. The outputs (emissions) are no burden for future generations.

No environmental system (hydrosphere, atmosphere, lithosphere) may be used as a sink for anthropogenic flows in such a way that the receiving system or parts of it becomes a "hazardous site" for the biosphere (see also the case study on designing waste management in chapter 5).

Four essential mass goods of the urban metabolism (i.e., water, biomass, construction materials, and energy carriers) were investigated in a region of the Swiss Lowlands (Baccini & Oswald, 1998). The results reflect the actual situation in a quasi steady-state situation. The following conclusions can be drawn:

• There are quantitative and qualitative indications that the use of groundwater does not meet the criteria of "sustainability."

• There is a large disparity between the ecological potential and the actual economic value of the two subsystems "agriculture" and "forestry." Within the urban system, agriculture and forestry have become a new type of commons (due to the high degree of subsidy) whose role has to be newly defined under the aspects of sustainability.

• Currently, the management system of construction materials exhibits a low efficiency of the material stock representing the infrastructure and a high dependence on nonrenewable resources. However, enough alternatives are available on a long-term scale.

• With regard to energy, the region strongly depends on imported fossil fuels. With respect to the criteria of regional sustainability, a new setup of energy transformation represents the greatest challenge. It is related to the transformation of the settlement area. A carbon flow analysis for a typical urban region is given in figure 2.31. The process "Settlement," using only about 12% of the total area of the region, is responsible for the predominant carbon flow (2000 kg carbon per capita and year, emitting about 7 Mg CO_2 per capita and year from nonrenewable carbon sources). The input stems mainly (90%) from imported fossil fuels. Approximately 60% of the demand for food is covered by the regional agriculture. The rest has to be imported. The carbon turnover of the process "Agriculture" (covering 55% of the regional area) is three to four times higher than the one in "Forestry" (33% of the total area). Their combined net contribution to the target process "Settlement" is only about 10% of the total. This is mainly due to the relatively high meat

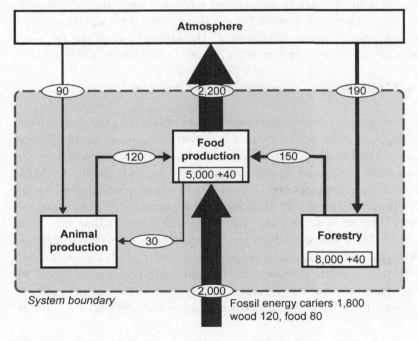

Figure 2.31
Carbon flow in an urban system (Swiss Lowland region; after Müller, Oehler & Baccini, 1995). Flow units in kg carbon per capita and year; stock units in kg per capita. "Agriculture" covers 55%, "Forestry" 33%, and the "Settlement" 12% of the total area.

production and consumption. The largest carbon stocks are found in the soils of the latter two processes. However, the concentration of the organic carbon in soils is relatively low (a few percent). The most concentrated forms are in the forest trees and in the wooden goods of the buildings. Both are in the same order of magnitude (10^3 kg/capita).

From a physiologic point of view, a DC urban system of the type "twentieth century," exemplified with the Swiss Lowlands, has the following main characteristics: The settlement stock and the type of energy transformation are the key factors to gain a "status of sustainability." The other three mass goods (water, biomass, and construction materials) are less critical for the region studied (i.e., less effort would have to be applied to reach their desired status in material management). From an engineering point of view, it follows that a transformation to a "sustainable status" is only possible by a reconstruction of the urban system (i.e., buildings and transportation network). In other words, it is not sufficient to construct only "new ecological cities" as additives to the existing stocks. However, the process of reconstruction is a cultural one and has to include all relevant properties of human society; namely, the political, socioeconomic, and ecological ones. It is a new challenge never encountered before in history. Former generations were mainly oriented toward growth processes in urban development. The greatest risk of the DC is to stay within a strategy sustaining the metabolic pattern of the current urban system.

An Overview of Methodological Approaches for Metabolic Studies

There are several approaches for investigations into and evaluations of metabolism. In this volume, MFA as described in chapter 3 is used to examine metabolic processes in the context of human activities, and MFA-based assessment methods are applied for evaluation, optimization, and design. Before focusing mainly on MFA and the activity concept, contemporary approaches to assess material turnover of societies are briefly summarized. The *total material requirement* (TMR) indicator (cf. Bringezu, 1997) is one of several aggregated physical measures that have been proposed as a metric for resource productivity and overall pressures on the environment. Other examples include (1) the "ecological footprint" and (2) the *sustainable process index* (SPI) that relate human activities and corresponding processes to the total area of land, including the hinterland that is utilized by a society, and (3) *life cycle assessment* (LCA) as a tool to assess environmental impacts of a good or a process.

TMR expresses the total mass of primary materials (without water and air) extracted from nature to support human activities. It is a highly aggregated indicator for the material turnover of an economy and is used to compare different economies. The TMR indicator includes materials used for processing (direct material input;

DMI) as well as "hidden flows" in the region and the hinterland that occur while resources are being mined and further processed such as tailings and wastes from primary industries inside and outside the system's boundaries. Data for TMR determination are collected from national statistics on industrial production, agriculture, forestry, and fisheries, as well as from foreign trade statistics for imported goods and specific information on hidden flows.

According to the authors, TMR indicates a generic pressure on the environment because the volume of resource requirements determines in general the scale of environmental disturbances by extraction as well as the throughput of the economy (DMI) and subsequent amounts of emissions and wastes. However, TMR does not indicate specific and localized environmental pressures and thus lacks operational power. Because TMR does not include water and air, it is difficult to apply the mass balance principle to verify the accumulated data.

Another aggregated measure relating human activities to the earth's ecosystems is the "ecological footprint." It compares human demand for materials and energy with planet Earth's ecological capacity to regenerate, more specifically the amount of area at land and sea needed to regenerate the resources a society consumes and to absorb the wastes produced. The approach is designed to assess how many "planets Earth" it would take to support mankind with a given lifestyle. For 2006, it was estimated that about 1.4 planets would be needed, indicating that the current lifestyle is not sustainable for longer time periods (Global Footprint Network, 2010).

A similar but more detailed and specific aggregated indicator is the SPI, which was developed to evaluate if given processes fulfill sustainable economic conditions (Narodoslawsky & Krotscheck, 1995). The concept is based on the assumption that a sustainable society can only be established on the flow of solar exergy. As for the ecological footprint, the ultimate unit is area, because the amount of solar radiation that can be harvested on Earth depends on the surface of the earth. Thus, the authors conclude that area is limiting any sustainable economy. The SPI is a measure for the area needed to provide raw materials and energy and to accommodate by-products, emissions, and wastes from a process. As a result, the per capita ratio between the area needed to supply a specific service and the area needed to supply all possible services is obtained. The data for calculating the SPI originates from process development. The SPI is as well suited as the ecological footprint for proficient visualization and communication of the results, but it has the advantage of being able to include all the necessary goods and substances that are relevant for a comprehensive evaluation of a process or system. The SPI has been applied to evaluate technological as well as organizational changes in processes and in consumption pattern.

LCA is a widely applied, standardized method to investigate the environmental impacts of goods, processes, and systems (ISO, 2006). The objective is to assess all environmentally relevant steps of a product or process in a life cycle perspective that is from cradle to grave (or recycling) in view of environmental loadings and resource consumption. An LCA comprises a life cycle inventory analysis followed by a life cycle impact assessment. Often, LCAs are used to compare and optimize the environmental performance of products and processes. Sometimes, LCAs are also performed on entire systems such as in a waste management scenario analysis.

All methods summarized above have their specific objectives, such as evaluation of environmental impacts, of resource consumption, of sustainability metrics, and so forth. However, they do not contribute to the understanding of metabolic processes in a way that allows designing the necessary policy measures on an operational level. For this purpose, anthropogenic systems that include metabolic systems as integrated parts are needed. Whereas data from LCA and TMR studies are accumulating fast, systematic work in the field of the metabolism of the anthroposphere is still rare. In the emerging science of industrial ecology (Ehrenfeld, 2009), authors are beginning to emphasize the need for new insights into and methods for the analysis, evaluation, and design of complex anthropogenic systems.

3

Analysis and Assessment of Metabolic Processes

A terminology and methodology for describing and evaluating the metabolism of the anthroposphere is proposed based on a classification of anthropogenic activities in the context of cultural values. Material flow analysis (MFA) is presented in detail as a base for all methods that are applied to assess anthropogenic systems. Examples of how to establish MFA and substance flow analysis (SFA) are given, and models and software products that support MFA/SFA are discussed. This chapter prepares us for the design of metabolic systems by answering the following questions:

- How to structure the anthroposphere for effective assessment and design of metabolic systems?
- How to choose a system for a given metabolic problem?
- Which of the system's properties do we have to measure and how do we measure them?
- What is to be calculated from the data gained?
- How do we interpret the resulting quantified metabolic system?

Structuring with Activities

The anthroposphere is a biological and cultural system. It comprises all human activities and the thereby integrated or connected geogenic entities. From the point of view of the natural sciences, all metabolic processes within this sphere have to follow the natural laws. The installation and the management of these processes (e.g., water supply, energy transformation, building construction and maintenance, transport infrastructure) are functions of the cultural properties of a society. These properties, their changes in time and space, are objects of the humanities and the social sciences. Therefore, it is evident that metabolic studies are to be developed in a transdisciplinary procedure (Hirsch-Hadorn et al., 2008).

Although the focus is on metabolic patterns (i.e., on physical, chemical and biological properties), the structure of the study object should allow linking it to

Table 3.1
Classification of cultural activities, corresponding phenomena, and scientific methods

Cultural Activity	Phenomena (Institutions, Materials)	Associated Scientific Methods
A	Family raising, education, art, religious service, military service	Friendship, marriage, language, music, church, war	Psychology, sociology, philosophy, theology to elucidate the origin of ethics and moral standards
B	Production of goods, trade, and consumption	Steel, buildings, cars, food, computers, cows	Engineering and economic sciences to construct, produce, and elucidate market mechanisms
C	Exploitation of resources to obtain primary materials and energy from the environment	Sun, water, air, soil, stones, forest, grass	Natural sciences to determine the laws that govern processes in natural systems

cultural properties. In table 3.1, three groups of cultural properties (A, B, C) are characterized with either social institutions or things (phenomenology) and with the scientific disciplines man has developed to understand and to control his own and nature's activities.

Such a simplified classification makes it immediately clear that a strict distinction of "things" in groups B and C, separate from A, is not correct. The reasons for building pyramids, cathedrals, or nuclear power plants, using primary energy and stones and applying certain techniques and labor, cannot be understood with only the help of the disciplines in B and C. Many things have at least a triple function for human beings: the first as a biological necessity (e.g., water as essential nourishment and for hygiene), the second as a physical tool (e.g., water as economic media to store energy in artificial lakes, to transport persons and goods on ships, or to dispose of wastes in sewers), the third as a religious symbol for purity (e.g., water to baptize).

In the phenomenology of water flows through the anthroposphere, different values are combined in various ways. Many of them can only be elucidated by the disciplines in group A, which give answers to the following question: Why have humans with a certain cultural background developed various activities (e.g., to reside, to transport, to play)? Not all activities are coupled inevitably with only one material. Let us take again water to illustrate this statement. It is a biological necessity to have water in the activity TO NOURISH. It is not necessary to transport goods on or with water. There are or could be other means. In the combinations of the disciplines in groups B and C, the activity TO TRANSPORT as a biological and cultural performance can be described. In addition, the following question is raised: Given a set of environmental, technical, and economic boundary conditions, what are the most efficient combinations of materials, goods, and processes to maintain and develop the activity TO TRANSPORT?

Activities and Values

An initial effort toward structuring of the anthroposphere to follow its metabolic processes is the selection of a very limited set of activities, which are introduced in table 3.2. The four activities comprise the basic needs of any societal entity independent of its cultural construct in space and time. The choice of activities as stimulators of metabolic systems is a consequence of the insight to consider the needs of *Homo sapiens* as a biological and cultural being. Whereas the first two activities, TO NOURISH and TO CLEAN, emerge more strongly from the biologically driven needs, the other two, TO RESIDE&WORK and TO TRANSPORT&COMMUNICATE, are driven mainly by socioeconomic needs and trigger more strongly the cultural evolution of settlements and trade (see chapter 4 for more details). In other words, the choice of activities for structuring the anthroposphere is a result of a transdisciplinary approach. Activities should eventually allow combining the various scientific disciplines for a holistic approach to the anthroposphere. In chapters 4 and 5, some illustrations are given to show the potentials. Up to now, this methodological approach is still at the beginning.

At first sight, the restriction to only four activities seems to be arbitrary and too simple. However, practical experience shows (see also chapter 4) that further refinement increases the complexity of description of the anthropogenic metabolism. Therefore, the number of activities should be as small as possible and as large as necessary. The values of a society are reflected in its setup of metabolic systems. Two examples are provided: (1) Religious beliefs codetermine the choice of nourishment of the adherents. One could extend the set of activities by introducing an activity TO BELIEVE. A metabolic system for TO NOURISH reflects the influence of the religiously induced diet on the flows and stocks and can be compared with other TO BELIEVE scenarios. (2) Politically given rights to regulate access to physical resources, be it a system of kleptocracy, meritocracy, democracy, and so forth, codetermine the distribution of a given stock among the households of the individuals. A special activity TO GOVERN could be chosen inducing different metabolic systems. However, practice shows that there is less hardship working comparatively with the four basic activities.

Activities generate metabolic systems. Contrarily, a given set of metabolic boundary conditions stimulates societies to install a pattern of values to secure the optimal use of resources. Two examples to this statement are provided: (1) In the Netherlands, the dyke reeve had the highest social status because the proper functioning

Table 3.2
Four societal activities and their corresponding metabolic processes

Activity	Definition	Remarks
TO NOURISH	Comprises all processes and goods: 1. To produce solid and liquid food for human beings (includes hunting, gathering, agricultural production) and to distribute it to the consumers. 2. To consume (cooking, eating, drinking). 3. To release the wastes of digested residues.	In producing and distributing, the processes "food conservation" and "food storing" have become important.
TO CLEAN	Comprises all processes and goods: 1. To care for human health (hygiene). 2. To maintain the quality (aesthetics and functioning) of any good. 3. To provide environmental protection (waste treatment and management).	"Cleaning" with water has become a dominant process such as washing human bodies, dishes, clothes, cars, windows, streets, and so forth.
TO RESIDE&WORK	Comprises all processes and goods: 1. To build residential units, work and recreation facilities. 2. To install and maintain all the equipment needed to run these facilities.	The separation of "to reside" and "to work" in the cultural evolution is relatively young. It is not applicable to every society.
TO TRANSPORT& COMMUNICATE	Comprises all processes and goods: 1. To transport persons and goods. 2. To transport information.	The processes range from road construction to education and administration. The goods range from railway tracks to cars, telephones, and personal computers.

of the water flow system was a matter of life and death for a community living below sea level. (2) In arid climates, hospitality gestures are initiated with the offer of drinking water, the most precious good because of its scarcity and a symbol of high esteem in welcoming a stranger.

Processes, Goods, and Substances

The notion *activity* asks for metabolic systems that consist of processes and goods. Processes and goods are linked together in a functional logic. In the anthroposphere, the linkage of process and goods is always driven by a human need (i.e., an activity). Goods are materials that are valued economically by trade. Goods consist of materials whose physical nature can be described by the chemical composition of the substances involved. A sheet of paper is a good that has a price. It is produced mainly for carrying information by print, handwriting, or drawing. In this function, it serves the activity TO TRANSPORT&COMMUNICATE. Paper is a material manufactured in a series of technical processes. Paper mainly consists of cellulose molecules (polymers extracted from wood) that are combined with small quantities of other substances (e.g., aluminum salts). On the level of substances, the material balance for the involved chemical elements for any process must obey the law of mass conservation. (The exemptions are processes with nuclear reactions.)

It follows that the structure of the metabolism of the anthroposphere is given by a series of activity-induced *material systems*. These systems consist of linked processes and goods through which material is flowing. If such a material system is extended by energy flows, it is called a *metabolic system*. Furthermore, it can be extended by flows of money and information. In this case, we speak of an *anthropogenic system*.

In table 3.3, the set of five essential notions is presented with definitions, scientific disciplines involved, and methods to be applied. They are the basics for the crucial method in metabolic studies, namely the MFA, which is introduced in this chapter.

Designing Material Systems for Metabolic Studies

In table 3.4, a few examples illustrate the connections between the four notions applied in material systems. It becomes obvious (see also table 3.2) that a clear and distinct assignment of a process or a good to one and only one activity is rarely possible. For example, it is reasonable to assign a tractor on a farm to the activity TO NOURISH. However, its contribution to the overall transport of goods is not necessarily restricted to the farming of products. If the same tractor serves other purposes, it has to be classified also under the activity TO TRANSPORT&COMMUNICATE. Another example: The food chain includes the intermediate storage of cereals; that is, storage buildings are necessary and are quantified as goods under the activity TO RESIDE&WORK, and the treatment of

Table 3.3
Notions and definitions for material management systems

Notion	Definition	Disciplines	Methods
Activity	Deed of human beings to satisfy their needs, inducing material management systems	Biology, medicine, philosophy, history, social sciences	Empirical and hermeneutic
Material system	Open system consisting of connected processes and goods through which substances are flowing	Mathematics, computer sciences	Computer-aided mathematical models based on physical and chemical principles
Process	Transport, transformation, storage of goods and substances	Engineering, economy, biology	Material balances
Good	Substances and substance mixtures that have valued functions	Economy	Market research econometrics
Substance	Chemical elements and chemical compounds	Physics, chemistry	Physical and chemical analysis

human feces (at the end of the food chain) in the sewage is in the activity TO CLEAN. It follows that decisions have to be made with regard to the precise definition of activities in order to omit a double- or triple-counting of material flows and stocks. This aspect will be illustrated in chapter 4.

The few examples, chosen to illustrate the relations between the four basic notions, make clear that affluent societies have a huge number and a very large variety of processes (about 10^4 to 10^5), goods (10^5 to 10^6), and substances (10^4 to 10^5) in use. At first sight, the resulting complexity of the anthropogenic metabolism seems to be out of reach for an analysis and consequently out of reach of an evaluation of the relevant characteristics of large-sized material management systems. The examples given in the section entitled "The Metabolism of a Region" of chapter 2 illustrate the strength of chemical element balances (phosphorus and lead). They serve as "indicator substances" and give, in a comparative presentation of their MFA, an initial overview of the relevant characteristics of a process and/or of a regional material management system and/or the metabolic role of a certain good. In table 3.5, a selection of 10 indicator substances is presented and commented on with regard to their indication potentials. The listed indications and the impact potentials are of exemplary character and do not cover, in an encyclopedic manner, the whole spectra of properties.

Table 3.4
Illustration of connected notions in material and metabolic systems

Activity	Process	Good	Substance
TO NOURISH	Cow	Grass	Nitrogen
		Water	Phosphorus
		Milk	Protein
		Fertilizer	
	Cooking	Stove	Iron
		Pan	Glass
		Broccoli	Vitamin
		Electricity	Copper
TO CLEAN	Wash machine	Shirt	Cotton
		Water	Water
		Detergent	Fatty acid ester
	Sewage treatment plant	Sewage water	Water
		Feces	Urea
		Sewage sludge	Zinc
TO RESIDE&WORK	Carpentry	Timber	Lignin
		Chair	Iron
	Cement manufacture	Limestone	Calcium
		Sandstone	Silicon
		Oil	Hydrocarbon
TO TRANSPORT& COMMUNICATE	Car driving	Automobile	Aluminum
		Gasoline	Polyester
	Telephoning	Mobile phone	Polyamide
		Battery	Nickel
			Cadmium

With the set of the five notions and the concept of their relations within a metabolic system, any metabolic property of the anthroposphere can be qualified and quantified. This statement, based on a large number of practical studies, is illustrated in the following with paper.

Exemplifying with Paper and its Metabolic Idiosyncrasies
The term *paper* has a long history beginning in antiquity when the papyrus plant was used as writing material. The name *paper* stayed in use although the material changed. Furthermore, it became a term for a diplomatic note and a scientific publication. The cellulose-based paper could conquer new applications in the hygiene market. One has to know the cultural subtleties in order to understand the meaning of the following letter to a journalist, which was written by an actor who was not amused by the criticisms he received: "Dear Sir, I sit in the smallest room of my

Table 3.5
Selection of indicator substances

Element	Symbol	Indications
Carbon	C	Carrier of chemical energy Carrier of nutrients Carrier of toxic substances
Nitrogen	N	As NO_3 an essential nutrient As NO_2 a potential air polluter
Fluorine	F	As F^- a strong inorganic ligand In incineration Hf is formed, a strong acid
Phosphorus	P	As PO_4^{3-} an essential nutrient Eutrophication of aquatic ecosystems
Chlorine	Cl	Forms as Cl^- very soluble salts Forms with organic molecules –C–Cl– bonds, leading to substances that can be very stable (e.g., Polychlorinated Byphenyls PCB) and toxic (e.g., dioxins and furans)
Iron	Fe	Forms as Fe^{3+} poorly soluble oxides and hydroxides As metal easily oxidized under atmospheric conditions (H_2O and CO_2) Recycling of metallic iron is economical
Copper	Cu	Forms as Cu^{2+} stable complexes with organic ligands As metal an important electrical conductor In small concentrations already toxic for unicellular organisms
Zinc	Zn	Forms as Zn^{2+} soluble salts Quantitatively important as anticorrosive and rubber additive
Cadmium	Cd	In many consumer goods as additive Nonessential element Toxic in higher concentrations
Mercury	Hg	Forms metallorganic and toxic compounds under reduced conditions Atmophilic element
Lead	Pb	Forms as Pb^{2+} very stable complexes with natural organic ligands Lithophilic element Was and still is an important additive in gasoline

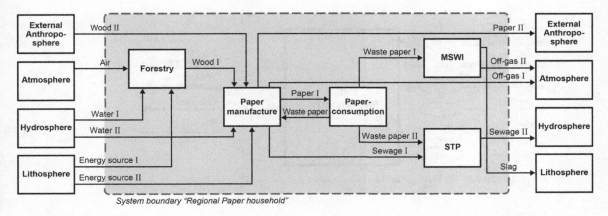

Figure 3.1
Regional paper management system (after Baccini & Bader, 1996).

house and have your paper before me. Very soon I shall have it behind me." It is a game with words (1) naming with the same material (paper) two goods in communication and cleaning, (2) moving the material from a process of information to a process for body cleaning and by doing so (3) lowering the value of the journalist's work to zero.

Two activities are mainly responsible for generating the "good" paper, namely to TRANSPORT&COMMUNICATE and TO CLEAN. For a regional scale, a system with five processes and 18 goods can be sketched to make a first approximation of the corresponding paper material system (figure 3.1). On the production side, the processes *Forestry* (wood production) and *Paper Manufacture* are connected in sequence to produce paper (for printing, packaging, and hygiene) of which a portion can be sold within the region and the rest is exported. Here, a first simplification is observed, namely the assumption that the regional consumers do not buy paper from extraregional sources. However, wood sources from external sources are considered. Paper consumption leads to two types of waste papers. The first (printed matter and packaging material) is transferred to the municipal solid waste and treated in a waste incineration plant (MSWI). The second is hygiene paper, which goes via sewage into a sewage treatment plant (STP). Incineration and sewage treatment plants belong to the activity TO CLEAN (see chapter 4). This is noteworthy because these installations, on a regional scale, are laid out for a complex mixture of substances, not exclusively for paper. The only recycling path within this system is the flow of used paper, from the process *Paper Consumption* (by separate collection) back to the process *Paper Manufacture*. This process, within the system, seems to have a key role, as it works with a variety of mass resources such as biomass, water, energy carriers, and recycled paper.

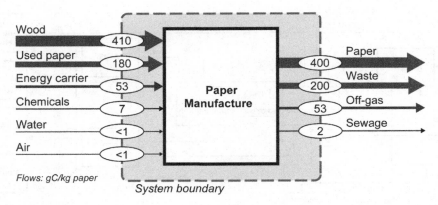

Figure 3.2
Carbon flow in the process "paper manufacture," in g C/kg paper (after Baccini & Bader, 1996).

An indispensable indicator substance would be carbon. Its flows for the process *Paper Manufacture* are given in figure 3.2. The relatively simple balance shows that the second-highest C flow of 200 g per kg paper is the solid waste, consisting of timber residues. It could serve as an energy carrier and replace the primary energy input with hydrocarbon from oil. Today, this is realized in most paper factories. It can also be postulated that use of more used paper instead of wood will mean less energy need, because used paper brings the already separated cellulose fibers into the process and does not have to pass an extraction process as does wood. Are such rough estimations of material and energy flows per mass unit of consumer goods relevant? The paper consumption per capita and year lies between 200 and 400 kg in developed countries and approximately 10 to 100 times less in developing countries. At present, the world average lies around 50 kg. For developed countries, the energy to produce the demanded paper flow is approximately 3 GJ per capita and year, or between 1% and 2% of the total energy demand. The water demand for the paper consumption is about 3 m^3 per capita and year, again a share of 1% to 2% of the total. The share of the total wood consumption lies between 10% and 20%. It follows that paper consumption with respect to the mass resources energy and water is not decisive, however is not negligible. With regard to the process *Forestry* (i.e., an important producer of renewable resources), paper plays an important role in the metabolic game of solar energy.

In summary, the metabolic phenomena of the anthroposphere show a broad variety of characteristics, on time and spatial scales differing by orders of magnitude. Technical processes, goods, and substances are interwoven in a complex way, obeying natural laws and socioeconomic rules. There are various methodological concepts and tools to grasp metabolic properties. In this volume, the method chosen to elucidate the essentials of the metabolism of the anthroposphere is based on the "activity approach." With a small set of four additional properties, namely *process*, *good*, *substance*, and *material system*, the main components of the structure of the anthroposphere are given. The example with paper given above helps to trigger the methodological questions stated at the beginning of this chapter. The answers are given in the sections that follow. The applications on a broad scale are shown in chapters 4 and 5.

Quantification and Assessment Using MFA

At present, there is a highly elaborated set of tools available to follow particular aspects of anthropogenic activities, mainly in the field of economics. An often used example is Leontief's input–output model, representing a national or regional economy as a matrix that links the main actors of the economy (Leontief, 1986). The set of life cycle analysis tools represents a methodology that is applied widely to evaluate the interaction of the anthroposphere with the environment. Methods for urban planning allow analyzing and optimizing challenges relating to transportation, housing, recreation, and the like. However, means to assess metabolic processes in a comprehensive and holistic way, linking the necessary disciplines in order to address all relevant issues in a common approach, are still scarce and just beginning to be developed (Oswald, Baccini & Michaeli, 2003). A multidisciplinary approach beyond natural and engineering science is necessary because decisions regarding metabolic processes are heavily influenced by social science issues, too. Such decisions affect all spheres and human society as a whole: households, communities, industry, and agriculture are concerned as well as environment and resource base.

In this chapter, the main emphasis is laid on MFA. The reasons for emphasizing MFA in this chapter are as follows: (1) flows and stocks of materials and energy are an objective, indispensable fundament of every society enabling human activities; (2) the three topics anthroposphere, activity concept, and MFA belong together allowing a comprehensive description, understanding, and comparison of metabolic systems; and (3) most assessment methods such as life cycle assessment (LCA), sustainable process index (SPI), and so forth, require a sound knowledge of the material base. Hence, it is important to acquire reproducible data about sources, pathways, and sinks of materials with known uncertainties that are derived from transparent methods with defined system boundaries.

MFA Methodology

MFA/SFA as described later is a consistent methodology that can—and should—form the base for any evaluation method. MFA/SFA is also value laden, as it requires subjective decisions about the selection of system boundaries, processes, goods, and substances that can deviate from one analyst to another. The main advantage of

MFA/SFA systems is that they must obey the law of conservation of matter, and thus they form a natural science base that can be cross-checked for consistency and probability.

As presented in chapter 2, modern societies have a high per capita turnover using a large variety of materials to fulfill their needs. It can be roughly estimated that about 10^5 to 10^6 different goods, 10^4 to 10^5 different chemical substances, and about 10^4 to 10^5 different technical processes are used to maintain today's metabolism (Baccini & Bader, 1996). To handle such a large amount of information seems to be a Herculean task, particularly when it is considered that the data sources are scattered and that numbers are often not available and—if available—changing rapidly. However, the task becomes less extraordinary when the purpose of data collection is taken into account: In most cases, individual, clearly defined metabolic issues are to be addressed, involving a limited range of materials and processes.

To structure and analyze the material part of the anthroposphere, a consistent, rigid, and transparent methodology is needed. The methodology must enable the user to focus on all material aspects relating to the anthroposphere, such as economic aspects, resource issues, and environmental considerations. It is thus necessary to distinguish between two kinds of "materials" introduced in table 3.3: (1) "goods," which stand for economic entities with an economic (positive or negative, in rare cases zero) value, and (2) "substances," which are defined as chemical elements or compounds, consisting either of uniform atoms (elements) or uniform molecules (compounds). To include both goods and substances is of primordial importance: Substances define the properties and functions of goods and thus control the value of goods. For instance, the function of a group of substances called brominated flame retardants is to prevent electronic devices such as radios or computers from catching fire. Phthalates [esters of phthalic acids such as di-2-ethyl hexyl phthalate (DEHP)] are highly efficient in softening plastic goods made of polyvinyl chloride (PVC). The price of scrap iron is determined, among others, by the content of nickel and copper: nickel increases and copper decreases the value. The content of metals or sulfur in fuel is important for energy generation: Fuel concentrated in heavy metals and sulfur requires advanced air pollution control devices, making energy generation comparatively expensive. The costs for treatment of the good "wastewater" depend on the concentration of the substances phosphorous, nitrogen, organic substances, and metals.

Information about goods is highly abundant: Because of their economic value, goods are contained in economic statistics and databases of enterprises, regions, nations, and even on a global level. To establish a Leontief input–output table for the flow of goods of a nation has become a routine technique because of the readily available economic data (a specific problem is that the units in economic statistics usually are currency units, whereas for studying metabolic processes, units such as mass or mass flows are needed, too). In fact, some authors make use of the abundant economic data and combine input–output techniques with MFA, for example, for waste input–output material flow analysis (WIO-MFA) (Nakamura & Kondo, 2009) or for linking economic systems with the environment (Xu, 2010).

On the other side, information about substances is scarce: To establish a national input–output analysis of lead or hexachlorobenzene is still a laborious task, and the results are associated with high uncertainties. In contrast to goods, most substances are invisible: When in the 1970s consumers bought floor liners made of PVC, they did not realize that PVC was stabilized by cadmium, which needs special care when the floor liner will be recycled in the future. And when consumers replace floor liners today, they most likely neither know if it is made of PVC nor if it contains zinc or any other metal that is used as a replacement for toxic cadmium. Both levels, goods and substances, require their individual set of toolboxes for analysis. To work on the level of substances is more challenging because of the fact that modern goods are complex mixtures of many substances that are not trivial to analyze. Procedures for representative sampling, sample pretreatment, and laboratory analysis are labor intensive and costly, hence other methods are sought for more cost-effective substance analysis. Examples for determining substance concentration in wastes are given in Brunner and Ernst (1986) and Morf and Brunner (1998).

It is important to note how crucial the level of substances is for decisions regarding metabolic processes: In most cases, the level of goods does not allow drawing conclusions about resource needs or environmental impacts without considering the substance level. When using coal, it is necessary to know how much carbon, sulfur, and mercury is contained in this coal, otherwise a coal-fired power plant cannot be designed and operated in an efficient and environmentally acceptable manner. For iron recycling (see also figure 4.12), the amount of copper and nickel in iron scrap determines the value of the scrap. When sustainable agriculture is discussed, a focus on the level of goods falls short without considering the level of substances (e.g., the nutrients phosphorous and nitrogen). Thus, a methodology to address metabolic processes must be able to take the two levels of goods and substances into account. This implies that information from two different disciplines must be used and merged, on the one hand from econometrics and statistics, and on the other hand from the field of chemistry. This challenging interdisciplinarity is a precondition for successful metabolic studies.

A highly useful feature of materials on both levels, goods and substances, is that they obey the law of conservation of mass, as stated by Antoine Lavoisier in 1789 (see also chapter 2). Mass cannot "disappear"; the mass of a closed system will remain constant over time. This allows checking the validity of a material flow system: All inputs into the system must be represented either by the outputs or by the changes in stock. In particular, for tracing materials on the substance level, the mass conservation principle is an excellent means to check if an observed system is incomplete.

Terms and Definitions

Substance

In general, a *substance* is defined as the physical matter of which a good consists. Substances are characterized by a unique and identical constitution and are homogenous. Chemists define substances as matter that is composed of uniform units. Because in chemistry there are two types of basic uniform units, namely atoms and molecules, *a substance is defined either as an element—which is by definition composed of the same kind of atoms—or as a chemical compound containing the same kind of molecules.* Examples of substances on the level of elements are the members of the periodic table of elements, such as hydrogen, carbon, iron, or selenium. Examples of substances on the level of compounds are carbon dioxide (CO_2), ammonia (NH_3), hexachlorobenzene (C_6Cl_6), cellulose (polycellobiose, $C_6H_{10}O_5$), iron chloride ($FeCl_2$), or calcium carbonate ($CaCO_3$, calcite or aragonite).

Often, a substance is meant to have a certain shape. The definition of substances as used here does not require this. A substance can be a trace within a matrix (e.g., cadmium as a pigment in a PVC polymer). For tracking a substance through a system, the physical form of the substance is usually of secondary importance. Hence, the definition of substances as used for SFA excludes shape.

The advantage of working on the level of elements is that they are "conservative": They cannot be destroyed by physical–chemical means (with the exception of nuclear reactions) and can easily be balanced if means for determination (e.g., methods for sampling and analysis or econometrics) are available. For compounds, this is different: Molecules can be transformed from one compound into a different one. If such transformations take place, mass balancing on the compound level is only possible as long as the molecule remains unchanged. If transformed, the mass flow ends in a sink, the "transformation process." An example is given in chapter 4 when the activity TO CLEAN is exemplified by a pathway analysis of the compound nonylphenol polyethoxylate.

Isotopes (atoms with varying amounts of neutrons) and ions (atoms or molecules with varying amounts of electrons) are usually not further differentiated in the definition of substances. In a carbon balance of a region, all carbon is included, not discriminating between carbon 12 or carbon 13 isotopes with 6 respectively 7 neutrons. However, MFA methodology can be applied to isotopes as well.

The number of substances on the level of elements is rather small; at present, the periodic table comprises about 110 elements. In contrast, the number of known substances on the molecular level amounts to about 10 million chemical compounds. Thus, whereas it seems possible to establish material balances of all elements (e.g., for a regional or national substance flow system), the task to determine comprehensive balances for all chemical compounds is out of reach because of the unacceptably high amount of resources required for such an undertaking.

SFA or substance flow and stock analysis designates the flows and stocks of a substance during the pathway through a predefined system. Other terms for such an analysis are *pathway analysis* or *mass balance analysis*. Typical applications are to quantify flows of potentially hazardous substances to the environment; to determine hazardous accumulations or depletions in specific material stocks such as soils or indoor environments; to reduce resource consumption by establishing a recycling system; and so forth. For these purposes, often mass flows of substances such as nutrients (C, N, P), heavy metals (Fe, Cu, Zn, Pb), or pesticides (hexachlorocyclohexane) are investigated.

In the general literature, the term *substance* is not always used in a strict and rigid manner as defined by chemists and as used here. Often, it is applied to materials that have similar specific properties such as color, density, solubility, or specific functions such as packaging materials, construction materials, and so forth. To be able to compare different studies, to foster transparency, and to communicate results in an impartial way, it is necessary to use general terms and to bring the different terminologies together.

Good

Economists define a *good* as a physical product that is useful to satisfy some desire or need. In MFA, goods are defined as economic entities of matter with a positive or negative economic value that comprise one or several substances. Goods can be contrasted with services, which are intangible, whereas a good is a tangible product with mass. More general terms used by economists for goods are *products*, *merchandise*, or *commodities*. Commonly, a product stands for the output but not the input of a process. Chemists use the notion "educt" to label the input of a process. Merchandise and commodity usually characterize goods with a positive economic value. Rarely, goods have a negative value like in feces or sewage sludge.

Examples for goods are steel, drinking water, copper ore, PVC, mobile phones, cars, municipal solid waste, and air. All goods mentioned contain a variety of different substances. Occasionally a good can be identical with a substance (e.g., the substance "pure gold" containing only gold atoms is also traded as a good). Goods can be measured in terms of turnover of value or mass flow. In rare specific cases, goods lack economic value, such as air for breathing or precipitation. Statistics are available on several levels that present detailed information about thousands of flows and stocks of goods: market research supplies details about the turnover of consumer goods, government statistics supply information about waste generation, the department of housing collects figures on the building stock, and so forth. Whereas some official figures on flows and stocks of goods are readily available, others are proprietary production or market research figures that are not free of charge. The availability of data about flows and stocks of goods is a prerequisite for establishing material balances.

Material

The term *material* is an overarching umbrella term to express both substances as well as goods. Thus, if metabolic processes are addressed in general, or if it is not decided yet if the level of substances or goods will be the focus of a study, the term *material* is appropriate. An MFA usually comprises the assessment of goods and substances, whereas the result of an SFA focuses exclusively on substances. Nevertheless, in many cases it is essential to include the flow and stocks of goods in an SFA for analytical reasons. Further terms are *inputs* and *outputs*, denominating flows of materials in and out of processes, and *imports/exports*, standing for flows across system boundaries.

Process

In MFA, a network of processes linked by flows is modeled. A *process* is defined as the transformation, transport, or stock change of materials. Processes fulfill the mass balance principle

$$S_t = \sum_{t_0}^{t} (\text{input}_t - \text{output}_t) + S_{t_0} , \tag{3.1}$$

where S_t is the stock after time step t, t_0 is the time of the initial time step, t is the current time step, and S_{t_0} is the existing stock at the initial time step.

Transformation Process

Of the three types of processes, the transformation of goods is the most important if material balances are to be examined. Through transformation, goods are changed into new products of new values, qualities, and often new chemical compositions. Transformations can be manufacturing processes, such as the transformation of steel into a car; physiologic processes ("digestion"), such as the conversion of biomass, water, and air into CO_2 and H_2O; or chemical processes, such as the synthesis of polymers from monomers. Transformations comprise not only technical processes but also natural transformations, such as photosynthesis, geochemical weathering, or stratospheric ozone depletion. Transformation processes can be balanced on a mass or molar basis. Balancing on a mass basis does not require a definition for

1 mol of a good X and thus is always possible as long as mass can be determined. Balancing on a mole level is the more suitable approach to understand the mechanisms behind a transformation because the chemical composition is given by the stoichiometric ratio of the individual elements involved in the transformation. Because it is difficult to define 1 mol of sewage sludge, municipal solid waste, or even milk, balancing on a mass level (e.g., in kg/d) is the method of choice for most purposes.

Transport Process

"Transportation" stands for the change of location of a good without transforming the properties of the good; materials entering a transport process subsequently leave it physically and chemically unchanged. In "transportation," material and energy fluxes that are needed to move a good are often included, thus the consequences of a transportation process can be that materials are transformed, too. Because the amounts of resources involved in transportation can be significant, it is often necessary to assess the contribution of these fluxes to the total system investigated. The methods used to determine the material and energy fluxes that are needed for a specific transport are the same as those used to determine the material and energy fluxes that are needed for the process transformation.

Storage Process

A process may host a stock of materials that often changes over time. Stocks can either be natural stocks such as iron ore deposits, the soil, ocean sediments, and the atmosphere or they can be anthropogenic stocks such as the building stocks, long-lived appliances in a personal household, the iron stock contained in a railway company, and so forth. If a stock is changed, the process becomes either a source (negative change of the stock) or a sink (positive change) of materials within this process. A typical example for a source process is iron ore mining, where the stock of iron ore within the process *Iron mine* decreases due to the exploitation of iron ore. A typical example for a sink process is a construction site where buildings are set up resulting in a growing stock of materials within the process *Construction*. In certain cases, it can be appropriate to define virtual stocks in a process: a municipal solid waste (MSW) incinerator converts organic substances such as cellulose mainly into CO_2 and water. Hence, in a flow and stock analysis of cellulose, the incinerator acts a final sink in the pathway of cellulose. Note that there is only a virtual addition of cellulose to the virtual stock of the incinerator. In reality, cellulose is decomposed. This example also shows that transformations and stock changes are often related. In general, a utilitarian approach is appropriate when defining processes: The definition of a process that is most useful for modeling metabolic system should be chosen.

Usually, the economic value of a good is changed by going through a process; for example, in manufacturing a car, value is added to the applied iron. During the activity TO CLEAN, drinking water with a positive economic value is converted to wastewater with negative economic value. In human digestion, the value of food is destroyed when highly valuable nourishment is transformed into less useful feces and urine. The value of food for the adult consumer is that this good renders energy when transformed from high-value food to low-value feces. The value discussion is different for babies and children when—in addition to energy gains—food is partly converted into living biomass, resulting in the new and valuable good human body. When it comes to evaluation and design of metabolic processes and systems, it is of key importance to observe the changes in value along the material pathway in order to understand the driving forces that control the material flows and stocks.

Depending on the issue to be investigated, processes can be rather complex topics, comprising many subprocesses, flows of goods, and substances. Even if it is advisable for reasons of feasibility to treat a process as a black box taking into account inputs and outputs only, a minimum understanding of the physical, chemical, and biological mechanisms within a process is required. The question of further dividing a process into subprocesses often determines costs and uncertainty of an investigation; hence, the deeper the comprehension about a process, the more efficient and accurate an MFA that can be established.

In MFA, certain symbols are used to display processes. Transformation and transport processes are represented by rectangular boxes and a stock within a process as a box within the process box (figure 3.3a).

Flow and Flux

A *material flow* is defined as the amount of mass that flows per time through a conductor. A *material flux* is defined as the amount of matter that flows per time unit through a unit area ("cross section"). Also, a flux is equal to the product of material density (mass per volume) and velocity of flow (distance per time). This latter definition is streamlined to pursue the goal of MFA; namely, to determine the density, transformation, and flow of a material in a given system over a certain period of time. In practice, the two terms *flow* and *flux* are not used in a rigid way. In fact, most authors prefer the term *flow* when they describe fluxes. This can be rationalized as follows: In MFA, a flow always refers to a certain system (e.g., a region, a household, a waste management company, etc.). This system can be looked at as the "cross section" mentioned above. Hence, the units that are applied to flows are actually "mass flow per time and system," thus representing a flux. Most often, the unit chosen for MFA is mass flow per time and capita, representing a true flux, with the "capita" as the cross section. In this book, the terms *material flow*,

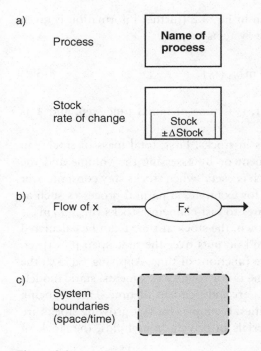

a)

Process   **Name of process**

Stock rate of change   Stock ±ΔStock

b)

Flow of x —————— F_x ⟶

c)

System boundaries (space/time)

Figure 3.3
Graphic representation of processes, stocks, flows, and systems: (a) symbols for processes, stocks within processes, and change of stock; (b) symbol for material flows and fluxes. The oval area may either contain the name of the good or the numerical value of the flow (in mass per time) or flux (mass per time and cross-section); (c) symbol representing system boundaries in time and space.

substance flow, and *flow of goods* are used without considering the term *flux* any further. The symbol representing both flows and fluxes is given in figure 3.3b.

Depending on specific characteristics of flows, the following additional attributes are used: Flows that cross system boundaries are either called imports (entering the system) or exports (leaving the system). Flows that enter processes are called inputs or inflows, and flows that leave a process are called outputs or outflows. When discussing material balances and metabolic processes, the differentiation between imports/inputs and exports/outputs is relevant because it serves to point out quickly whether a specific flow crosses system boundaries or not.

Stock
In general, a *stock* of a material is the mass of a material residing in a process or system during the balancing period of the system. This stock, which is likely to change with time, has been presented in Eq. 3.1. for the quasi steady-state. For the

dynamic case, it can be expressed as given in Eq. 3.2 (further information is given in the section on "Modeling of Metabolic Systems"):

$$m_{stock}(t) = \int_{t_0}^{t} \dot{m}_{input}(\tau)\,d\tau - \int_{t_0}^{t} \dot{m}_{output}(\tau)\,d\tau + m_{stock}(t_0),\qquad(3.2)$$

where $m_{stock}(t)$ is mass of stock at time t, t_0 is time of initial time step, and t is current time step.

There are two ways to assess materials in stocks. First, total mass of stock can be determined either by direct measurement or by assessing the volume and the material density of the stock. This approach is useful when stocks stay constant over long time periods ($\dot{m}_{storage} / \dot{m}_{stock} < 0.01$), for example, in natural processes such as soils or oceans. The second approach relates to fast-changing stocks ($\dot{m}_{storage} / \dot{m}_{stock} > 0.05$). If the size of the stock at t_0 is known, the stock at time t can be calculated based on the difference between inputs and outputs over the time span $[t_0 - t]$ (cf. figure 2.19). Usually, $\dot{m}_{input}$ and $\dot{m}_{output}$ are functions of time. Applying Eq. 3.2, the stock (m_{Stock}) can be calculated for any time t. For rough assessments, static models are used, assuming that $\dot{m}_{input}$ and $\dot{m}_{output}$ are independent of time. Fast-changing stocks are typical for anthropogenic activities. Examples for fast-growing stocks are metals in urban settlements, plastic materials in private households, the stock of mobile phones, and wastes in landfills.

Transfer Coefficient

The result of the partitioning of materials in processes can be expressed by transfer coefficients k. These coefficients define the fraction of a material that ends up in one of several products of a process. If a process has several input goods, k is defined as the fraction contained in one output in relation to the total input of all goods. For substances, k determines the fraction of a specific substance in one output in relation to the total input of the specific substance. In some particular cases, it may be appropriate that the transfer coefficient denominates the fraction of a substance contained in a good in relation to a specific (and not the total) input. An example is the emission of nitrogen oxides from MSW incinerators when the question addressed is how much nitrogen oxides are being released to the atmosphere per unit of MSW, not taking into account nitrogen oxides that are produced in the furnace by the input of N_2 contained in air.

Although partitioning is sometimes presented as percentage, transfer coefficients k are given as dimensionless values equal to or smaller than 1. They depend on the kind of process, process parameters, and on the properties of the materials. If all transfer coefficients of a system are known, the system properties are completely determined. As presented in figure 3.4, the sum of all k_i of a process equals 1.

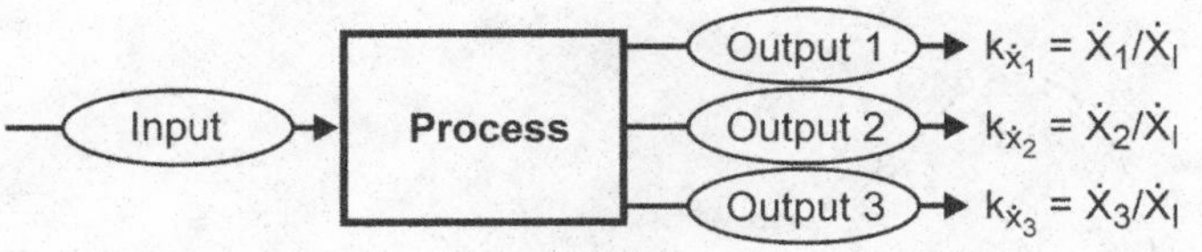

The diagram shows: Input → Process → Output 1, Output 2, Output 3 with:

$$k_{\dot{X}_1} = \dot{X}_1/\dot{X}_I$$

$$k_{\dot{X}_2} = \dot{X}_2/\dot{X}_I$$

$$k_{\dot{X}_3} = \dot{X}_3/\dot{X}_I$$

$$k_{X_i} = \frac{\dot{X}_{O,i}}{\sum_{i=1}^{n_I} \dot{X}_{I,i}}$$

$$\sum_{i=1}^{n_I} k_{X_i} = 1$$

Figure 3.4
The partitioning of a material X in an input by a process between the outputs 1, 2, and 3. The transfer coefficients k_x of a material X are defined as $X_{product}/X_{educt}$. X is usually given as a flow value (mass per time). The sum of all k_{xi} equals 1. n_I respectively n_O = number of input respectively output flows.

Transfer coefficients are useful when modeling entire MFA systems. Because transfer coefficients are technology dependent, they can be used to evaluate the impact of alternative or new technologies on the flow of materials through a system. An example is the comparison of two scenarios for waste management; namely, separate collection versus improvement of air pollution control technology for MSW incinerators. Because of the high load of cadmium and other heavy metals in the surroundings of MSW incinerators in the 1970s, it became urgent to introduce new measures in waste treatment to control heavy metals. Advanced air pollution control devices for MSW incinerators are capable of reducing emissions by a factor of 140, hence decreasing $k_{Cd\ emissions}$ from 0.6 to <0.004. The introduction of new, harsh waste collection schemes with source separation can lower the input of Cd into MSW incinerators by 80% to 90%, resulting in a 10-fold reduction of emissions ($k_{Cd\ collected\ separately} = 0.9$, $k_{Cd\ collected\ with\ MSW} = 0.1$). Thus, the management decision to invest in advanced air pollution control instead of separate collection reduces air pollution by MSW incineration by one order of magnitude more than that by separate collection of cadmium-containing wastes. Of course, the most effective but also most costly measure is to do both, emission abatement and separate collection.

Although transfer coefficients are often regarded as constants, they can also vary with time and other parameters such as process conditions (e.g., temperature, pressure) and input composition. In some cases, they are constant within a certain range (see figure 3.5). This is of considerable practical use in sensitivity analysis and scenario analysis.

System and System Boundary
An MFA embraces a material system that comprises two elements; namely, processes and flows of materials within a defined system boundary. In principle it is an open system (i.e., the system allows imports and exports). When designing an MFA system, the two key steps are (1) to decide which elements are inside

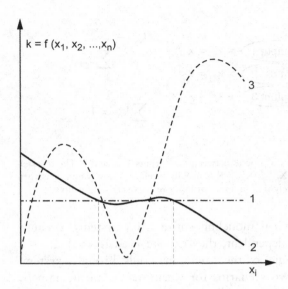

Figure 3.5
Different types of transfer coefficients: (1) k_1 is constant and independent of parameter x_i; (2) k_2 varies with x but can be regarded as constant within a certain range of x_i; (3) k_3 is highly sensitive to variations in parameter x_i and cannot be assumed as being constant (Brunner & Rechberger, 2004).

and which are outside the boundaries and (2) to define how the various elements within a system are linked and related to each other. A system can be composed of one or many processes and flows and may be disaggregated into subsystems.

The system boundary has to be defined in time and space: For both limitations, the choice depends on the scope of the project, the system investigated, and the questions to be addressed. System boundaries in time can range from 1 second for a study on combustion processes to 1000 years for balancing inputs and outputs of MSW landfills including aftercare. For anthropogenic systems such as a private household, a town, or a company, a temporal boundary of 1 year is often selected because financial accounting and reporting is done on an annual basis, and thus information that covers periods of 1 year is usually easier to retrieve than data for shorter or longer time periods.

As a spatial system boundary, usually a fixed geographical area is chosen that includes all the relevant processes. The area can be the location of a particular household or an enterprise, a region specified by, for example, hydrologic characteristics (uniform watershed), or the area of a town, province, or country delimited by administrative borders. In addition to this type of real spatial system boundaries including all processes and flows of goods, a second type of a more abstract system boundary in space may be defined: all households of a region, or the whole waste management sector of a country, or all processes related to the activity TO NOURISH. This second type of boundary includes not all processes within the defined area but only certain processes (e.g., households, or waste management companies, or retailers) that are relevant for the issue to be investigated. Material systems for natural systems usually include neighboring spheres, too. For example, forest ecosystems may include also the lower layers of the atmosphere. Material transport and transformations in atmosphere, soil and subsoil, and groundwater are important processes. Depending on the problem to be solved, they require selection of a wider system boundary that includes, for example, the planetary boundary layer where the main exchange of air and air pollutants between regions takes place, or the aquifer where also large amounts of materials are transported and transformed by groundwater. The system boundary is symbolized graphically by a rectangle with broken lines and rounded corners (figure 3.3c).

The systems boundary that discriminates between internal and external processes has an important function because processes within the boundaries have to be mass balanced, whereas processes outside the system such as source or sink processes for imports and exports are not balanced in an MFA. Hence, if a process cannot be balanced due to missing information, it may be helpful to consider drawing an alternative system boundary and to position the process in question outside the boundary.

System boundaries and so-called functional units are linked to each other. During the process of evaluation, the question arises how to compare the data generated (e.g., for two alternative systems). A functional unit is the basis on which a certain service or good is offered and provides a reference for the material flows and stocks of a system. Two examples are provided: (A) the functional unit to relate a 1-year study of the activity TO NOURISH may be an adult person or the household of a family of four (parents and two small children). It is clear that the results of the MFA are different for the two functional units, as the children will accumulate about 4 kg/person and year, something the adults usually try to avoid. (B) Decision making in waste management, comparing MSW land-filling with MSW incineration: The functional unit can be the service that is supplied for citizens or the amount of waste treated and disposed of per time. In all cases, a rigid and comprehensive definition of the systems boundary resolves also the issue of the functional unit: When the systems boundaries "in space" are defined, it must be specified what kind of system it comprises. For example in case of example A, the boundaries can be defined for a specific single person, a specific household with defined characteristics, an "average" national private household, and so forth. For example B, the same applies for boundary in space. In addition, the boundary in time deserves special attention, too, because the mass flows due to incineration last from hours (operational balance) to 25 years (total balance including construction and demolition of the plant), and material balances and thus boundaries in time for a landfill must include very long time spans exceeding 100 years (see chapter 4 "TO NOURISH and TO CLEAN"). For meaningful comparisons, it is of paramount importance to choose a correct functional unit represented by appropriate system boundaries.

Activity

The term *activity* has already been introduced in the discussion of the anthroposphere in chapter 2. It stands for a basic human need such as TO NOURISH and is represented by a set of processes and flows and stocks of goods and substances required to fulfill these basic needs. This set usually comprises a very large amount of goods and processes. For instance, when the activity TO NOURISH is investigated, the system boundary comprises everything from the seed-corn, fertilizer, agricultural soil, farming equipment, and plant production to food industry, food trade, grocery store, private households, and the consumer. The activity concept is based on the idea that whereas basic human needs do not change over time, the cultural, social, technological, and economic conditions change constantly, allowing new and more effective ways of meeting the requirements of mankind. Also, the activity concept allows early recognition of constraints regarding resource availability and environmental protection, and it can be used for designing activities to meet certain goals such as "sustainability" or "autonomy."

The problem in defining fundamental activities and in establishing a hierarchy of activities is to identify relevance and to filter out as few as possible and as many as necessary. The four main activities TO NOURISH, TO CLEAN, TO RESIDE&WORK, and TO TRANSPORT&COMMUNICATE are summarized in table 3.4. These four key activities comprise most of the material and energy turnover of a society. They are crucial for understanding anthropogenic metabolism and to comprehend all metabolically relevant properties, independent of the cultural idiosyncrasies in space and time of an anthropogenic system. However, the activity concept is an open concept and allows inclusion of additional activities if needed: There are other human activities, such as LEISURE, with a certain material turnover. It should also be noted that some human activities like REPRODUCTION are associated with hardly any material turnover and thus are not of primary interest for metabolic studies. However, such activities might be of relevance when the control of material flows is in focus: Comparing different religious denominations such as the Amish versus baroque Catholic believers or Buddhist vegetarians versus Christian carnivores, it becomes obvious that an activity TO BELIEVE can have a large impact on the control of metabolic processes, material flows, and activities. The power of the activity concept is further elucidated in chapter 4 when case studies are introduced.

Procedure to Establish MFA and SFA

In this chapter, methods to analyze flows and stocks on both levels—goods and substances—are introduced. The term *material flow analysis* (MFA) is used here in accordance with the terminology chosen by the International Society for Industry Ecology (ISIE) and applied by the *Journal of Industrial Ecology* and denotes the analysis of flows and stocks of goods. Note the difference with the definition of materials given in this chapter, which comprises both substances and goods. In accordance with the ISIE, *substance flow analysis* (SFA) stands for the analysis of flows and stocks of substances. Although most data published to date has been acquired by MFA and concerns issues such as resource efficiency (Moll, Bringezu & Schütz, 2005; OECD, 2007b), the substance level has received much attention recently by the studies of the Yale School of Forestry on regional and global metal stocks and flows (Gordon, Bertram & Graedel, 2006; Gerst & Graedel, 2008).

The establishment of MFA/SFA is a heuristic procedure based on experience. Besides the definitions given above and the principle about the conservation of mass, there are no theories available to perform MFA/SFA. However, there are computer models such as STAN and SYMBOX to support the establishment of MFA/SFA, and they are listed in the following section on "Modeling of Metabolic Systems." There are many groups active in MFA/SFA using their individual techniques and accumulating their specific experience and data. The procedure presented below to establish an MFA/SFA is thus not the only possible approach, but it is one that has been applied on numerous occasions by various groups from America to Asia. Also, despite the many approaches, there are only small differences between the *methods* of the individual schools of MFA/SFA. The main divergence is the *focus*: One group of authors concentrates on the level of goods promoting a general reduction of material turnover in order to conserve resources and protect the environment ("Factor 10, 5, or 4": von Weizsäcker, Lovins & Lovins, 1995; Schmidt-Bleek, 1997; OECD, 2002; von Weizsäcker, Hargroves & Smith, 2010). Another group prefers to include substances for a more specific assessment, acknowledging that the quality of resources and the environment is determined by substances, too.

With respect to metabolic processes, MFA/SFA is performed for analytical reasons (early recognition, monitoring), for decision support, for problem solving, and for designing new or improved processes and systems. Thus, in most cases, an MFA/SFA has a distinct goal that is given by the objective of the issue in question. However, supporting decisions and problem solving with the MFA/SFA method is an iterative process with a moving target that requires first to define the objective in a clear and transparent manner. It may even be that a proper objective can only be developed after an initial rough mass balance, giving some hints where the important key issues are. In any case, a proper, clearly defined objective is a prerequisite, be it defined beforehand or during the elaboration of an MFA/SFA.

A flow diagram for the establishment of an MFA/SFA is given in figure 3.6. It involves eight steps: (1) definition of the task, (2) setting of objectives, (3) definition of the MFA system, (4) rough balancing for testing if the chosen system is appropriate and efficient to reach the objectives, (5) mass balancing of goods (MFA) and substances (MFA/SFA), (6) evaluation and interpretation of the results, (7) drawing conclusions, and (8) reporting. There are two crucial steps in every MFA/SFA: the definition of the system and the collection of data for the mass balance. When defining the system, momentous decisions about the choice of processes, flows and stocks of goods, as well as about system boundaries in time and space are required. Also, complex processes might have to be broken down into subsystems. Defining a system is a step that requires experience, creativity, and effectiveness; it is a highly innovative process. Usually, a system is not defined in one approach but rather by an iterative process that is optimized along the pathway toward the MFA/SFA. A key step is the selection of relevant flows, stocks, and processes. The art of performing an MFA/SFA consists of choosing as few elements as necessary to render the system as simple as possible but still representing the essence that is necessary to reach the objective. The effort to create an appropriate and elegant MFA/SFA system is often underestimated because addressees of MFA/SFA reports see only the final product, which—if properly presented—looks simple and straightforward. Because the establishment of an MFA/SFA is a heuristic process, it is obvious that experience is a necessary prerequisite and that with increasing experience, system definition will improve.

The process of establishing an MFA/SFA is seldom linear but often iterative and with loops: The rough balance may show that the objectives cannot be reached efficiently with the chosen system and thus a new system has to be designed. Sometimes, data are missing to balance a process that lies within the system boundary. In such cases, it may be beneficial to place the process outside of the boundaries, thus eliminating the need for balancing and reducing the requirements for data. Budgetary restrictions may not allow analyzing all flows and stocks, hence aggregation may be appropriate for reducing the analytical effort.

An important step is the presentation of the results with text and figures, too. Often, MFA is used as a basis for decision making. Decision makers are not familiar with MFA, and they usually have neither the time nor the desire to get engaged in MFA methodology. Hence, the results of an MFA have to be presented in a way that allows clients to comprehend quickly the main message of the MFA. The visualization of the MFA results is thus a didactic process that must be performed with great care and skill. Some software programs like STAN (cf. Cencic & Rechberger, 2008) allow presenting detailed MFA results graphically. However, it may be necessary first to aggregate results in order to make them easily understandable and then to extract the key message from the combined results.

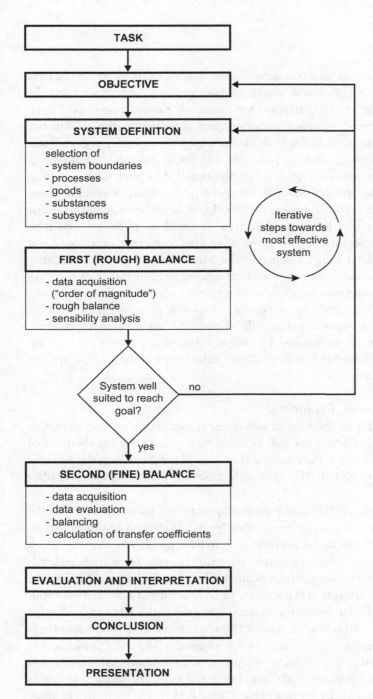

Figure 3.6
Procedure for establishing an MFA.

Because goods are entities that are defined economically ("positive or negative value . . ."), data about flows and stocks of goods are abundant and can be found in economic statistics (Organization for Economic Co-operation and Development (OECD) Factbook; national import–export statistics; national statistics about production, consumption, disposal of wastes or wastewater; regional statistics; statistics by various economic branches and large individual companies). Nevertheless, it may be necessary to acquire additional data that have not been published before (e.g., by market research, econometric methods, or specific investigations into particular inputs, outputs, and stocks of households, enterprises, or regions). To keep the effort for data collection as small as possible—extensive data collection or even actual analysis is the most expensive part of an MFA—it is suggested to first make a rough balance of the MFA system in order to identify the key flows and stocks. Based on this first assessment, further investigation can be focused on the relevant elements, with emphasis on determining the sensitive flows and stocks as accurately as possible. It is not necessary to determine all flows belonging to a process individually. Using the mass balance principle, missing flows can often be determined by simple calculations. Indeed, the redundancy of information in an MFA system allows calculation of flows that are not accessible by other means.

MFA on the Level of Goods, Exemplified

As an example of an MFA on the level of goods, the consumption of food in private households is presented. The results will be used to discuss MFA on the level of substances. In chapters 4 and 5, the example is expanded for describing the material system of the activity TO NOURISH and for the case study on designing phosphorous management.

The private household (PHH) is the most important metabolic process of the anthroposphere because it is the consumer who "pulls" at the end of the consumption pathway, and hence—with his decision to purchase goods—he decides which goods have to be produced, which resources are required, and, for a given technology, which environmental loadings are generated. Thus, for decisions regarding the design of anthropogenic systems, it is necessary to have sufficient information about the process PHH. For a first assessment, material flows through PHH can be divided into fast-flowing goods with residence times <1 year (consumption goods) and long-lived goods with residence times >1 year (investment goods). The main consumption goods from a mass flow point of view are water, air, food, printed matter, and products for cleaning and hygiene. Both mass flows and stocks of investment goods outweigh consumption goods by far (see table 2.2 for different activities). The most important category is construction material (concrete, steel, wood, tiles, etc.), followed by interior finishing, appliances, and others.

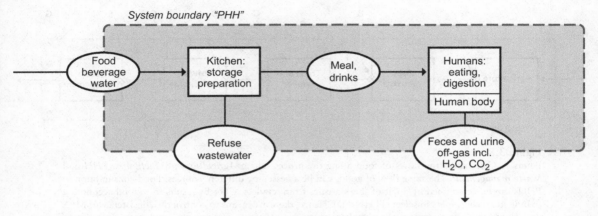

Figure 3.7
Scheme of flows and stocks of materials for the activity TO NOURISH in PHHs. Appropriate system limits comprise the spatial boundaries delimiting the PHH and 1 year in time. Also, a multitude of PHHs as well as the lifetime of consumers or a histogram including different PHHs in terms of, for example, purchasing power or household size (persons per PHH) can be taken into account. The choice of the system boundaries must be based on the issue to be investigated ("goal-oriented selection of system boundaries"); there is no general "most appropriate choice."

Food fulfills a special function in a PHH because it is essential for humans as displayed by the activity TO NOURISH (figure 3.7). Food supplies the necessary energy, liquid, nutrients, vitamins, and trace substances to sustain human life. The agricultural production of food requires large amounts of energy, nutrients, man-power, and capital. It also exerts pressures on the environment (e.g., by releasing nutrients into waterways causing eutrophication or by emitting veterinary medication and food additives into soil and water). Thus, food production today is a widely debated topic, with human health, environment, and resource availability (phosphorous) as the main issues.

A food balance of the PHH allows quantifying dietary intakes and outflows of consumers. It is thus a base regarding decisions to optimize the activities TO NOURISH and TO CLEAN; for example, in view of overloads or deficiencies of phosphorous (P) and nitrogen (N) in the diet of consumers, for efficient recycling of postconsumer nutrients, or for removing P and N in wastewater treatment plants. The emphasis on the process PHH does not mean that the material fluxes associated with the production in agriculture and the processing in the food industry are less important (figure 3.8): On the contrary, for example, agricultural wastes represent the largest amount of any waste material produced in most advanced societies. The priority has been placed here because the consumption of food in PHHs is the key variable for the production of food: If there is a large demand for and a high income to spend on food, the activity TO NOURISH will appear quite different than that

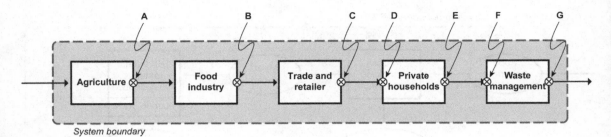

Figure 3.8
Points A–G to assess the flows of food along the process chain *Agricultural production* to *PHH* and *Waste management*. The same flow of goods can be assessed by different sources: Food consumption in PHHs can be determined by point-of-sale statistics from retailers (C) or by results from food accounting of individual consumers in selected PHHs (D). Hence, there are means to control the consistency of MFA data. Note that the system is not complete; there are additional imports and exports such as water, chemicals, wastes, wastewater, and off-gas not included in the figure.

of another society where only limited resources are available. Note that in highly regulated economies, the production of food is controlled not only by the demand of the PHHs but also by agricultural policy or structural boundary conditions (economic interventions such as subsidies and taxes; support for "green" agriculture recognizing the importance of prudent farming for environmental protection).

If the total activity TO NOURISH is to be investigated, all floor space, appliances, materials, and energy carriers needed for food preparation, storage, and disposal have to be included, too. For simplicity, this is omitted here, because the purpose of this section is to show how to establish a balance on the level of goods, and this can be well exemplified by focusing on food only.

The example discussed here is structured in eight steps according to figure 3.6.

1. The task is to establish a database on the level of goods for the activity TO NOURISH of a typical PHH.

2. The objective is to determine all flows of food goods entering a PHH over a representative period in order to give an average figure.

3. The system is defined as follows: The system boundary in time is chosen as 1 year. This ensures that annual variations in dietary habits are taken into account. As a boundary in space, the physical boundaries of the PHH are considered. In a first approach, only one process is taken into account: The process *Kitchen* includes preparation and storage of food (figure 3.7). The import goods such as meat, milk, vegetables, and the like (table 3.6) are selected according to available literature on household consumption of food (BAS, 1987). The export goods are meals and drinks. The choice of substances is discussed later in the section about SFA. No subsystems are defined.

Table 3.6
Rough assessment of food consumption in PHHs based on statistical data about main food items from selected European countries

Country	Cereals	Potatoes	Milk	Meat	Sugar	Total
	(kg/capita and year)					
Switzerland[a, b]	74	50	160	80	42	361
Poland[c]	120	160	260	72	42	654
Hungary[d]	118	60	150	72	36	436
EEG countries[e]	104	67	NA	80	39	NA
Average food consumption (used for rough balance)	120	80	180	80	40	500
Estimated consumption of beverages and water	—	—	—	—	—	1000

NA, not available.
[a]BAS (1987).
[b]BAL (1983).
[c]OECD (1981a).
[d]OECD (1981b).
[e]de Haen et al. (1982).

4. In figure 3.8, the access points for data acquisition about food consumption are given. Each access point has its own advantages and disadvantages. For example, the measurement at point D by individuals accounting for all their expenses for food gives highly accurate individual data. This data can only be generalized and used for extrapolation to a larger amount of households if either very many PHHs are included in the survey or if the effect of household size, purchasing power, demographics, and the like are considered. In contrast, if points B or C are used to collect data, a very large quantity of PHHs is taken into account because the flows of B and C reach many households. Note that the figures derived from B and C may be different because of losses during transportation and storage. Access point A is also convenient for cross-checking data, for example, an excellent database on the production of milk exists in most countries, and imports/exports as well as losses are well known, so that A can be properly compared with data collected at D.

The rough data for food input of the first-draft MFA system was collected from literature about different countries and is summarized in table 3.6. This information was not sufficient for balancing the process "kitchen" because there is no data about the main product of kitchen activities: the meal. Thus, to be able to verify the input data, a new MFA system was designed. This system is presented in figure 3.7 and includes the process *Humans* producing the export flows *feces and urine, off-gas including H_2O and CO_2*. The advantage of this new system design is that the data about import into the PHH (table 3.6) can be cross-checked by data about the export of the PHH (table 3.7). Hence, a rough balance of flows of food into the household and flows of off-materials leaving the household can show if the chosen system is more or less complete or if some major goods have been missed in the system design.

For this balancing attempt, data at the back end of the PHH has been assessed from the medical literature (average daily intakes versus average daily outputs of feces, urine, respiratory air), from waste management (amount of compostable biomass collected), and from wastewater management (dry matter per person and day arriving at a sewage treatment plant). This second rough balance based on food consumption and off-products from households showed (i) quite good agreement between imports and exports and (ii) pointed out the main flows due to the activity TO NOURISH in a PHH. Because water turned out to be a key good for this activity, it is necessary to look into the non-water fraction of food, too. Hence, in addition to total mass flow of goods, the flow of so-called *dry matter* has been investigated, too (dry matter is defined as the matter that remains when a good is heated up to 65°C for a period of 24 hours; other specifications of temperature and time are common, too).

5. Hence, because the second rough balance shows that the revised system is well suited to reach the objectives given, a more detailed study was performed resulting in table 3.8. The basis for this table is a comprehensive survey on the consumption

Table 3.7
Production of food wastes in PHHs

	Mass Flow	Dry Matter Flow	
Waste Product	(kg/c.y)	(%)	(kg/c.y)
Kitchen			
Food, garbage[a] (to MSW)	100	20	20
Liquid wastes	400	5	20
Humans			
Feces[b]	45	23	10
Urine[b]	440	5–7	22
Respiration and transpiration[c]	440–2600		70
			(C in CO_2)
Total wastes	1430–3600		142

[a]Twenty-five percent of 400 kg, which is the average MSW generation rate for Europe.
[b]Lentner (1981).
[c]Calculated by the food input and the excretion rate (Kraut, 1981; Lentner, 1981).

of food in Swiss households (BAS, 1987). For a period of 1 year, close to 500 Swiss PHHs accounted daily for their purchase of consumables, resulting in a large database about food input into PHHs. The investigated households consisted of various sizes from one to more than six members. Thus, by using results from each group size and knowing the regional distribution of household size, it becomes possible to "synthesize" the consumption of an average regional PHH. Table 3.8 contains data on a per capita consumption base for an "average" Swiss PHH of 2.3 persons per household.

With the exception of water, values for column A of table 3.8 originate directly from federal statistics of the Swiss Bundesamt für Statistik (BAS, 1987). Water consumption was calculated (i) assuming 1 L of drinking water consumption per day, (ii) based on the purchase of coffee (2.8 kg/c.y) and tea (0.5 kg/c.y), assuming that consumers use the ratio of water to ingredient recommended by the suppliers of the ingredient, and (iii) by adding water used for cooking and subsequently contained in dry food such as cereals, rice, and farinaceous food. Values for dry matter content in column B were taken from Lentner (1981) as well as from nutritional information of products. The per capita amount of dry matter was calculated by multiplying column A with the percentage of dry matter given in column B. Values for phosphorous and nitrogen are discussed in the next section.

When the values of table 3.8 are compared with other values for food consumption, it should be kept in mind that in addition to eating in PHHs, most persons also eat out. For the Swiss households presented here, it is assumed that about 10% of food has to be added to the total value for food consumption given in the bottom line of table 3.8. In other areas, or during other times, it may well be that this amount is smaller or larger (e.g., larger because a considerable percentage of the population is working out of house and enjoys food at a canteen or restaurant at or near the workplace).

The export side of the PHH is summarized in table 3.9. The partitioning of goods in the processes *Kitchen* and *Humans* is assessed as follows. All food purchased is stored and prepared in the kitchen (subprocesses *Storage* and *Preparation*). Although because of today's refrigeration technologies wastes resulting from storage are negligible, preparation of food results per definition in residues that are discarded, such as fruit peels, bones from meat, and others. In table 3.7, the category waste from preparation includes the fraction of food, too, that is purchased but not consumed because of surplus and leftover. According to recent literature, this fraction is increasing in affluent countries and has reached a considerable amount of about 36 kg/c.y or 12% by weight of the residual waste from households (Salhofer et al., 2008). Based on analysis from waste composition and on household practice, it can be estimated that altogether on an as-received basis, about 25% of the 400 kg of nonliquid food is discarded in the kitchen. On a dry matter basis, this loss amounts

Table 3.8
Food products consumed by an average Swiss household in 1986 (Lentner, 1981; BAS, 1987). Values are averages of 484 households of various sizes (one to more than six persons per household, with 15% greater than four-person households)

A	B	C		D		E	
	Mass Flow	Dry Matter[a]		Phosphorous		Nitrogen	
Food	(kg/c.y)	(%)	(kg/c.y)	(g/kg)	(g/c.y)	(g/kg)	(g/c.y)
Water[b]	1100	0.5	5.5	0.00001	0.01	0.005	5.5
Beverages (without alcohol)	76.2	10	7.6	0.1	7.6	0.8	60
Wine	16.3	10	1.6	0.1	1.6	0	0
Beer	14.0	10	1.4	0.15	2.1	1	14
Fruits	61.0	15	9.2	0.15	9.2	1	60
Vegetables	45.3	15	6.8	0.5	23	4	180
Milk	92.5	12	11.1	0.9	84	5	460
Cheese	12.5	60	7.5	6	75	40	500
Butter	3.8	83	3.2	0.16	0.6	1	4
Bread	23.8	65	15.5	2	48	10	240
Cereals (flour, com, eats, barley, semolina)	8.3	88	7.3	2	17	15	120
Farinaceous foods	5.3	90	4.8	2	11	22	110
Rice	3.0	88	2.6	1	3	12	36
Potatoes	23.3	35	8.2	0.4	10	3	70
Meat	19.4	35	6.8	2	39	35	680
Sausages	9.7	45	4.4	2	19	20	190
Poultry	4.2	30	1.3	2	8	35	150
Fish	2.2	20	0.4	2	4	30	70
Eggs	6.5	25	1.6	2	13	20	130
Sugar	7.3	99	7.2	0	0	0	0
Sweet foods (chocolate, cacao foods, honey, jam)	7.2	90	6.5	1.5	11	15	110
Oil	3.6	99	3.6	0	0	0	0
Margarine	2.2	80	1.8	0.15	0.3	1	2
Vegetable fats	0.3	99	0.3	0	0	1	0.3
Coffee	2.8	98	2.7	2	6	30	84
Tea	0.5	98	0.5	1	0.5	10	5
Total[c]	1551	8.3[d]	129.4	0.25	393	2.1	3280

[a]Dry matter content when purchased (the values when harvested may vary by more than 10%); "dry matter" is defined here as 100% minus water content, which implies that liquids such as alcohol are included.
[b]"Water" denotes drinking water and water used for cooking (e.g., coffee, tea, etc.).
[c]To calculate the total food consumption per person, approximately 10% has to be added to the "total" for out-of-house food intake (restaurants, canteens, etc.).
[d]Calculated from values for "Total" in columns A and C.

Table 3.9
Waste products associated with the activity TO NOURISH of the process *Household* (Brunner, Ernst & Sigel, 1983; Obrist, 1987)

A	B	C		D		E	
	Mass Flow	Dry Matter[a]		Phosphorous		Nitrogen	
Good	(kg/c.y)	(%)	(kg/c.y)	(g/kg)	(g/c.y)	(g/kg)	(g/c.y)
From kitchen							
Food, garbage (to MSW)	100	20	20	0.45	45	3	300
Liquid wastes	400	5	20		20[f]		200[f]
From humans							
Feces[b]	45	23	10		180		660
Feces[c]							610
Feces[d]							360
Urine[b]	440	5–7	22		230–440	80	4400
Urine[c]							1200–4400
Urine[d]						320	2790
Respiration and ranspiration	440–2600		70 (C in CO_2)	0.0002	0.1	0.2	120
Respiration and transpiration[d]					0		130
Total wastes	1430–3600		142		370–690		2400–5700
(Total wastes including out- of-house eating)[e]	1700		140		430		3700

[a] Twenty-five percent of 400 kg.
[b] Lentner (1981).
[c] Kraut (1981).
[d] Calculated by the food input and the excretion rate (Kraut, 1981; Lentner, 1981). Phosphorus: feces 20%, urine 80% for intake of P of 400 g/capita and year. Nitrogen: feces 11%, urine 85%, and transpiration (including epidermis erosion) 4%.
[e] Including 10% of out-of-house eating.
[f] Estimated.

to 15%. The same amount of dry matter is lost by liquid wastes from cooking: dishes such as pasta or vegetables require a large amount of water for cooking, which is discarded. This liquid waste contains not only the minerals contained in drinking water but also other constituents such as salt and ingredients that have been dissolved from food; thus, the dry matter content of this wastewater is quite high (50 g/kg). The mass of wastewater is estimated based on the water needed for cooking (~1 L/d).

For a first assessment of the process *Humans*, the following assumptions were taken: It is assumed that the annual change of a person in weight, dry matter, and P and N content is negligible when balancing a PHH. This assumption seems reasonable if one considers that the average yearly increase of the body mass of an adolescent does not exceed 5 kg, which is less than 0.5% of the food consumed in 1 year. The reason for this small fraction is that a human being requires food mainly to maintain his or her energy metabolism and not to produce body mass. The remaining data about human excreta and other losses from the body surface are taken from the physiology literature (Kraut, 1981; Lentner, 1981).

6. The results presented in table 3.9 and figure 3.9 show that with respect to mass, the most important export path for food consumed in the PHH is the air: More than one third of the food (including drinking and cooking water) brought into a household ends up in the exhaled air of respiration and, to a lesser extent, in transpiration. This balance is mainly determined by the water budget of man: the amount of water contained in respired air is usually similar or larger than the amount of urine excreted per day. In addition, a large fraction of the dry matter taken up with food is mineralized to CO_2, resulting in a considerable transfer of food dry matter to the waste product *human off-gas* (respired air). In fact, this pathway of dry matter appears to be larger than any other pathway such as garbage, feces, or urine. It is interesting to note that the dry matter flux by means of urine is larger than by means of the feces. Of course, the composition of dry matter of the two goods is quite different; water-soluble salts make up the bulk of urine, whereas in feces the ash content is only 20% and the bulk is composed of organic substances such as fats, organic acids, bacterial biomass, and undigested organic material like cellulose, and so forth.

7. The following conclusions can be drawn based on the results given in tables 3.8 and 3.9:

   a. With respect to the activity TO NOURISH, two processes are important in a PHH: the human body transforms food into three different new goods that are released to the atmosphere and sewer; and the sewer acts as the main collector for off-products.

b. About one fourth of the food input (including drinks) into a PHH never reaches man but leaves the kitchen as liquid (sewage) and solid wastes (municipal solid waste or compost).

c. About 50% of the total food-related mass flows leave the PHH (kitchen and humans) through the sewage system. For dry matter, the sewer accounts for one third of the mass flows.

d. About three fourths of the 1500 kg of food consumed annually by a human being is beverages and water contained in tea, coffee, and the like, and only about one fourth is solid food products.

e. Because of the large consumption of water and beverages, feces are waste products of minor importance compared with other products of the process household. Mass flows and dry matter flows by means of urine are both larger than those by means of feces.

These five points have implications for many aspects of the anthroposphere (e.g., waste and wastewater management and resource management). For example, they clearly show that the food-related mass flow by means of sewage is nearly 10 times larger than by means of MSW or compost. On a dry matter basis, the flow in the sewer is still more than double the flow in MSW. The results of the MFA allow dimensioning management and treatment systems; for example, by identifying the maximum potential for composting of biogenic organic material deriving from households (about 17% of the mass of the non-water food entering a household). Some of these implications are discussed in the case studies of chapter 5.

8. The final results are condensed and presented in figure 3.9. Parts (a) and (b) of figure 3.9 clearly convey the main message about the importance of the two processes *Kitchen* and *Humans* summarized in step 7. They show the main flows, show that stocks are of minor importance, and allow differentiation between total mass flows and dry matter mass flows. Decision makers are able to grasp the conclusions quickly.

MFA on the Level of Substances (SFA)

Basically, the path to establish an SFA is identical to that for an MFA (figure 3.6). In addition, substances have to be selected in step 3, and data about substance concentrations and substance flows have to be acquired and balanced in steps 4 and 5. Information about substance contents in goods can be taken from the literature as well as from sampling campaigns and measurements. Measurements of substance concentrations in goods are usually expensive, prohibiting the application of large-scale measuring campaigns. Generally, the assessment of substance flows is more expensive and less straightforward than the assessment of flows of goods. The reason is that economic information on the level of goods is usually available due

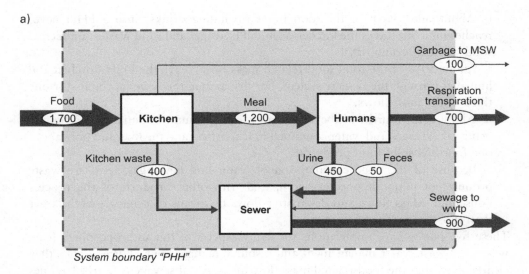

Figure 3.9
(a) Mass flow of food and food wastes through PHH, in kg/person and year. (b) Dry matter flow of food and food wastes through PHH, in kg/person and year. MSW, municipal solid waste; wwtp, wastewater treatment plant.

to reasons of bookkeeping and financial accounting. Thus, if combined, such information allows mass balancing of goods along the product life cycle chain. In contrast, chemical information about substances is less abundant and only punctually available when such knowledge is required for product quality or for protection of safety, health, and environment.

Flows and stocks of substances are calculated by multiplying flows and stocks of goods with the corresponding substance concentrations. Again, as in the case of MFA, the redundancy of information in an SFA can be used to optimize the SFA system: There are several sources for similar information along the substance flow pathway as displayed in figure 3.8 with the information access points A to G. Whereas the bulk flows of substances in and out of a system can be determined at points A and G, points in between allow assessment of more individual flows. For instance, nutrient flows are often known at point A for reasons of subsidizing or controlling agricultural practice and at point G for reasons of waste and wastewater management. Thus, it can be an efficient means for rough balancing to compare A and G in step 4 of an SFA.

Selection of Indicator Substances

The purpose of SFA is to support analysis, evaluation, and design of metabolic processes. As discussed in chapter 2, the substance level serves to include resource

b)

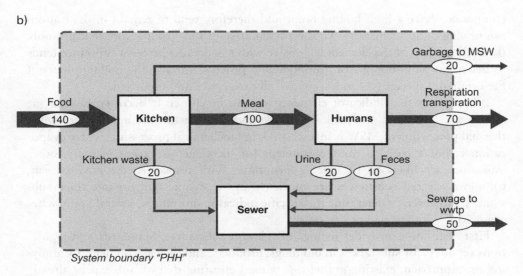

Figure 3.9
(continued)

as well as environmental aspects. Individual substances make up resource and environmental qualities, be it by their high concentration in an ore (e.g., gold) or by their unwanted presence in a secondary raw material (copper in iron scrap). Because several millions of substances have been synthesized and described to date, the question arises which substances to select for a particular investigation; how to reduce as much as possible the number of substances that have to be taken into account for evaluation.

To achieve this goal, so-called indicator substances summarized in table 3.5 are selected. This is an important means to achieve maximum information with minimum resources. An indicator substance represents a group of substances. It shows a characteristic physical, biochemical, and/or chemical behavior that is a specific property of all members of the group. Indicator substances can therefore be used to predict the behavior of other substances. For instance, metals can be separated into *atmophile* and *lithophile* elements. (In geochemistry, elements are grouped as follows: siderophile elements are those elements that are concentrated in the earth's iron core; chalcophile elements tend to combine with sulfur in sulfide minerals; lithophile elements generally occur in or with silicates; and atmophile elements are prominent in air and other natural gases.) Atmophile elements have a lower boiling point and thus are transferred during incineration into off-gas. Examples for such atmophile elements are Cd, Zn, Sb, Tl, and Sn. Hence, for example, Cd serves well as an indicator for this group. Lithophile elements such as Ti, V, Cr, Fe, Co, and Ni and their compounds have a high boiling point and therefore tend to remain in the bottom ash or slag of an incinerator. Fe is representative of this group. Note that not only the boiling point of the element is decisive with respect to whether a substance tends to behave as an atmophile or lithophile in a process, but also the boiling points of the occurring species (chlorides, oxides, silicates, etc.) are important.

It is obvious that indicator elements have to be chosen in accordance with the system and processes analyzed. The above indicator metals are appropriate when thermal processing of MSW is investigated. In biochemical processing, where evaporation is not a path to the environment for most metals except mercury, other indicators are likely to be more appropriate. With respect to the environment, the most selected substances are nutrients and inorganic and organic trace substances. However, to determine the specific indicator substances, several approaches are used.

First, legislation provides listings of relevant substances. Standards and regulations set limits for substances in buildings, products, and the environment. By applying this approach, existing knowledge is used ensuring that all substances already selected by the respective authorities are considered.

Second, in absence of legislation or standards, the relevance of a substance has to be evaluated based on specific evaluation methods. A workable rule is as follows: All import and export flows of goods of the system are grouped into solids, liquids, and gases. From each group of imports and exports, those flows are selected that cover 90% of the total mass flow. This yields a set of important flows of goods for the system. Next, the ratio of substance concentrations in the selected anthropogenic flows of goods is compared with the geogenic reference materials "Earth crust," "water bodies," and "atmosphere." Substances with an anthropogenic to geogenic ratio of >10 are favorite indicators for environmental pollution. If all or most ratios are <10, those substances with the highest ratios should be further looked at. This rule of thumb can only assist the selection process; it is necessary to check for plausibility and consistency during the entire evaluation process.

Third, a substance is already in the focus and must not be selected anymore. Studies such as the global metal flows and stocks project at Yale University (Gerst & Graedel, 2008) or nutrient balances of the River Danube basin (Somlyódy, Brunner & Kroiß, 1999) focus on data acquisition about individual metals and nutrients in order to have a database for improved decisions regarding future metals and nutrient management. The selection of substances is part of the project definition.

In practice, the selection of substances depends on the scope and the resources (financial and human) that are available for an MFA study. Hence, indicator substances are usually determined pragmatically by the objectives and the corresponding problems to be solved. Experience with SFA shows that many metabolic systems can be roughly characterized by a comparatively small number of substances, such as about 5 to 10 elements. In table 3.5 a list of frequently used indicator elements is presented. For example, if the activity TO NOURISH needs to be optimized, *resource-related* questions concern nutrients, water, and energy. *Environment-related* issues focus on nutrients, soil salination, greenhouse gas emissions, and, in addition, trace chemicals. Hence, an appropriate choice of indicator substances includes phosphorous, nitrogen, carbon, H_2O, and possibly some individual refractory organic substances from the range of pesticides and agrochemicals.

For an assessment and optimization of waste management, the selection of indicator substances must be based on the objectives of waste management; namely, protection of men and the environment, conservation of resources, and long-term aftercare-free waste management practice (see case study "Designing Waste Management" in chapter 5). Each category of goals has to be operationalized by different, quantifiable sets of indicators; for example, protection of men and environment entails that the key substances are selected that pose the greatest risk for the environment. As displayed in figure 3.10, MSW is an important "conveyor belt" for the two heavy metals cadmium and mercury: Nearly 50% of all cadmium imported into Switzerland in the 1990s appears in MSW. Hence, processes that treat MSW

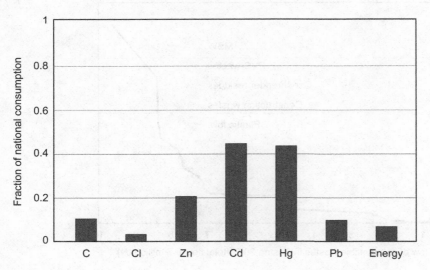

Figure 3.10
Selection of key substances for optimization of waste management regarding the objective "protection of men and environment." The fraction of nationally imported cadmium and mercury that ends up in MSW is much higher than the fraction of carbon and energy. Thus, waste management processes such as incineration must focus on removing these heavy metals from the anthroposphere and on safe storage in the environment.

must be able either to concentrate Cd and Hg for recycling or to isolate the two heavy metals in an inert and long-term immobile form that is suitable for final storage (underground or above-ground landfilling). In both cases, the emissions of Cd and Hg from the treatment must remain low and comply with standards. The approach described here allows setting priorities when indicators are selected: Because the potential contribution of MSW to the national energy demand is relatively low (~5%) and the flow of heavy metals through these plants is relatively high, it becomes clear that MSW incineration must primarily concentrate on solving the Cd and Hg problem first and then focus on energy-related issues.

Because Cd has been used extensively as an additive to stabilize PVC, the highest load of Cd is found in plastic wastes, MSW, and car shredder residue (figure 3.11). Whenever processes handling plastic materials and wastes are assessed, it is important to select Cd as an indicator for environmental protection. The fact that in many countries Cd is not used anymore for stabilizing polymers is no reason for dismissing Cd as an indicator: There are still large Cd reservoirs in long-lived products that have to be removed before significant effects can be observed on the output side of the anthropogenic system. It is notable that despite the decreased use of Cd by the

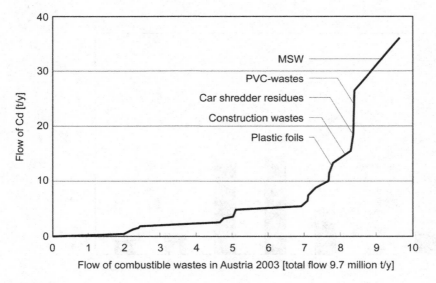

Figure 3.11
Cadmium in combustible wastes in Austria 1999: 20% (2×10^6 Mg) of all combustible wastes contain little Cd and thus do not need advanced air pollution control when incinerated. Plastic-containing wastes such as MSW, car shredder residues, and organic residues from construction wastes contain large amounts of Cd due to its past use as a plastics additive. Thermal treatment of such wastes deserves careful and advanced incineration and air pollution control. Thus, Cd serves as a key indicator substance for waste incineration.

polymer industry, the mass flow in MSW is not decreasing: It appears that the increase in consumption of NiCd batteries more than compensates the reduced Cd flow in plastic wastes (Skutan, Vanzetta & Brunner, 2009). For further examples, see also Brunner and Rechberger (2004).

Example of Establishing an SFA

To demonstrate how to establish an SFA, the same example TO NOURISH in PHH is used as in the case of the MFA on goods. This example is further expanded to a case study on regional phosphorous management in chapter 5. According to figure 3.6, the first two steps of the SFA are identical to those of the MFA. In the third step, as a part of the system definition, indicator substances have to be selected. Because the main reason for food consumption is the supply with energy and nutrients, the two nutrients phosphorous and nitrogen have been selected as indicators. Other substances may have been chosen as well, such as carbon, trace metals, or vitamins. The procedure to establish an SFA is the same for all of them.

Because step 4 of the MFA has resulted in a new system, as a first approach it is assumed that this new system is also appropriate for SFA. Thus, the SFA proceeds at step 5. Flows of nitrogen and phosphorous are calculated by multiplying the mass flows of food and wastes (column B in tables 3.8 and 3.9) by the respective substance concentrations (columns D and E in tables 3.8 and 3.9). Concentrations of N and P in individual food items and waste products are taken from the literature, where such information is abundant (cf., e.g., Lentner, 1981; European Communities, 2001; BUWAL, 2003). Care must be taken to ensure that the various data sources contain compatible information: sometimes nutrient contents are given on a "as received" basis and sometimes on a "dry matter" basis. Also, the state of the food has to be taken into account: by preparing, peeling, or cooking, nutrients can be lost. Literature values originating from medical sources, where only the nutrient intake per person is concerned, do not include losses that are caused before consumption takes place. On the other side, values about nutrient content in wastewater from PHHs are often based on complete analysis of sewage, including also nutrients from other sources such as detergents. Thus, it is necessary to specify and define in detail the system to be investigated and to scrutinize cautiously the values taken from the literature to ensure that they are compatible with the setup of the SFA.

The results of the calculations of nutrient flows and the balances are given in figure 3.12. They demonstrate the metabolic power of the human body: While the body is converting biomass into energy, most of dry matter is mineralized producing CO_2 and H_2O (c.f. figure 3.9b). Nitrogen, taken up mainly as protein in meat and milk products, leaves humans preferentially as urea dissolved in urine. Nitrogen remaining in solid products (feces) does not exceed 20%. Similarly, about 70% of

a)

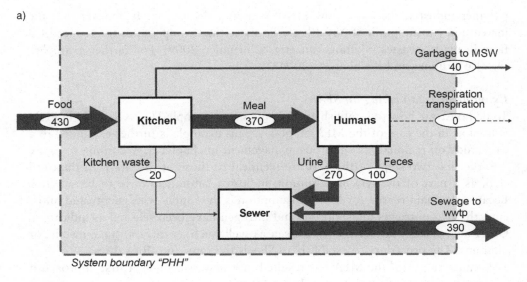

Figure 3.12
(a) Phosphorous flow of food and food wastes through PHH, in g/person and year. (b) Nitrogen flow of food and food wastes through PHH, in g/person and year.

phosphorus contained in food leaves the body by urine, and only 30% of P is contained in feces.

The evaluation step 6 of the SFA reveals that urine is the main carrier of nutrient flows. Both feces and kitchen residues are of less importance. This result opens new possibilities for future design of alternative wastewater and nutrient management systems (cf. chapter 5).

In summary, SFA of food consumption in a PHH allows the following conclusions:

1. As in the case of MFA, with respect to the activity TO NOURISH, two processes are important: the human body that transfers most of the nutrients from food into urine, and the sewer that collects close to 90% of all food-related nutrients entering a PHH.

2. Feces are waste products of less importance compared with urine and other products of the process household. Mass, dry matter, and P and N flows by means of urine are all larger than by means of feces. More than half of the phosphorus and nitrogen that enter a household leave it by means of urine. Nevertheless, for certain substances such as metals, feces can be an important carrier.

3. The amount of P and N in solid kitchen waste is only about 10% of the total food input. Thus, if nutrients are to be recycled, the focus needs to be on sewage and not MSW.

b)

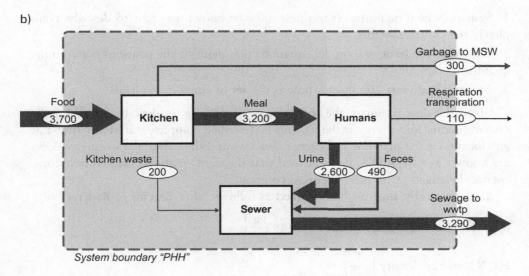

Figure 3.12
(continued)

4. About 14% of the nutrients contained in purchased food never reach man but leave the kitchen as liquid (sewage) and solid wastes (MSW or compost).

As in the case of MFA, the results of the SFA have implications for several fields (waste management, nutrient management, nutritional economics, etc.). For example, they point out the goods that are to be selected for most efficient recycling of N and P (urine and sewage). For further applications and discussion of the results, see the case study on P in chapter 5.

Modeling of Metabolic Systems

Physical and Mathematical Formulation of Metabolic Systems

The general goal of a physical–mathematical model is a quantitative description of the metabolic properties of an anthropogenic system. An anthropogenic system is a selected sector of the anthroposphere (see chapter 2). Within this sector the metabolic processes, defined by their physical and chemical properties, are to be simulated with tailor-made mathematical models. In the following, a short extract of an extended presentation of the theoretical background and of various applications (Baccini & Bader, 1996) is given and, for didactic reasons, illustrated with one case study, the handling of glass bottles. A mathematical model for the MFA method asks for answering the following four questions:

1. System: Which quantities (variables) must be known in order to describe completely the chosen system?

2. Which are the basic and model equations that describe the relations between the quantities (variables)?

3. At given boundary conditions, how is the set of equations solved?

Given the description in the foregoing sections, it is evident that a metabolic system (matter and energy management) is described completely if at any time t at any location x the material and energy densities and the material and energy flows are known. For MFA/SFA, it is assumed that the term "material" stands here for a chemical element or an inert chemical compound.

Mathematically, this can be described as follows (after Baccini & Bader 1996):

$\rho_i(t, \vec{X})$: material density of substance I $\left[\dfrac{kg}{m^3}\right]$

$e(t, \vec{X})$: energy density $\left[\dfrac{J}{m^3}\right]$

$\vec{m}_i(t, \vec{X})$: material flow (a flux density) $\left[\dfrac{kg}{m^3 \cdot s}\right]$

$\vec{s}(t, \vec{X})$: energy flow (a flux density) $\left[\dfrac{J}{m^3 \cdot s}\right]$.

These four variables build a complete set for a metabolic system. Densities are scalar field functions, and flow densities (or fluxes) are vector field functions.

Physically described, the densities reflect the current state of the system (i.e., the current distribution of material and energy within the system). The flow densities describe the change of material and energy of the system. The metabolic system is a continuous (mechanical) system, analogous to a hydrodynamic system.

The basic relations between densities and flow densities are given as balance equations, based on the law of mass and energy conservation:

$$\frac{dM_i^{(j)}}{dt} = \Phi_i^{(j)} + R_i^{(j)} \tag{3.3}$$

$$\frac{dE^{(j)}}{dt} = S^{(j)} + Q^{(j)} \tag{3.4}$$

$M_i^{(j)} = \int_{V_j} \rho_i dv$: material stock in V_j

$\Phi_i^{(j)} = \oint_{\partial V_j} \vec{m}_i \cdot d\vec{f}$: net material flux into V_j

$R_i^{(j)} = \int_{V_j} r_i dv$: material production rate

$$E^{(j)} = \int_{V_j} e\,dv: \text{energy stock in } V_j$$

$$S^{(j)} = \oint_{\partial V_j} s_i' \cdot d\vec{f}: \text{net energy flux in } V_j$$

$$Q^{(j)} = \int_{V_j} q\,dv: \text{energy production rate in } V_j.$$

The physical meaning of the two equations is as follows: The left term in the equations gives the change of material and energy in volume j per time unit. The first term on the right side is the net flux of material (energy) into V_j and the second term the net production of material (energy) within V_j. If these second terms, as source terms, are zero, the two basic equations are reduced to "conservation equations." However, they do not yet describe fully the system. Additional system equations are needed, namely the specific model equations.

MFA/SFA Model

In table 3.10, the terms of MFA/SFA are interpreted in their physical and mathematical meaning.

Table 3.10
Synopsis of key terms from the MFA method and from mathematical description

MFA/SFAa	Mathematical Descriptionb
Metabolic system	Spatial and temporal unit with boundaries, where material and energy are investigated with regard to space and time.
Material system	A metabolic system comprising only material densities and flow densities.
Process	A distinct volume within the system for which a material (energy) balance is made.
Good	Carrier of specific material (energy) for which flows and stocks are determined. In general, goods do not obey the law of conservation of matter.
Substance	Chemical element and chemical compound.
Material	All physical components contained in goods.
Activity	A mixture of biological and cultural deeds to satisfy basic needs within the anthroposphere, generating functional anthropogenic systems that contain metabolic systems.

[a]Baccini and Brunner (1991).
[b]Baccini and Bader (1996).

There are other approaches to model metabolic systems on the basis of statistical mathematics and computer sciences. They are strictly descriptive. The mathematical MFA model MMFA (after Baccini & Bader, 1996) serves basically to help understand the essence of a metabolic (material) system within the anthroposphere on the basis of the principles of the natural sciences. In the following, a brief example for applying the MFA model is given, first for a steady-state case (time independent) and then for a dynamic case (time dependent).

Time-Independent Material Systems: The Steady State or Quasi Steady State

A classical input–output case is chosen illustrating a two-process system. Let us simplify the flow of glass bottles within a region that produces and distributes glass bottles (as containers for liquids that are consumed). *Production & Distribution* is the first process. The buyers of liquids in glass bottles form the second process, named *Consumption*. Consumers give the empty bottles either to the waste, which is exported, or return it to the first process (recoil). Thus, the first process has two glass sources, the input by importing and the input from recycling. It is also assumed that within this system, the material glass stays chemically unchanged. This system choice leads to the MFA material system given in figure 3.13.

For a mathematical description, the corresponding set of six system variables is as follows:

Material: $M^{(1)}(t)$, $M^{(2)}(t)$

Flows: $A_{12}(t)$, $A_{21}(t)$, $A_{31}(t)$, $A_{23}(t)$.

The index 3 stands for all external processes. Obviously the flows A_{13} and A_{32} do not exist.

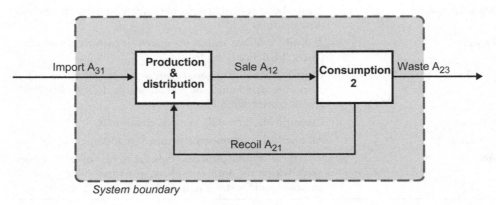

Figure 3.13
Material system for a simplified regional glass bottle management.

The balance equations are as follows:

$$\frac{dM^{(1)}}{dt} = A_{21} + A_{31} - A_{12} \qquad (3.5)$$

$$\frac{dM^{(2)}}{dt} = A_{12} - A_{21} - A_{23}. \qquad (3.6)$$

Further assumptions are needed, because we need six equations. For a classical input–output situation in a steady state, we choose a given input. We assume further that the stocks in the two processes are zero and there is, consequently for a steady state, no change in stocks. Because of these assumptions, the set of unknown variables reduces to three, namely A_{12}, A_{21}, and A_{23}. Now the set of six equations is

$$M^{(1)} = 0 \qquad (3.7)$$

$$M^{(2)} = 0 \qquad (3.8)$$

$$A_{31} = 1 \text{ given input} \qquad (3.9)$$

$$A_{12} = k_{12}(A_{21} + A_{31}) \text{ given transfer coefficient } k_{12} \qquad (3.10)$$

$$A_{21} = k_{21}A_{12} \qquad (3.11)$$

$$A_{23} = k_{23}A_{12} \text{ given transfer coefficient } k_{12}. \qquad (3.12)$$

Transfer coefficient k_{12} is equal to 1 because stock changes take place. The third transfer coefficient k_{23} is also given due to the relation $k_{23} = 1 - k_{21}$. As already introduced at the beginning of this chapter, the transfer coefficients of a process reflect its distribution properties for the outgoing material to connected processes. Consequently, we can give for the still unknown three flows a set of three equations as follows:

$$A_{12} = \frac{I}{I - k_{21}} \qquad (3.13)$$

$$A_{21} = \frac{k_{21}I}{I - k_{21}} \qquad (3.14)$$

$$A_{23} = I. \qquad (3.15)$$

Choosing the numeric example

$A_{31} = I = 19 \text{ kg/c.y}$

$k_{12} = 1$

$k_{21} = 0.87$

$k_{23} = 0.13,$

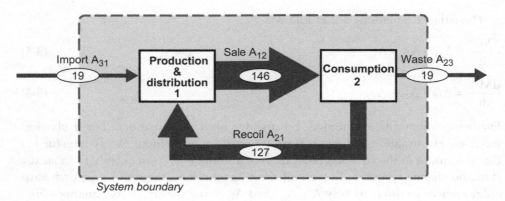

Figure 3.14
Input–output example for a simplified regional glass bottle management. The input and the recycling coefficient are given. Consumers recycle 87% of the bought bottles. The thickness of the flow arrows is proportional to flow quantities in kg/c.y.

the unknowns can be determined. The corresponding scheme of the material system shows the following flows (figure 3.14).

The largest flows are the two inner flows between the processes. For a material with short residence times in the processes, the model illustrates well the consequences for the metabolic system supporting such a (simplified) glass bottle management. The glass bottles are mainly "on the road" and therefore need a suitable transport capacity on vehicles within the region (see also chapter 4 with regard to the activity TO TRANSPORT&COMMUNICATE). If we increase the recycling rate (transfer coefficient k_{21} has a value close to 1) and do not reduce the input, the two inner flows grow hyperbolically (figure 3.15).

It is evident that the first assumption with a given input is not suitable for a market economy. More adequate is a given consumption rate of glass bottles (flow of sold liquids in glass bottles per capita and year). We set the sale flow $A_{12} = V_0$. Thus, the input A_{31} is a function of the sale flow and the recycling flow. The corresponding set of flow equations is as follows:

$$A_{12} = V_0 \tag{3.16}$$

$$A_{23} = k_{23} V_0 \tag{3.17}$$

$$A_{21} = k_{21} V_0 \tag{3.18}$$

$$A_{31} = \frac{V_0 (1 - k_{12} k_{21})}{k_{12}}. \tag{3.19}$$

The following numeric example is chosen:

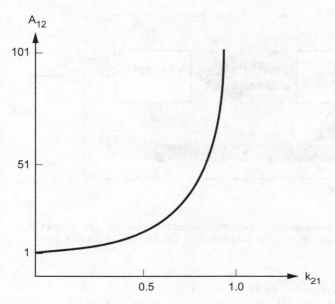

Figure 3.15
The sale flow of glass bottles A_{12} as a function of the recycling transfer coefficient in consumption k_{21} (see equation 3.13 for A_{12}) with a given constant glass bottle input into the system.

$V_0 = 146$ kg (see result with a given constant input in figure 3.14)

$K_{12} = 1$

$K_{21} = 0.95$

$K_{23} = 0.05$

The resulting flow scheme is given in figure 3.16. In comparison with figure 3.14, we can observe that an increase of glass recycling of roughly 10% (transfer coefficient k_{21} from 0.87 to 0.95) leads to reduction of the input (import) and output (waste) of roughly 60% (from 19 to 7 kg/c.y).

It follows that because of political considerations with regard to waste minimization, namely to increase the recycling proportion at a relatively high level (see also chapter 5, "Designing Waste Management"), recycling decisions can have severe economic effects for the suppliers of import goods.

For further descriptions of systems in a quasi steady-state situation (i.e., only the stocks M of the processes are time dependent), we recommend consulting Baccini and Bader (1996). Because field measurements or statistics are always accompanied by errors, the same mathematical formulation can be extended by error propagation calculation and sensitivity analysis.

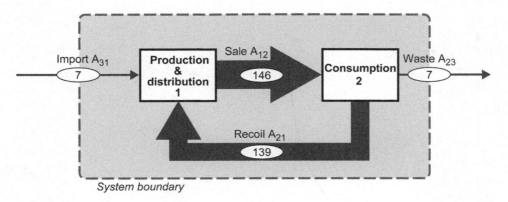

Figure 3.16
Input–output example for a simplified regional glass bottle management. The sale and the recycling coefficient are given. Consumers recycle 95% of the bought bottles. The thickness of the flow arrows is proportional to flow quantities in kg/c.y.

Time-Dependent Material Systems: The Dynamic State

In reality, stocks and flows of material systems are time dependent. In addition, real material systems are more complex with regard to the number of processes and flows. Again, the glass bottle example is applied to illustrate the essentials of a dynamic system. The statistical data stem from the Swiss situation in the years between 1960 and 1990. The concrete situation in this country asks for a differentiation between *Production* of glass bottles (from the imported *raw materials*, silicon oxides and carbonate salts, and from *recycling* glass) and *Distribution* of glass bottles with the liquid. This is necessary because the system offers two ways of reusing the consumed bottles: The distributor takes back the reusable bottle (this path is called the bottle *multiuse*) or the producer takes back the glass to make new bottles (glass *recycling*). In addition, distributors import and sell various liquids in glass bottles ("import bottles"). The extended system has now three processes and eight flows (figure 3.17), or 11 variables. Looking over several decades, the flows change within this period due to changes in technology, logistics, marketing, and lifestyle.

From statistical data there is information on three flows; namely, imported raw materials, recycling, and imported bottles. They are presented in figure 3.18. Recycling (bottle glass back to the producer) started in the 1970s, grew continually, and led to a corresponding reduction of the imported raw material. The multiuse path was already installed and functioning before the 1960s.

Model Formulation

1. The changes of stocks are neglected. Therefore, the three balance equations are as follows:

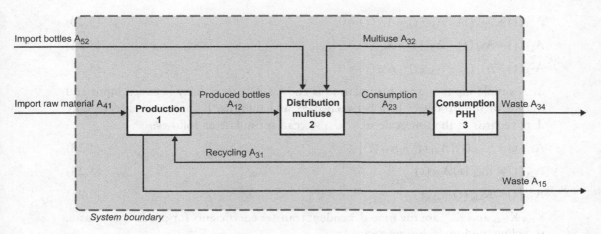

Figure 3.17
Material system for an extended regional glass bottle management (compare with figure 3.13).

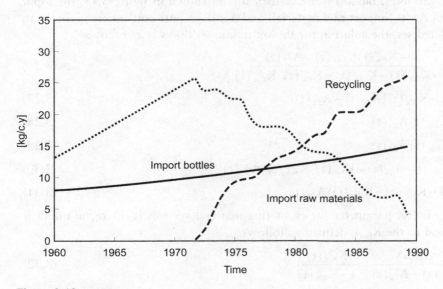

Figure 3.18
Glass bottle flows in Switzerland: graphic presentation of statistical data (after Baccini & Bader, 1996).

$$A_{41}(t) + A_{31}(t) - A_{12}(t) - A_{15}(t) = 0 \tag{3.20}$$

$$A_{12}(t) + A_{32}(t) + A_{52}(t) - A_{23}(t) = 0 \tag{3.21}$$

$$A_{23}(t) - A_{31}(t) - A_{32}(t) - A_{34}(t) = 0. \tag{3.22}$$

2. A simple input–output formulation is chosen. The time delay between input and output of the processes lays between days and months. Because we observe only data records in the timescale of years, we can formulate as follows:

$$A_{15}(t) = K_{Pw}(t)[A_{41}(t) + A_{31}(t)] \tag{3.23}$$

$$A_{31}(t) = K_{Rec}(t) A_{23}(t) \tag{3.24}$$

$$A_{32}(t) = K_{Mu}(t) A_{23}(t). \tag{3.25}$$

K_{Pw}, K_{Rec}, and K_{Mu} are the time-dependent transfer coefficients for production waste, recycling, and multiuse, respectively.

3. With the already chosen model formulations, we have obtained nine equations for the 11 variables. Based on the statistical data shown in figure 3.18, the input functions of $A_{41}(t)$ (import raw material) and $A_{52}(t)$ (import bottles) are given. With this completed set, the solution for the six unknown flows is as follows:

$$A_{23}(t) = \frac{[1 - K_{Pw}(t)] A_{41}(t) + A_{52}(t)}{1 - K_{Rec}(t) - K_{Mw}(t) + K_{Pw}(t) \cdot K_{Rec}(t)} \tag{3.26}$$

$$A_{12}(t) = [1 - K_{Mu}(t)] A_{23}(t) - A_{52}(t) \tag{3.27}$$

$$A_{32}(t) = K_{Mu}(t) A_{23}(t) \tag{3.28}$$

$$A_{31}(t) = K_{Rec}(t) A_{23}(t) \tag{3.29}$$

$$A_{15}(t) = K_{Pw}(t) \cdot A_{41}(t) + K_{Pw}(t) \cdot K_{Rec}(t) \cdot A_{23}(t) \tag{3.30}$$

$$A_{34}(t) = (1 - K_{Mu}(t) - K_{Rec}(t)) A_{23}(t). \tag{3.31}$$

A generally useful parameter for evaluating material systems is the recoil quota K (not identical to the K_{Rec}), defined as follows:

$$K(t) = \frac{A_{31}(t)}{A_{52}(t) + A_{12}(t)} = \frac{K_{Rec}(t)}{1 - K_{Mu}(t)}. \tag{3.32}$$

This parameter indicates the ratio between the glass (from bottles) that flows back (within the system) to the new production (within the system) to the sum of bottles flowing to the distributor. It is a measure for the "system's ability" to keep the material glass (from bottles) within the system. K is directly proportional to K_{Rec} but in a nonlinear relation to K_{Mu}. In other words, the Mu path has, quantitatively seen, a much stronger influence on the recoil quota of the whole system.

From statistical data the values for $K_{Rec}(t)$ and $K_{Mu}(t)$ can be estimated and are presented in figure 3.19. The values for K(t) were calculated with Eq. 3.32.

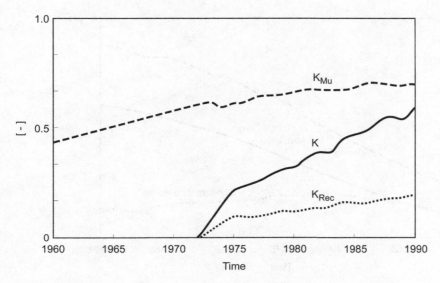

Figure 3.19
Estimated transfer functions for $K_{Rec}(t)$ and $K_{Mu}(t)$ and calculated values for the recoil quota $K(t)$ (after Baccini & Bader, 1996).

Based on this model, the dynamic development of the regional material system "glass bottles" can be reconstructed and is illustrated with three flows Sale $A_{23}(t)$, Multiuse $A_{32}(t)$, and Consumption waste $A_{34}(t)$ in figure 3.20.

For the end of the observed period (in 1990), a flow scheme is given in figure 3.21a. It shows, in a three-process system, the same sale flow of 146 kg/c.y, already given for the steady-state case with the given sale flow (see figure 3.16). Differentiating between the two reuse paths, the multiuse flow $A_{32}(t)$ shows the dominant role in the recoil quota. The hypothetical effects of eliminating only the recycling path A_{31} (figure 3.21b) or cutting also the multiuse path (figure 3.21c), calculated with the elaborated mathematical model, illustrates the importance of the path multiuse.

The glass material system reacts relatively quickly to technological changes (e.g., the introduction of the recycling path A_{31}). This has to do with the fact that glass bottles have very short residence times (days to months). However, the material glass is a "long-living" substance and can be kept within the system.

The mathematical model helps not only to simulate scenarios for material systems, built up with the MFA method, but also to elucidate the relevant properties of the system. It can support evaluation of metabolic effects within the system if one changes just one or several parameters. The glass system's quick reaction has to do with the fact that the stocks play practically no role in the metabolic game. The glass is on the road. If we choose systems with large dynamic stocks, the system's reactions

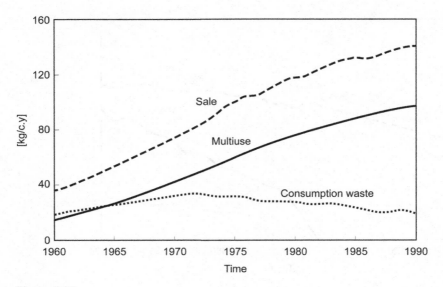

Figure 3.20
Model calculations for the dynamic regional material system "glass bottles." Sale, multiuse, and waste flows as functions of time, in kg/c.y.

are much different. An illustration for such cases will be given in chapter 5 with the design study on "Urban Mining." Above all, this model approach deepens the understanding of the metabolic system and gives also insights into possible effects stimulated by regulatory decisions of the overriding anthropogenic system. It does not, as discussed below in this chapter, answer specific questions on ecological impacts and resource demands. It does not, applying again the glass example, give an answer to questions on the ecological impact using glass bottles. However, it is of crucial importance to have sound metabolic models to test recommended changes that should be inserted in a metabolic system based on impact evaluations.

Examples of Software to Model MFA/SFA

To support investigations into and the understanding of material systems, models and corresponding software have been developed for calculating mass balances in a static as well as dynamic way. The main purpose of these models is to facilitate the establishment and calculation of MFA/SFA systems by setting a strict methodological framework for the user. The software allows us to expand considerably the work done earlier by pencil and paper or by a simple handheld calculator. It allows the user to convert his subjective view of a material system into a predefined formalism: to draw systems boundaries, to define processes with and without stocks, to link the processes with flows of materials, to insert available data, and finally to

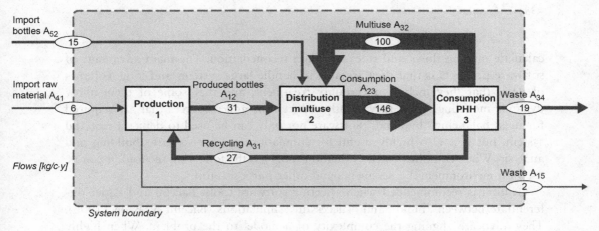

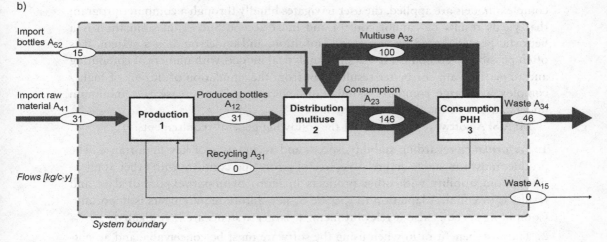

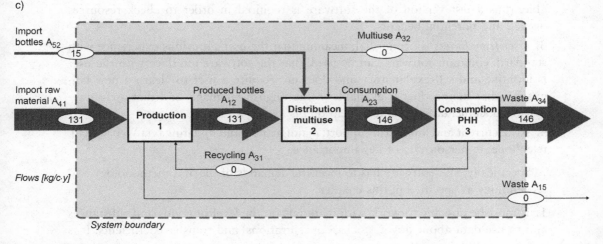

Figure 3.21
Synopsis of three situations for a regional material system "glass bottles" (Switzerland, after Baccini & Bader, 1996). (a) Status in the year 1990; (b) hypothetical elimination of recycling; (c) elimination of recycling and multiuse.

calculate missing flows and stocks by data reconciliation. The main advantage of software products is that they are able to handle large systems and large volumes of data, that they include uncertainty treatment, and that some of them allow dynamic modeling. Thus, they greatly reduce time and resources that are required for mass balancing "by hand." Software not only can be used to depict a material system, but it is also highly useful for simulation and for scenario building and analysis. When combined with evaluation tools (cf. the following section), it can be used for environmental assessments and other purposes, too.

Caution is recommended when selecting software tools. Baccini and Bader differentiate between "blind" and "successful" simulations (Baccini & Bader, 1996). They advocate aligning the complexity of a model to the problem. When highly complex models are applied, the user navigates blindly through a computer program that yields results he cannot control and must accept. Successful simulation is a heuristic process based on experience, intuition, and understanding a system. It is often possible to combine a few simple analytical models with numerical computing and to get the same or better results than from the simulation of dozens of highly complex and large computer simulations. Necessary in any case is a minimum understanding of the underlying system.

MFA/SFA software should fulfill the following general requirements:

1. *Performance* regarding stability, speed, and accuracy is of key importance. The product must run stable and reliable, free of errors and conflicts with other applications. Compatibility with other products in terms of import/export of data and results is essential. Adaptation to specific or new future requirements is important. Computing speed is also a relevant factor if larger systems are investigated.

2. The cost–benefit ratio when using the software must be conceivable and acceptable, thus a test version of the software is required in order to check resources needed and potential benefits.

3. *User friendliness* is primordial, meaning that the user's previous experience with standard, common software can be used, that the software interface with the user is intuitive and self-explanatory and does not require a user to "learn a new language," that the product is available in the English language, and that a comprehensive documentation (manual, help file) is available.

4. Also, current and long-term support, maintenance, and updating (via Web, e-mail, telephone, in person) are of key importance.

Specifically, SFA software has to meet the demands of the SFA methodology and terminology as specified in this chapter.

1. It must be possible to work with materials on the level of goods and substances and to use data about flows, stocks, concentrations, and transfer coefficients. It is

a prerequisite that the software applies the principle of conservation of mass for processes and systems on both the material and the substance levels. Application of standard symbols as presented in figure 3.3 facilitates the communication with other authors. In order to keep an overview of complex systems comprising many processes, it must be possible to aggregate processes to subsystems and to define subsystems. Also, the software must be able to cope with stocks, with feedback loops (materials flowing back to a process they have been in before), and storage with time-retarded outputs.

2. Regarding *data*, the software should have the following features: the possibility to input data locally and remotely in the form of values, functions, or graphs from external databases (e.g., Microsoft Access, Microsoft Excel, DBase, Oracle) or via scripting languages. Efficiency of data input is of prime importance because this is a time-consuming process. It must be possible to cross-check easily input data for plausibility and correction of errors. The output data (results) can be presented in different numerical and graphical ways (tables, figures, Sankey diagrams).

3. The software should be able to model both the static as well as the dynamic behavior of a system. It must be able to manage uncertainties applying different kinds of statistical distributions and to reconcile an overdetermined system. The propagation of uncertainties can be taken into account by using Gauss' law of error propagation and/or Monte Carlo simulation methods.

4. By the application of the software, three results are expected: (i) a "model," that is, a description of the actual state of a given material system; (ii) the possibility of using this model for simulation of scenarios with new processes, new materials, or new input data; (iii) scenario analysis for comparing the original system with alternative systems.

As of today, many software products have been developed in the field of MFA, SFA, and LCA, such as SIMBOX (Baccini & Bader, 1996), STAN (Rechberger & Cencic, 2010), SFINX (Van der Voet, 1996), ORWARE (Dalemo et al., 1997), FLUX (Olsthoorn & Boelens, 1998), DYNFLOW (Elshkaki, 2000), GaBi (PE International, 2010), and UMBERTO (IFU Hamburg GmbH, 2010). A review of available software is given in Elshkaki (2007).

There are two software products that fulfill almost completely the requirements stated above and that are tailor-made for MFA/SFA methodology as presented in this volume: the commercial product SIMBOX (not to be mistaken for SIMbox or SIM BOX) developed by Bader at Swiss Federal Institute of Aquatic Science and Technology (Eawag), Switzerland (Bader & Scheidegger, 1995), and the freeware STAN by Cencic and Rechberger at the Vienna University of Technology, Austria (Rechberger & Cencic, 2010). Both adhere strictly to MFA terminology as presented in this chapter, and both are particularly well suited to investigate material systems.

The results of SIMBOX and STAN are simplified and idealized numerical images of the material system investigated. In the first place, these models promote a better understanding of the material system. This is in contrast to other software products on the market that have specific purposes as assessment tools, such as a focus on the environment, on sustainability reporting, on eco-design, and so forth. With some of these products, the substance level can be addressed, too. However, their methodological structure is not based on the MFA/SFA approach presented in this volume. In addition to SIMBOX and STAN, the two commercial products GaBi and UMBERTO are briefly presented. The reason is that they have been widely applied for LCA and environmental assessment and that they can also be used for analyzing material systems on the level of goods and substances. They are based on different calculation models, are well established, supported, and maintained, are available in the English language, and are accessible as test versions. For a detailed comparison of the application of GaBi, UMBERTO, and MS Excel, including a case study of MFA, see Cencic (2004).

MFA/SFA-Based Software Products

SIMBOX
The software SIMBOX (Baccini & Bader, 1996; Bader, 2010) has been developed for simulating material, substance, energy, and monetary flows in anthropogenic systems. It is programmed in HT Basic and designed to run on a standard Windows PC (Windows 95 up to Windows XP). The user has the choice to define and adapt the investigated system either manually graphic-interactively or automatically (using a system generator). Data can be entered either by assigning data graphically to flows and processes or by using a data-input interface. The user defines model equations; both stationary as well as dynamic systems can be modeled by SIMBOX, which calculates the flows and stocks for a current state (stationary system) or as a function of time (dynamic system). For standard input–output approaches, predefined built-in equations can be used. A particular feature is that the software offers calculation modules for specific purposes, such as the calculation of total substance flows, of transfer coefficients, and of missing flows, including corresponding uncertainty. Also, best estimates for a given set of data can be calculated by a least squares fit. The software is designed to evaluate credibility of the output by means of uncertainty analysis and uses sensitivity analysis to determine key model parameters. It is well suited for simulations of parameter variation and for scenario analysis. The results of SIMBOX consist of either numerical readouts (data export to other programs) or graphs representing key flows and stocks with their uncertainties. Recently, monetary flows have been combined with SIMBOX for investigations into economic aspects. Expert judgments can be included as probability functions in order to include subjective knowledge.

However powerful SIMBOX is as a dynamic SFA modeling tool, access to the software remains somewhat limited and requires—in contrast to the other commercial products—contacting and purchasing the software from the authors.

STAN

STAN—the acronym stands for sub<u>st</u>ance flow <u>an</u>alysis—is a user-friendly freeware that supports MFA according to the Austrian standard ONORM S 2096. It runs under Windows and was programmed in C#. The system requirements are Windows XP with Service Pack 1 or higher; Microsoft .Net Framework 2.0 or higher; Intel Pentium III 1 GHz with 512 MB RAM or more; and 20 MB free disk space. STAN fulfills the software requirements given earlier. It draws on three types of equations to describe the complete material system in a mathematical way:

1. Mass conservation equations (e.g., balance equation and stock equation):

Σ inputs = Σ outputs + change in stocks

$stock_{Period\ i+1} = stock_{Period\ i} +$ change in $stock_{Period\ i}$ with $i \ldots$ period.

2. Linear relations equations (e.g., transfer coefficient equation or additional linear relation equation between similar quantities):

$output_x =$ transfer coefficient $_{to\ output\ x} \cdot \Sigma$ inputs

flow x = factor $\cdot$ flow y.

3. Concentration equations:

$mass_{substance} = mass_{good} \cdot concentration_{substance} \cdot mass_{substance} = volume_{good} \cdot$ $concentration_{substance} \cdot mass_{good} = volume_{good} \cdot density_{good}$.

The input data into these equations may comprise unknown, measured, and/or exact (constant) variables. If through elementary transformation of the three equations (equality constraints) at least one equation can be found that contains no unknown and at least one measured variable, the data set can be reconciled in order to improve the accuracy of the results. Some of the equality constraints used may be nonlinear. Hence, for solving the problem of nonlinear data reconciliation, successive linear data reconciliation is applied iteratively. Statistical tests are used to identify sources of errors such as random or gross errors, thus improving values that are subsequently used to calculate unknown quantities. Corresponding uncertainties are determined by error propagation.

The user builds an MFA/SFA system by using a graphical toolbox of the STAN desktop with predefined components for processes, flows, subsystems, system boundaries, and text fields. Subsystems offer the opportunity to model the inner structure of a process in more detail by disaggregating into subprocesses. Data about mass flows, stocks, concentrations, and/or transfer coefficients is inserted by hand or imported from databases. In addition to the two levels of goods and substances, data for several time periods can be included, too. Although not designed for dynamic modeling like SIMBOX, STAN is also suitable for dynamic modeling if the time period feature is used. Instead of applying continuous mathematical input functions, STAN offers the possibility to assign single values of discretized functions or time series to the corresponding time steps (periods). Additional levels such as energy or economic units are not yet standard but are also possible as inputs. For default values of input data, mean values of a standard distribution are assumed. However, if uncertainties are known, values including standard uncertainties of the mean value can be inserted. If sufficient data about a system is available, calculation algorithms of STAN allow one to make use of redundant information to reconcile uncertain data and subsequently to compute unknown variables including their uncertainties (error propagation). Gross errors in a given data set can be detected by statistical tests integrated in the software. STAN results comprise both numerical outputs (tables for export into other common formats) and graphical outputs (mass balances and freely scalable Sankey diagrams) on the level of goods and substances. Graphs can be exported in various common formats and resolutions. If multiple periods have been investigated, it is possible to display the results of a single flow as a time series. Resulting data can be rendered anonymous by referring to an arbitrary value (e.g., in percent or fraction of total import or export).

One of the main features of STAN is its user-friendliness. The system design allows straightforward composing and changing of systems components. The hierarchical levels of goods and substances are easy to administer, and additional substances or other attributes can be added if needed. The system boundary and values for total import/export and stock changes are automatically displayed. Inconsistencies in data can be reconciled, and best fits can be found without guessing.

Other Software for Modeling MFA/SFA

GaBi

The acronym of the GaBi 4 software system for life cycle engineering originates from the German *GAnzheitliche BIlanzierung*, which means "comprehensive balancing." It was designed by the Institute for Polymer Testing and Polymer Science at the University of Stuttgart and is managed today by PE Europe GmbH (PE International, 2010). It runs on a standard Windows PC equipped with Windows XP or higher operating system. GaBi is a software tool for modeling products, processes, and systems from a life cycle perspective. It allows modeling emissions from products and processes, balancing inputs and outputs of materials and energy, aggregating results for graphic display, and features an add-on to create interactive reports.

GaBi 4 is a package consisting of software and life cycle database covering a broad range of topics.

GaBi is based on linear equation systems and has the following functionalities: Like STAN, it allows the user to define the system graphically by an intuitive user interface with drag-and-drop features. Data are entered via a hierarchical database manager allowing the use of default or customer-supplied data. Default data input is supported by an embedded database documentation set up according to the International Reference Life Cycle Data System ILCD published by the European Union (European Union, 2010). Customer data can be imported in many formats using a data exchange tool. Data input is not restricted to materials and energy but encompasses also life cycle costing (LCC) and life cycle working environment (LCWE) data, and includes data quality indicators, too. It is possible to consider data uncertainties and their propagation within the system and to include normalization and customer-defined weighting factors if needed. GaBi allows advanced modeling techniques such as scenario analysis of material systems, enabling support for development and optimization of products and processes.

The outcome of GaBi is presented in a balance window that is designed to view life cycle inventory (LCI) and life cycle analysis (LCA) results. Features important for LCA are incorporated, such as switching between different impact categories of LCAs (e.g. Eco-Indicator 99; CML 2001 by the Institute of Environmental Sciences, Leiden University; EDIP 2003 (Environmental Development of Industrial Products) by the Institute for Product Development (IPU) at the Technical University of Denmark; TRACI (Tool for the Reduction and Assessment of Chemical and Other Environmental Impacts) by the U.S. Environmental Protection Agency). Individual impacts categories and specific emissions of a product in an LCA can be identified and quantified with filters. Customizable graphs are available for presenting specific results or for transferring results into other programs for creating own tables and graphs. GaBi offers special tools for sensitivity analysis, Monte Carlo simulation, and interpretation of results.

With all its advanced features, GaBi is clearly directed toward LCA and the corporate world for environmental reporting and sustainability reporting. Also, GaBi is well suited for analyzing and improving products and processes. Even if it has not been designed as a software to investigate metabolic systems, it has the potential to fulfill this task, too. However, because terms and definitions of GaBi differ significantly from MFA/SFA methodology, a considerable effort is needed to apply the sophisticated and elaborated GaBi software for investigations into metabolic systems. Also, in contrast to, for example, SIMBOX, GaBi is not focused on stocks and stock balancing and is not designed as a dynamic model, and thus dynamic simulations are difficult to perform.

UMBERTO

This software has been developed by the Institute for Energy and Environmental Research Heidelberg Ltd. in cooperation with the Institute for Environmental Informatics Hamburg Ltd. It is designed as a software tool to support process optimization during manufacturing (company scale) or along a product life cycle (product scale). UMBERTO visualizes a process model that can be assessed according to different evaluation criteria and allows subsequent optimization of the process. It is available in four versions for consultants, business, education, and as a test version. The basis of the software is so-called Petri networks, which model the network of material and energy flows. Petri networks originate from informatics; they facilitate definition and calculation of complex systems of material flows and inventories. As in the case of GaBi, UMBERTO methodology and MFA/SFA terms and definitions presented in this chapter are not compatible.

As in the case of STAN, a graphical user interface (GUI) allows comfortable defining of even complex systems. System components can be chosen freely and are referred to as products, raw materials, pollutants, forms of energy, and so forth; they are organized in a hierarchically structured material list. Components are defined either by the user or by recalling modules from the UMBERTO library, which contains data about numerous standard processes. Data about production processes, material flows, and inventories can be entered from such libraries, from external information systems, or from own, internal calculations. In the case of establishing an LCA, a life cycle inventory based on the Ecoinvent database is available to support data input. The software calculates material flow networks including cycles (loops) sequentially and locally, independent of flow directions. Besides materials, also information about energy and costs can be entered into the model. UMBERTO accepts change of the level of detail when needed. Thus, the user can start with the highest aggregation stage (e.g., a company) and later focus on a specific production line or a process if needed. Because causal relationships are modeled, the consequences of changing parameters remain transparent.

The results about flows and stocks calculated by UMBERTO can be evaluated using standard or individual environmental and economic performance indicators. Both cost accounting and cost allocation are possible, as well as scaling per unit of products or per period. The software is suited to identify relevant flows and environmental effects of production processes and products. It can be combined with scenario analysis, characterized by a graphical network structure, and thus allow optimization of production processes. Time periods for balancing can be defined for each scenario (e.g., the fiscal year or a month), thus enabling time-dependent calculations. The main application areas of UMBERTO are to model and optimize production processes and to perform LCAs and LCC. As in the case of GaBi, terminology and methodology do not correspond with those presented in this chapter.

Thus, whereas both software products are well suited to and have extensive experience with environmental assessments and similar challenges, they are not designed to investigate anthropogenic systems. Up to now, it is the privilege of SIMBOX and STAN to be tailor-made for exactly that purpose.

Evaluation of Metabolic Processes

The system "anthroposphere" is created by human beings to fulfill their needs. Because external circumstances (environment, first-order resources) may change as well as internal boundary conditions (technology, societies, and other second-order resources), the anthroposphere and its metabolic processes are transformed continuously with varying rate of change. To recognize early and to understand beneficial and harmful developments induced by these transformations, it is necessary to evaluate ongoing changes. Also, to prepare for the future, a careful assessment of potential future directions is necessary. In short, methods to evaluate metabolic processes are required. Such methods must be based on two pillars: first, a sound natural science backbone that forms a solid numerical framework for the second pillar, the evaluation step. This step is crucial: In contrast to the objective measurements of flows and stocks of materials, the interpretation and evaluation of MFA results is a subjective process that is based on values of individuals and societies. It is important to note that—in contrast to natural science principles such as the law of conservation of matter—values change over time, and hence the evaluation step is a dynamic process that is also subject to transformation over time.

There are two different categories of evaluation methods: The first group such as LCA, ecological footprint, and SPI aims at determining the impact that a process, material flow, or human action has on a target such as people, environment, resource base, or the like. The objective of the second group of MFA-based evaluation methods is to understand the metabolic processes in the context of the triptych anthroposphere, activities, and material flows: What are the relevant flows and stocks in view of basic human needs (activities), how do they develop over time, and what controls them? The main difference is that in order to reach their goals, methods of the first category focus on flows at interfaces, such as anthroposphere/environment, whereas methods of the second category concentrate on the metabolic, sometimes anthropogenic (to control!) system, taking into account key flows and stocks of relevant goods and substances according to the mass balance principle. This chapter serves to test the hypotheses that (1) only methods of the second category are appropriate to investigate metabolic systems, and (2) if methods of the first category are based on mass balancing by MFA/SFA, their significance is highly enhanced because of the possibility of cross-checking and linking with metabolic investigations.

In the following, we address three questions: "What is the main objective of the method?" "What are the evaluation indicators used, and how have they been selected?" "What is the physical basis of the evaluation system?" These questions center on the key issue involved in the assessment of any system: What is the purpose of evaluation, and which evaluation methodology is appropriate to fulfill this purpose? Objective and methodology belong together. Behind these questions is the matter of values, because each assessment method summarized in the overview of methodical approaches for metabolic studies of chapter 2 and discussed here in more detail has a general objective based on values that are shared by the inventor and the user of the method. Whereas the following paragraphs serve to answer the above questions and to give an overview of current assessment approaches, the text is not intended to compare or score the various methods such as LCA, SPI, and so forth.

SFA-Based Assessment Methods
In the following, three SFA-based evaluation methods are briefly introduced and discussed in view of the questions stated earlier. The methods have been selected (1) because they require SFA results founded on the law of conservation of mass, and (2) they have been successfully applied to investigate metabolic processes. For a detailed description of these methods, see Brunner and Rechberger (2004).

Anthropogenic versus Geogenic Flows
The principle of this evaluation method first mentioned in Güttinger and Stumm is to compare anthropogenic substance flows and stocks with geogenic substance flows and stocks and to apply the precautionary principle to this comparison (Güttinger & Stumm, 1990). As a normative code derived from the precautionary principle, the following two rules have been postulated by SUSTAIN and cited in Narodoslawsky and Krotscheck (SUSTAIN, 1994): (1) Anthropogenic substance flows should not exceed local assimilation capacity and should be smaller than natural fluctuations of geogenic flows. They should not alter natural stocks. (2) Anthropogenic substance flows should not alter the quantity and quality of global substance flows and stocks.

For an anthropogenic versus geogenic flows (A/G) evaluation, a regional SFA is required. For a given process, metabolic system, or activity, the anthropogenic flows from and to the environment are determined by establishing mass balances of relevant substances. The mass balance approach is crucial in order to consider all possible flows and stocks of a substance. If not all effluents are taken into account or if stocks are neglected, the system cannot be evaluated comprehensively: It is likely that a fraction of a substance is hidden in processes like multiple recycling or in the stock of a landfill, and thus this fraction might bypass the A/G comparison.

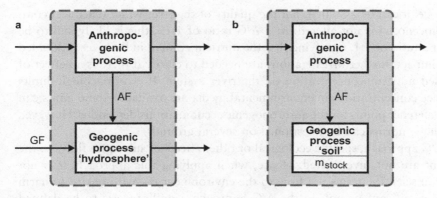

Figure 3.22
Exemplification of evaluation based on comparing (a) anthropogenic flows (AF) and geogenic flows (GF), respectively (b) anthropogenic flows and geogenic stocks. The ratios AF/GF respectively AF/m_{stock} serve as a measure for environmental pressure of an anthropogenic process. For AF/GF, a ratio of approximately <0.01 signifies no pressure.

The actual sources and recipients of the flows from and to the environment are defined (specific compartments in water, soil, and air such as groundwater or forest soils of particular regional qualities), and the geogenic substance concentrations in flows and stocks of these compartments are determined. It may be difficult to find genuine geogenic concentrations because many of today's ecosystems show traces of anthropogenic activities. Also, natural geogenic concentrations vary considerably between regions; hence, it may require considerable efforts to find or estimate regional geochemical data about substance concentrations in unpolluted soil, air, surface water, and groundwater. Next, the contribution of anthropogenic substance flows to geogenic substance flows is calculated for each particular compartment. Also, the accumulation—or, for example, in the case of resource extraction, the depletion—of substances in natural stocks is analyzed. This comparison allows evaluation of whether a man-made action induces substance flows that alter geogenic flows and whether natural stocks are likely to be changed in a harmful or beneficial way (cf. figure 3.22).

Main Objectives
There are three main objectives of the A/G approach: (1) to foster understanding and to provide physiologic information about material systems, (2) to compare geogenic and anthropogenic flows and stocks, and (3) to use this information to test the hypothesis whether the flows at the interface anthroposphere–environment are likely to change environmental and resource qualities.

As an example: Assume that the flow of nutrients from PHHs as displayed in figure 2.9 remains below 1% of the geogenic nutrient flow in a river system. Based on experience, it can be assumed that the quality of the river water (chemical composition, biocenosis) is not altered. At a A/G ratio of 1:10, this is likely still to be the case. At a ratio of 1:1, signifying that the nutrient content in the river is doubled by human impact, further investigations are needed to clarify if there is an effect of the increased nutrient concentration on the river system. If eco-toxicologic limits for substance concentrations in environmental media are available, these limits can be used as reference points for the anthropogenic accumulations/depletions. However, eco-toxicologic norms can be questioned on several grounds.

In the A/G approach, specific ecological or other effects of substance flows to the environment are not investigated. Hence, when applying this approach, it is not discernible in scientific terms if a flow to the environment is—with regard to harm or benefit—significant or not. In the A/G approach, significance is to be defined rather by heuristic and precautionary arguments. This may be regarded as a weak point for the A/G approach. However, this reflects the inherent problem of evaluation as a subjective process.

Physical Basis of Evaluation
The starting point of an A/G evaluation is a mass balance on the level of substances (SFA) according to the law of the conservation of matter. Terms and definitions meet the specifications of SFA methodology. A second basis consists of the flows and stocks of the specified substances in the regional environment. The normative element of the method comes into play when a certain ratio of A/G is set as a limiting value.

Indicators
The main indicator is the quotient A/G, representing the ratio of an anthropogenic substance flow to the corresponding flow in water, air, or soil of the environment. This main indicator is based on individual substances that are chosen because of their properties to act as indicators for various purposes. These indicators are selected in accordance with the objectives: They are on one hand key substances that characterize processes, metabolic systems, or activities investigated (e.g., the nutrients phosphorous and nitrogen for the activity TO NOURISH or carbon for the process *Room heating*). On the other hand, for the purpose of evaluation, indicator substances are selected that might be of relevance as resources or as potential pollutants for investigations about material systems, resource conservation, and environmental protection. In the ideal case, the two kinds of indicators coincide, such as N and P, which are crucial for both the activity TO NOURISH in PHHs and eutrophication of aquatic systems. In other cases, the set of indicators used in mass balancing has to be expanded to include not only substances yielding physiologic information but also additional substances that act as indicators for resources

and environment. Such a case represents the system "waste management," where often trace substances have to be considered that are of little relevance to the physiologic information (cf. chapter 5).

Statistical Entropy Analysis

Statistical entropy analysis (SEA) is a quantitative method to assess the power of a process or system to concentrate or dilute substances (Rechberger, 1999). The results of an SEA are relative statistical entropies (RSEs) of inputs and outputs of a process or system and substance concentrating efficiency (SCE), which is calculated from RSEs (see below). SCE is of relevance because it is a basic characteristic of any metabolic process or system and because concentration and dilution are important processes: If substances are highly accumulated in a good, the good may become a resource; if they are dissipated (that means dispersed to a point that they cannot be recovered with a finite amount of energy), they are lost for reuse and possibly pollute the environment. Traditional metrics to assess environmental pollution focus on emission loads and on substance concentrations in the environment. They are not designed to and cannot take into account the distribution of substances in processes or systems. This is a drawback when it comes to the evaluation of processes such as, for example, recycling or waste treatment. Let us assume that two waste management systems A and B handle the same waste and have the same emissions and that both comply with all environmental standards. System A produces three fractions: emissions that comply with environmental standards, a small mass fraction that is highly concentrated in hazardous substances, and a bulk fraction that is nearly clean of hazardous substances. System B also generates three fractions: the same emissions that comply with environmental standards, and two fractions of equal mass that both contain slightly elevated concentration of hazardous substances. System A is preferable from a metabolic, environmental, and resource point of view: It yields a concentrate of hazardous substances that possibly can be recycled or that can be treated properly before final storage, and it "purifies" the bulk of waste enabling recycling or further treatment. SEA is designed to recognize and quantify the differences of the two systems. Today, especially in waste management, the importance of concentrating substances as resources and dissipating pollutants in sinks is not yet fully appreciated. One reason may be the earlier lack of an appropriate metric to quantify these effects.

SEA is based on rigid MFA/SFA and requires little additional computing. Because information on both goods and substances is required, neither MFA nor SFA alone satisfies the data need of SEA, and it is necessary to develop a complete mass balance on both levels of goods and substances. The method is able to quantify statistical entropy of single-process systems as well as that of systems containing several processes. In SEA, a system is regarded as a procedure that transforms an input set of concentrations into an output set of concentrations; the same applies to the mass flows. Each system is a unit that concentrates, dilutes, or leaves unchanged its throughput of substances. To measure this transformation, an appropriate function that quantifies the various sets is required. The transformation can be defined as the difference between the quantities for the input (X) and the output (Y). This allows determination of whether a system concentrates (X − Y > 0) or dilutes (X − Y < 0) substances.

To calculate X and Y, a mathematical function developed in information theory is used. C.E. Shannon measured the loss or gain of information by a system. He applied a function that originates from Boltzmann's statistical description of entropy and developed the so-called Shannon entropy. Note that Boltzmann's thermodynamic entropy, termed S and measured in J/(mol·K), and Shannon's statistical entropy, termed H and measured in bit, are formally identical. However, their background is different, and there is no physical relationship between the two: the former applies to thermodynamics and the latter to information sciences.

The statistical entropy H of a finite probability distribution is defined by Eq. 3.33

$$H(P_i) = -\lambda \cdot \sum_{i=1}^{k} P_i \cdot \ln(P_i) \geq 0, \tag{3.33}$$

where P_i is the probability that event i happens.

For illustration, let us assume a probability distribution for a case where one of three events can happen. In principle, there are three different distributions possible: (1) the probability of an event, for example, 2, is unity ($P_2 = 1$). The statistical entropy of such a distribution is zero. (2) The probabilities for all three events are equal. The entropy of such a distribution becomes a maximum. (3) All other possible combinations of probabilities must yield an H value between 0 and maximum. In order to be applied to material systems with known sets of substance concentrations and mass flows, the statistical entropy function is transformed in three steps (for a detailed description, see Brunner & Rechberger, 2004): (1) The statistical entropy function is applied to both the input and output of the investigated system. (2) The resulting equation for H^I is modified to quantify the attribute "mass flow." The mean concentration in a good (c_{ij}) is weighted with the mass flow ($\dot{m}_i$) of this good; $\dot{m}_i$ can be regarded as the frequency of "occurrence" of the concentration c_{ij}. The entropy H^{II} of a mass-weighted set of concentrations is deduced from equation H^I. (3) For a final adjustment of the initial equation, the gaseous and aqueous emissions are taken into account, which are diluted in atmosphere and receiving waters, resulting in an increase in entropy. The resulting H^{III} allows quantification of the distribution of a substance. The maximum of H^{III} is reached when the total flow of a substance is directed to the environmental compartment with the lowest geogenic

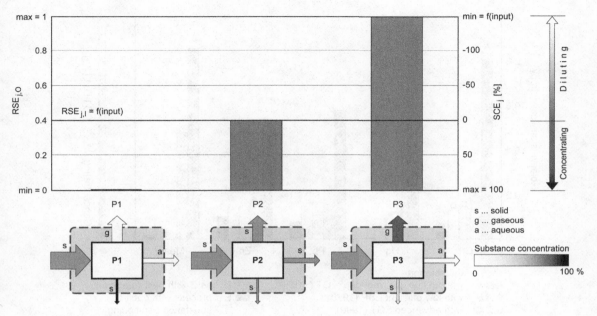

Figure 3.23
Example of relative statistical entropy (RSE) and substance concentrating efficiency (SCE) of three processes: Process P1 concentrates substance j at the maximum, producing a pure solid residue that consists only of substance j; all other products contain no substance j. Process P3 transfers substance j entirely into the atmosphere, which is equivalent to maximum dilution. Process P2 does not discriminate between the different outputs; all concentrations in inputs and outputs are the same (reprinted with permission from Rechberger & Brunner, 2002).

concentration. Using equations for H^I to H^{III}, the RSE of inputs and outputs of a material system can be calculated.

The SCE is defined as the difference in the RSE of inputs and outputs of a system. It is given as percentage and ranges between a negative value and 100%. An SCE value of 100% for a substance means that this substance is transferred completely to one "pure" output good. SCE equals zero if RSE values of input and output are identical. Such a system neither concentrates nor dilutes substances. However, this does not imply identical sets of mass flows and concentrations in inputs and outputs. The SCE_j equals a minimum if all of substance j is emitted into that environmental compartment that allows for maximum dilution (in general, the atmosphere). These relationships and an example are illustrated in figure 3.23.

SEA has been applied to MFA results of various systems such as waste treatment facilities and multiprocess systems such as the European copper cycle (Brunner & Rechberger, 2004). SEA considers all information of an MFA except the magnitude of stocks. Only inputs into and outputs from a stock are included in SEA. The relevance

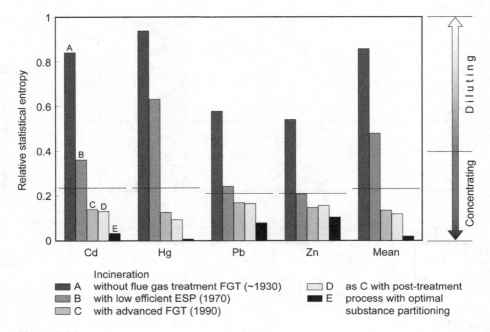

Figure 3.24
RSE values for the partitioning of Cd, Hg, Pb, Zn, and their weighted mean values for five MSW incineration technologies A–E of increasing technological sophistication. The horizontal lines indicate RSE
values for MSW. Incinerators without advanced flue gas treatment (A and B) dissipate heavy metals into
the environment. Modern incinerators (C and D) concentrate these substances. New technologies C and
D (including after-treatment of solid residues) are nearly as effective in partitioning Pb and Zn as the
most favorable process E, which is designed to produce environmentally compatible emissions as well
as Earth crust or ore-like residues (reprinted with permission from Rechberger & Brunner, 2002).

of a stock has to be assessed by comparison with other geogenic or anthropogenic
reference reservoirs (cf. figure 2.21). Figure 3.24 gives an example of how SEA quantifies the power of waste treatment processes to concentrate heavy metals.

Main Objectives
(1) To understand and quantify material systems in view of their power to concentrate and dilute substances. (2) To fill the gap of existing evaluation methods regarding substance flows that are neither contained in emissions nor evaluated in economic
metrics. (3) To support the design of processes and systems that deliberately concentrate substances for reuse instead of diluting and wasting them.

Physical Base of Evaluation
SEA methodology is completely built on MFA/SFA information: mass balances on
the levels of goods and substances are a prerequisite. It is indispensable to have

sufficient information about flows of goods and concentrations of the selected substances in these goods for the process or system evaluated. These mass balances are transformed into distributions that characterize the partitioning of goods and substances within a material system. It is important to note that SEA is based on Shannon entropy, a term from information science, and not on thermodynamic entropy.

Indicators
The two indicators are RSE and SCE. They are substance-specific characteristics of a particular process respectively system. To calculate these indicators, relevant substances have to be selected. The selection process for these substances must be based on preexisting knowledge about the fate of substances in particular treatments or systems. For example, to determine the SCE of waste incineration, the analyst must have sufficient knowledge about the partitioning of substances between the products of incineration (Hg $\rightarrow$ off-gas $\rightarrow$ scrubber sludge, Cd $\rightarrow$ filter ash, Fe $\rightarrow$ magnetic separation of bottom ash, Cl $\rightarrow$ scrubber water, C $\rightarrow$ off-gas, etc.). Only this knowledge enables him to select the appropriate set of indicator substances (here: Hg, Fe, Cd, Cl, C) that have to be included in the SFA. Without this information, an analyst might choose the substances Cr, V, Co, and Ni, which all behave quite similar in an MSW incinerator and are not representative of the process of incineration as a whole.

Exergy Analysis
In exergy analysis, material and energy flow data are used to produce a single aggregated measure of exergy changes within a system. The results are used to compare and improve the efficiency of industrial production systems as well as of entire national economies.

In thermodynamics, the exergy of a system is the maximum useful work that can theoretically be gained by bringing a resource into equilibrium with its surroundings through a reversible process. Resource in this context includes both energy and materials: If materials are considered, the chemical composition must be known. Reversible process stands, for example, for a process working without losses such as friction, waste heat, and so forth. The surroundings are defined as a reference state, for instance a reference environment, and must be specified by temperature, pressure, and so forth. For metabolic studies, the surroundings usually is the environment comprising atmosphere, oceans, and Earth crust.

Exergy is also referred to as "technical working capacity," "available work," and "essergy" (= essence of energy). It is an extensive property with the same unit as energy (e.g., J/g). In contrast to energy, there is no conservation law for exergy, and exergy is consumed or destroyed due to the irreversibility of any real process. The exergy content of a solid substance j can be taken from standard chemical exergy values $e^0_{ch,j}$ as introduced by Wall and listed in Ayres and Ayres (Wall, 1993; Ayres & Ayres, 1999). $e^0_{ch,j}$ are substance-specific values that are calculated for the standard state T_0, p_0. They relate to the mean concentration of reference species of substance j in the environment, assuming there is only one reference species for each element. Applying the standard chemical exergy values and having information about the chemical composition of materials, the exergy of materials can be calculated, and exergy balances for combined materials/energy systems can be established. A detailed description with examples is given in Brunner and Rechberger (2004).

Because exergy can be calculated theoretically for all materials and energy flows, it can be applied to any materials balance, too. Hence, it is predestined for aggregation of material flows and energy flows to one final indicator, namely exergy. For instance, industrial production and manufacturing can be described as a system that uses exergy in the form of fossil fuels and raw materials to produce consumer goods and wastes of less exergy. The technical efficiency of any system can be expressed as exergy efficiency. Exergy studies for single branches of industry as well as for entire national economies have been carried out (Wall, 1977; Wall, Sciubba & Naso, 1994; Michaelis & Jackson, 2000; Costa, Schaeffer & Worrell, 2001). Exergy is also used as an indicator for environmental effects of emissions and wastes (Wall, 1993; Rosen & Dincer, 2001). The rationale for this application is as follows: The higher the exergy flow in a material or energy, the more the flow deviates from the thermodynamic and chemical state of the environment and the higher is the potential to impair the environment.

A drawback of the method is that exergy and environmental impact do not strongly correlate. For instance, exergy values of substances emitted to the atmosphere are not proportional to their toxicity. While exergy values for PCDD/F (polychlorinated dibenzodioxins and polychlorinated dibenzofurans) and for carbon monoxide are quite similar (13.0 kJ/g versus 9.8 kJ/g) (Costa, Schaeffer & Worrell, 2001), their emission limits differ by eight orders of magnitude (MSW incinerators 0.1 ng/m^3 for PCDD/F and 50 mg/m^3 for CO). Exergy balances are often dominated by energy flows; materials such as wastes and emissions are of minor importance as displayed by the following example taken from Brunner and Rechberger (2004). The emission of 1 kg PCDD/F, which is about the order of magnitude of the total emissions of dioxins and furans in Germany in 1990, equals an exergy value of 13 MJ equivalent to the release of 500 L of warm water. Hence, caution is needed when the exergy approach is applied to material and metabolic systems. For detailed information on the application of exergy in resource and environmental accounting, see Wall (1977), Szargut, Morris, and Steward (1988), Baehr (1989), and Ayres and Ayres (1999).

Main Objectives
To define one single aggregated indicator that serves as a metric to quantify energy and material flow data in view of their effects on environment and energy use. The chosen indicator exergy is meant to compare and improve the efficiency of production systems as well as of entire national economies. Whereas exergy analysis yields values for each substance and energy involved in a process, it is not suited for fostering understanding of metabolic processes because it does not cover the level of goods.

Physical Base of Evaluation
Exergy analysis, a measure for the maximum amount of work that can be obtained by bringing a substance into equilibrium with its surroundings, is based on thermodynamics and is measured in energy units such as J/g. For exergy analysis, balances for energy and substances, and thus SFA, are indispensable. Exergies of all inputs and outputs of energy and substance are determined for a metabolic system. The normative character of the exergy analysis is given by specifying the term "surroundings." As "surrounding," the environment or any other specified state can be defined and used for comparison.

Indicators
The main indicator is exergy, a substance-specific, extensive property. To establish an exergy analysis of a metabolic process, exergies of all flows of substances and energy have to be selected as indicators. Hence, it may be that a large amount of substances has to be selected and included in the evaluation. There is no possibility to select an indicator as a representative for other substances, hence all substances pertinent for a process must be chosen.

Methods Including Both MFA and SFA level
The following methods (LCA, SPI, and ecological footprint) are based on both MFA and SFA results. Although none of the methods requires rigid mass balancing, it is possible to apply these methods to mass balances; in fact, such a practice would greatly enhance the validity of the methods.

Life Cycle Assessment
According to the Society of Environmental Toxicology and Chemistry (SETAC), life cycle assessment (LCA) is a standardized (ISO, 2006) method for

• evaluating the environmental burdens associated with a product, process, or action by identifying and quantifying energy and materials used and wastes released to the environment;

• assessing the impact of energy and materials used and released to the environment; and

• identifying and evaluating opportunities to effect environmental improvements.

LCA does not evaluate if a product or service is environmentally safe, but it provides indicators regarding impact assessment scores of the relative contributions of entire or partial product life cycles to specified impact categories.

The assessment includes the entire life cycle encompassing extracting and processing raw materials; manufacturing, transportation and distribution; use, re-use, maintenance; and recycling and final disposal (European Union, 2010). LCAs are useful in two ways: first for estimating the total environmental impact of a system (product, process, or action) and alternative scenarios with the aim of comparing them. The results, which usually are given in emissions per "functional unit," allow improving the design process and to choose the alternative system with the least environmental impacts. Second, LCA facilitates identification of the key sources for the biggest environmental impact of a system. The results point out where to focus on first in order to improve the environmental performance of the system. Typical areas for application of LCA are packaging materials (glass, paper, polyethylenterephthalate (PET), wood), consumer goods (batteries, food, diapers), transportation vehicles, construction sector (steel, machinery, floor liners), and waste management (incineration, recycling).

An LCA consists of four steps (Guinée et al., 2001):

1. Definition of goal and scope in terms of time, space, technology, and level of detail. This step includes also the selection of a "functional unit," which is related to the function that a product or service will deliver. The definition of the functional unit is determined by the objective of the LCA. Thus, there are several functional units possible for a single process or service.

2. The basis of an LCA is a life cycle inventory (LCI), a cradle-to-grave accounting of the energy and material flows into and out of the environment that are associated with a product, process, or action. For an LCI, system boundaries and relevant processes have to be selected. A LCI requires usually a large data set about raw materials and products, which may come from different parts of the world and from various origins of manufacturing. Data about the use phase and about waste disposal are also indispensable. A special issue is the allocation of emissions to functional units in case of multifunctional processes (e.g., the allocation of greenhouse gases resulting from a power grid receiving electricity from different power plants serving many customers). There are several databases available to support LCI; for example, the ecoinvent database (ecoinvent Centre, 2010) with more than 4000 consistent and transparent up-to-date data sets in the areas of agriculture, energy, transportation, chemicals, construction, waste treatment, and others. The generic

LCI data sets are based on industrial data and have been compiled by research institutions and LCA consultants. SFA can also be used as a method to establish the inventory for an LCA. The advantage of such a database is that because of the mass balance principle, outflows of hazardous materials of processes or systems become visible that in a traditional LCI cannot be discerned.

3. In the life cycle impact assessment (LCIA), data from the LCI and societal preferences are used to calculate and interpret the environmental impacts per functional unit. Emissions are classified according to preselected impact categories such as acidification, eutrophication, ozone depletion, greenhouse gas emissions, depletion of resources, eco-toxicity, noise, and others. In the "characterization" step, the impacts are quantified by a specific unit for each category (e.g., kg CO_2 equivalents for climate change). This enables aggregation of the results of one category to a single score, the category indicator result. An additional, optional step is to "normalize" and "weigh" impact categories leading to a single final score, called the eco-indicator score (Goedkoop & Spriensma, 2001). The higher the indicator, the larger the environmental impact. Based on LCI databases, eco-indicators have been calculated for many materials and processes.

4. In the last step, the results are interpreted in terms of completeness, consistency, robustness, plausibility, and conclusions, and recommendations are deduced.

In addition to this general procedure, other variations have been developed, such as the economic input–output life cycle analysis (EIO-LCA) by Hendrickson, Lave, and Matthews (2005). It is based on aggregate sector-level data and allows allocating environmental impacts to economic sectors. EIO-LCA has been applied successfully to extended production and manufacturing chains.

Main Objectives
Life cycle analysis aims at determining the environmental impact of a product or service by key indicator emission flows that are aggregated to emission categories. The objective is to compare and improve the environmental performance of man-made systems by identifying stages along the life cycle where large emissions occur. LCA is intended as a tool for eco-design, supporting the green design of products and services.

Physical Basis of Evaluation
The main fundamental of an LCA is the LCI, a comprehensive database of emission factors for processes and products. LCIs list several thousand processes and associated emission factors from every branch of an economy. The data of the inventory are flow and concentration data on the level of processes, goods, and substances. The LCI does not depend on mass balances; it is also possible that emission data

only are given for a particular substance, without a mass-balanced data set of all inputs and outputs.

Indicators
In principle, all substances can be used as indicators in an LCA. However, in practice, a limited and predefined set of classical indicators for emissions such as CO_2, N, P, SO_2, ozone precursors, heavy metals, and so forth that is aggregated to impaction categories (e.g., global warming potential, acidification, eutrophication, etc.) is used. The reason for this is as follows: LCI databases focus on an extended but still limited (due to the enormous amount of processes and emitted substances) set of indicators. To perform an LCA on the basis of these LCI data sets is a comparatively straightforward task. As soon as unfamiliar emissions are chosen as indicators, LCI data are missing, and the effort to perform an LCA becomes much more demanding.

Sustainable Process Index
The rationale behind the sustainable process index (SPI), which was developed by Krotscheck and Narodoslawsky, is that ultimately the basis of any sustainable economy is solar exergy (Krotscheck & Narodoslawsky, 1996). Because area is required to harvest solar exergy, area becomes the limiting factor for the sum of anthropogenic activities. To determine the SPI, the area is quantified that is required for a given process or service to provide materials and energy and to accommodate by-products, wastes, and emissions. The SPI relates this area to the total area available for a person in a given geographical space. The lower the area demand, the less is the impact on environment and resources. The result of an SPI analysis is the ratio between the area required to supply a particular service per person and the total area a person needs for all his products and services.

SPI analysis has been applied to a wide field: products such as aluminum, pulp and paper, and steel and services such as transportation, energy from biomass, and entire regional economies have been evaluated by the SPI method. Advantages of the SPI are aggregation to one single indicator (area), universal applicability, scientific basis, and the possibility for application to analyze and optimize industrial processes and systems. The SPI has been applied to processes and systems of production and consumption in order to evaluate changes in technology as well as organization.

In an SPI analysis, to calculate the total area demand for a product or service, the following subjects are taken into account: materials (including raw materials) and energy for consumption (A_R and A_E); materials, energy, and area for infrastructure (A_I); and area necessary to assimilate the products, wastes, and emissions of the process (A_P). For labor-intensive processes and systems, area for personnel has to be considered, too (A_{ST}). The total area is calculated as the sum of all areas (Eq. 3.34):

$$A_{tot} = A_R + A_E + A_I + A_P + (A_{ST}). \tag{3.34}$$

To determine the SPI, three types of raw materials are differentiated: renewable materials, fossil materials, and mineral materials. For each type of material, the area required is determined by dividing the mass flow of the material used by the "yield" of the corresponding renewable respective fossil or mineral material. For the case of renewable materials, this corresponds with Eq. 3.35:

$$A_{RR} = \frac{F_R \cdot (1 + f_R)}{Y_R}, \tag{3.35}$$

where F_R is renewable material flow required by the process or system, f_R is fraction of "gray" material required downstream to provide F_R (also called cumulative expenditure or "rucksack"), and Y_R is biogenic yield of renewable material in mass per time and area.

For fossil materials, the calculation is formally the same as for renewable materials. In contrast to Y_R, Y_F stands for the "yield" of sedimentation of carbon in the sea [about 2 g/(m^2·y)]. The rationale behind this "sedimentation rate" is that the global carbon cycle is not changed significantly as long as carbon emissions can be immobilized by sedimentation.

For mineral materials, the area A_{MR} is defined by equation 3.36:

$$A_{MR} = \frac{F_M \cdot e_D}{Y_E}, \tag{3.36}$$

where F_M is mineral raw material flow required by the process, e_D is energy demand to provide one mass unit of the considered mineral, and Y_E is yield for industrial energy.

Y_E depends on the energy mix (coal, hydro, fossil, or nuclear sources). For a sustainable energy system, Y_E is approximately 0.16 kWh/(m^2·y). The area for electricity consumption A_E equals the demand for electricity (kWh/y) divided by the yield for industrial energy Y_E. The contribution of area for infrastructure A_I to A_{tot} is usually marginal, thus a rough assessment of A_I suffices. In contrast, the area A_P required to assimilate and/or dissipate products, wastes, and emissions is usually quite important. To determine A_P, the SPI follows a similar concept as the A/G evaluation method: A renewal rate is assumed for the assimilation capacity of every environmental compartment. Sustainability is accomplished if emissions are smaller than the renewal rate of the environmental compartments respectively if the substance concentrations in the compartments are not changed. Thus, the area for assimilation/dissipation A_P is determined according to Eq. 3.37:

$$A_{Pci} = \frac{F_{P_i}}{R_c \cdot c_{ci}}, \tag{3.37}$$

where Index ci is compartment c and substance i, F_{P_i} is product/emission/waste flow P of substance I, R_c is renewal rate of compartment c (mass/area and year), and c_{ci} is geogenic concentration of substance i in compartment c. Finally, A_{tot} is calculated and related to the product or service investigated.

SFA and SPI analysis can be combined and yield good results if data about the various yield factors and other nonspecific SFA data (energy demand) are available. SPI is superior to many other evaluation methods because resource consumption is considered distinctively, and specific emissions and wastes are included, too.

Main Objectives

The general aim of SPI methodology is to supply a single indicator for sustainability. More specifically, the objective of an SPI analysis is to calculate comprehensively the area that is necessary to supply a product or service under the constraints of limited solar exergy. The goal is to reduce all aspects of a product or service to aspects of "area," be it for raw material extraction, greenhouse gas production, or the dissipation of emissions and wastes. The rationale for using area as the normative value is that in a sustainable economy, the only real input that can be used over an indefinite period of time is solar energy, which is connected to the surface of the planet.

Physical Basis of Evaluation

The physical basis of an SPI analysis consists of data about anthropogenic flows of goods and selected substances and of data about the flows and stocks of corresponding substances in the environment (water, air, and soil). Mass balances are not explicitly required. However, because information about inputs (resources and raw materials) as well as outputs (products, wastes, emissions) are required for the calculation of area, a mass balance approach will strongly facilitate establishing consistent SPI results.

Indicators

The key indicator is area. All human activities (products and services) are traced back to the area they ultimately need. The "indicator behind the indicator" is solar exergy, which is a given quantity per area. Additional indicators are chosen along the analysis, such as raw materials extracted, fossil fuels consumed, greenhouse gas emitted, or heavy metal emissions.

Ecological Footprint

The ecological footprint (EF) approach developed by Wackernagel and Rees has a similar objective as that of the SPI: The aim is to determine the total area a person or a product or a service requires for resource extraction, consumption, and waste disposal and to relate this area to the entire ecologically productive area (soil, hydrosphere, forest) of the planet (Wackernagel & Rees, 1996). The result shows the ratio of anthropogenic pressure to planetary carrying capacity and allows evaluation of whether a certain process or service is exhausting the planet's capacity if extrapolated to the whole world. The EF approach has first been applied to countries, cities, and regions, and recently also to products and services such as transportation, housing, food consumption, and others.

The EF is determined as follows: The direct uses of area for housing, transportation, production, and for agriculture as well as the indirect requirements for area are added. The areas are grouped in eight land-use categories, such as agriculture, forest, urban land, unproductive land, oceans, (forest) land required to absorb CO_2 from fossil energy consumption, and so forth. For nonrenewable resources, the footprint is calculated by considering the energy and CO_2 expenditure for production, but also the loss of renewable resources through extraction (e.g., agricultural soil). Some studies take only direct consumption and CO_2 emissions into account, whereas others include upstream processes. Because in many studies the largest part of the EF is due to the (forest or biomass) area that is needed to sequester CO_2 from fossil energy utilization, energy issues often dominate the calculation of the EF. As a consequence, sometimes only the CO_2 emissions are taken into account, leading to the "Carbon Footprint."

Main Objectives
As in the case of the SPI, the aim of the EF approach is to supply a single indicator for sustainability. The main objective is to compare the area required for a process or service with the total available productive area on the planet. The objective is to judge if changes are needed in order to use this process or system ubiquitously. Another objective is to coin a term ("footprint") that is straightforward to understand, that has potential for the political debate, and that easily conveys the message that product A is better than B, and that a "change is needed."

Physical Basis of Evaluation
To establish an EF, flows of goods and energy caused by a process or service have to be assessed. Sometimes, substances are taken into account, too (e.g., CO_2 for the calculation of area required for sequestering carbon from fossil fuels). Neither material balances nor stocks are investigated. The material and energy flows are transformed into direct and indirect requirements for area. Hence, the physical basis comprises direct determinations of area used for a process or service and indirect determinations of area required for production, consumption, and disposal of materials and energy.

Indicators
The key indicator is area. To determine area utilization, processes and materials, which require directly and indirectly area, have to be selected. This selection is done in a goal-oriented way based on the objectives of the specific EF study.

MFA-Based Assessment Methods
The following metrics (material input per service unit, total material requirement) are based on MFA results and do not take into account the substance level. Both methods do not require mass balancing, but it is possible to apply the mass balance principle.

Material Input per Service Unit
The rationale for the first group of metrics developed and promoted by the Wuppertal Institute for Climate, Environment and Energy is that a large input of materials causes more environmental impacts and requires more resources than a small turnover. Thus, according to this rationale, measures to reduce pressures regarding the environment and resources should be directed toward decreasing the amount of materials that are required for a single service unit. The material input per service unit (MIPS) concept has been introduced as an aggregate measure to estimate and compare the environmental impacts caused by products or services. It represents a metric to determine the input into a national economy, an enterprise, or a production process and to calculate the resource use associated with the services and products of this economy or enterprise. Recipes for application of the MIPS concept are given in Ritthoff, Rohn, and Liedtke (2002).

The normative part of the MIPS concept is the idea that less material turnover is better. Schmidt-Bleek and von Weizsäcker, who developed the MIPS concept, have proposed reductions of MIPS by a factor of 10, 5, and 4 (von Weizsäcker, Lovins & Lovins, 1995; Schmidt-Bleek, 1997; von Weizsäcker, Hargroves & Smith, 2010). They advocate dematerialization, eco-efficiency, and factor X reduction of material flows based on MIPS methodology. They see MIPS as the main tool to develop instruments, concepts, and strategies for improving resource productivity and sustainable resource management on the level of companies, regions, and nations.

In a MIPS analysis of a certain product or service, the total mass flow of goods from cradle to grave is accounted for, including raw material extraction, manufacturing, consumption and use, and waste management and recycling. Objects of investigation can be products such as steel, cars, beverage containers, building materials, as well as services such as food supply, transportation of passengers or freight, and dish washing. In a MIPS analysis, all material flows associated with a product or service are considered. Some of these flows happen outside of the system boundaries of the MIPS analysis, in the so-called hinterland, and thus—because they are not visible within the system boundaries—are called "hidden flows" or "ecological rucksack." For example, in order to produce 1 Mg of primary copper, about 700 Mg of materials has to be turned over in the hinterland (500 Mg of abiotic materials, 260 Mg of water, and 2 Mg of air); for gold it is more than 3000 Mg because of the even smaller concentration of gold in the ore.

In the MIPS approach, inputs are grouped into the five categories "earth movements," "biotic materials" (biomass), "abiotic (inorganic) materials," "water," and "air." Sometimes, energy (fuel, electricity) is chosen as a sixth category. All materials within the five groups have the same weight or relevance, there is no distinction between different goods such as 1 kg of iron and 1 kg of arsenic. The authors have chosen this concept making a case that it is an impossible task to determine the eco-toxicity of thousands of substances due to unknown long-term effects and unknown synergistic and antagonistic effects (Hinterberger, Luks & Schmidt-Bleek, 1997; Cleveland & Ruth, 1998).

Main Objectives
The starting point of the MIPS concept is the understanding that—if similar chances for all countries are a goal—the developed economies that consume 80% of the world's resources have to cut down their turnover of materials and energy. Because resource use and environmental loadings are linked, and because both today are too large for a sustainable economy, resource use per service unit must be decreased. Thus, the specific objectives are (1) to measure and monitor the total turnover of materials associated with a certain product or service as an approximation of the overall environmental burden and (2) to apply the MIPS concept to decrease resource extraction, improve resource efficiency, and increase recycling. Individual substances and mass balances are not addressed in the MIPS concept.

Physical Basis of Evaluation
The physical basis is materials accounting on the level of goods within a predefined service or product-oriented system boundary: Statistical econometric data are used to account for all economic imports, domestic production, consumption, and exports of an economy. Goods used or wasted in the hinterland can be taken into account, too. Complete mass balances are not attempted, and some stocks and minor flows are not taken into account.

Indicators
The indicators used in the MIPS concept are flows of the five material categories "earth movements," "biotic materials," "abiotic materials," "water," and "air." For each good necessary to provide the service or product, so-called material intensities are determined that specify how many Mg of each of the five material categories is

used to produce one ton of the good. Indicators are input oriented, and neither stocks nor outputs and emissions are taken into account.

Total Material Requirement

Total material requirement (TMR), which is related to the MIPS concept, has also been developed by the Wuppertal Institute for Climate, Environment and Energy (Bringezu, 1997). It serves as a metric for resource productivity and overall pressures on the environment and measures the total mass of primary materials extracted from nature abroad and domestically that are used to support human needs. Sometimes, total domestic output (TDO) is used as a measure, too. The authors of TMR and TDO regard them as highly aggregated indicators for the material basis of an economy. As displayed in figure 3.25, TMR comprises all extracted materials in and outside the system boundary: (1) imports and associated hidden flows outside of the system boundary and (2) domestic extraction and associated hidden flows. TMR does not include water and air. Domestic material input (DMI) comprises all flows used for direct processing—that is, both foreign and domestic material flows—but does not include any hidden flows or water and air. Both TMR and DMI are

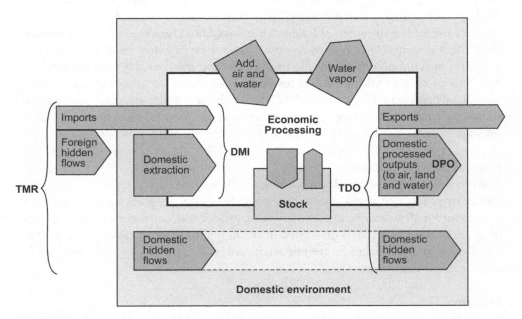

Figure 3.25
Definition of terms used for TMR methodology: TMR, total material requirement; DMI, domestic material input; TDO, total domestic output; DPO, domestic processed output. All units are measured in Mg per year.

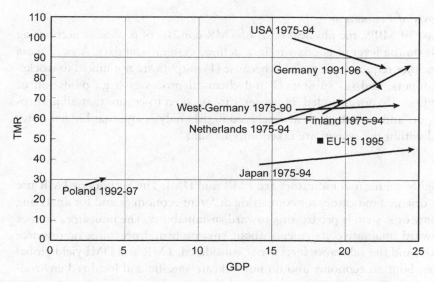

Figure 3.26
Total material requirement (TMR; in Mg per capita and year) versus gross domestic product (GDP; in 1000 $ per capita and year) for selected countries and periods. The data suggest that TMR for mature economies (United States, Germany) is starting to decrease and that a first trend in decoupling of economic growth and total throughput of processed materials can be discerned. EU-15 comprises the member states of the European Union in 1995 (reprinted with permission from Bringezu et al., 2004).

measured in Mg per year. Data to determine TMR is taken from national and foreign statistics on imports, industrial production, agriculture, forestry, and fisheries and specific information about hidden flows.

TMR has recently been used to give an overview of the material throughput and the eco-efficiency of various economies (cf. figure 3.26), as a basis for demateria-lization, to gain insight into the material metabolism of economic sectors by comparing different countries, and for looking into the selection of indicators for sustainability.

Main Objectives
The major aim of the TMR concept is to find a single, aggregated, and measurable indicator pointing out generic pressures on the environment. This aim is based on the assumption that the volume of resource requirements determines in general the scale of environmental pollution and wastes. Additional objectives are to get to know better the structural properties of the anthropogenic metabolism (e.g., by analyzing the ratio of renewable to nonrenewable inputs) and to supply instruments (TMR analysis) for monitoring economic activities.

Physical Basis of Evaluation
As in the case of MIPS, the physical basis of TMR consists of economic accounting of materials on the level of goods within a defined system boundary. A rigid mass balance is neither required nor possible because (1) outputs are not linked to stocks-in-use and inputs, and (2) substances and chemical processes (e.g., oxidation of biomass or fuel) are not included. However, care is taken to ensure that all anthropogenic import and export goods are included in the analysis and that hidden flows outside and within the system are taken into account.

Indicators
The two highly aggregated indicators are TMR and DMI. The authors of TMR use these input-oriented indicators for comparing different economies and for analyzing if an economy or system is progressing toward sustainability. The indicators are not directed toward qualitative statements about environmental pressures or resource consumption, and the substance level is not considered. TMR and DMI yield global information about an economy and do not indicate specific and localized environmental pressures. Thus, they lack operational power and are rather thought to be complementary to an SFA approach that delivers specific information about resources and the environment. The authors of TMR recommend combining TMR and SFA approaches for more comprehensive assessments.

Conclusions Regarding Evaluation Methods
Evaluation methods are based on a variety of objectives, resulting in different methodologies and in diverse results. Table 3.11 summarizes the objectives, physical basis, and indicators of the selected evaluation methods.

Nearly all of the methods focus on environmental protection. The reason may be that during the time when these methods were developed, issues related to the environment were of first relevance. Two of the eight methods (MIPS, TMR) are based on the assumption that anthropogenic material turnover is too complex to take individual substances into account, hence they focus exclusively on goods. EF is similarly based on a materials approach but takes CO_2 into account. The general objective of these three methods is to decrease consumption of materials and energy because they are strongly linked to environmental pressures and, in addition, to conserve resources.

Five evaluation systems include the substance level and thus are well suited for operational decision making on the level of environmental protection and resource use. However, for two of the methods (LCA and SPI), obeying the mass balance principle is not mandatory because they focus mainly on the topics emissions and resources. The main drawback of such an approach is that it is likely that the fraction of a substance that is not contained in either an emission or resource flow is

Table 3.11
Objectives, physical basis, and indicators of key evaluation methods

Method	Objectives	Physical Basis	Indicators
A/G	Environmental protection, understanding material systems	MFA/SFA mass balanced	Substances
SEA	Understanding and designing metabolic processes and systems	MFA/SFA mass balanced	RSE, SCE, substances
Exergy	Environmental protection, improving efficiency of systems	SFA and energy balance	Exergy, substances
LCA	Environmental protection eco-design	LCI, flows of substances and goods	Substances
SPI	Improve sustainability, eco-design	Flows of goods, substances, and energy	Area (solar exergy), goods respectively substances
EF	Improve sustainability	Flows of goods and energy	Area, processes, and goods
MIPS	Environmental protection, improvement of resource efficiency	MFA on level of goods	Goods
TMR	Environmental protection, understanding structural properties of economies	MFA on level of goods	Goods

not detected. Examples are available from waste management, where hazardous substances in gaseous emissions are heavily regulated and hence included in an LCA, whereas the same substance diverted to a recycling product is not evaluated. Only closed mass balances force evaluators to take the total flow and stocks (which are responsible for future emissions!) of hazardous and valuable substances into account. Hence, all evaluation methods should be based on the mass balance principle. This does not just apply to those methods including the substance level but also to those focusing on the level of goods. Without balances, system analysts have no warranty that all relevant flows and stocks of a system are taken into account. Even if such balances seem time consuming and costly, the value that they generate in terms of systems understanding and security makes up for the effort.

Two methods (SPI, EF) consider the finite space on planet Earth and use area as the main indicator. However, the rationale for using area is different: In the SPI approach, area is important as a means to capture valuable solar exergy, whereas in the EF method, area is considered as land that can supply various services, from raw materials supply to CO_2 sequestration by forests. SPI and the A/G approach have in common that they use geogenic flows as a reference for anthropogenic flows.

The goal to aggregate the results to one final value seems attractive for designers of evaluation methods. However, it is more difficult to derive operational support for decision making by a highly aggregated single indicator. In fact, it is impossible to design effective and efficient measures to protect the environment without taking the substance level into account. There is a balance needed between the attractive and easy to understand presentation of a single indicator and the need to investigate the full complexity of real-world anthropogenic systems.

It is surprising to see that none of the eight assessment metrics focuses on evaluating activities as a whole, including resource extraction, materials use, and environmental protection. Given the importance of activities (basic human needs!), one would expect to find methods to foster understanding of the challenges and opportunities associated with activities within anthropogenic systems. MFA and SFA and the activity concept as proposed in this volume are important parts of such a new methodology. They form a rigid natural science base that can be cross-checked for validity due to the fact that the investigated systems observe the mass balance principle. They allow detection of the key flows and stocks of useful as well as hazardous materials and are instrumental to identify the relevant processes within a material system or activity. MFA/SFA are per se not evaluation tools, but they must form the basis for all such assessment methods. Three methods are particularly well suited to complement MFA/SFA for evaluation purposes: SEA can play an important part because it points out which processes are able to concentrate and which dilute substances, thus destroying or creating values within the anthroposphere but also in the environment. Exergy analysis serves to improve the overall efficiency of a metabolic system. And A/G analysis is a premier tool to assess whether overloading of environmental compartments will be caused by an anthropogenic system.

In short, three kinds of methods can be discerned: (1) methods that focus on a single or highly aggregated indicator that conveys a straightforward message (EF, MIPS, TMR); (2) methods with in-depth analysis of material flows and the corresponding pressures on resources and environment (LCA, SPI); and (3) methods that foster understanding and improvement of anthropogenic systems (SEA, A/G, exergy). As mentioned before, in order to improve existing evaluation tools, it is recommended to base all such methods on MFA or SFA observing the mass balance principle. Such a procedure guarantees that all relevant flows and stocks of processes and services are taken into account in an assessment. In particular, the enclosure of stocks and stock changes is important because stocks determine the potential for future emissions and resources. Without stocks, the metabolic picture is not complete, and decisions regarding metabolism are prone to lead to inefficient solutions.

Outlook

Status quo of MFA/SFA

Today, MFA and SFA are well-established tools that are widely used in the scientific community. They are applied for life cycle analysis, for calculating resource efficiencies and MIPS, for optimizing waste management, for accounting of greenhouse gases, for regional planning, and for environmental and resource management. The International Society for Industrial Ecology (ISIE) and ConAccount are two groups that foster MFA/SFA by supporting information exchange between scientists and users, pointing out research needs and necessary developments, and promoting the application of MFA for resource conservation and environmental protection. ISIE and the corresponding *Journal of Industrial Ecology*, featuring regular columns on both MFA and SFA, are also active with regard to terminology: It is their merit that more and more, the terms MFA and SFA are used in a way that complies with the definitions given in this volume. In addition to application, both methodologies are further developed at renowned institutions such as Yale University [e.g., the STAF project on regional and global flows of metals of the group of T. Graedel (Lifset et al., 2002; Gordon et al., 2004)]. A large stock of data about flows and stocks of goods and substances is being built up worldwide. On the level of goods, the main drivers are the Wuppertal Institute for Climate, Environmant and Energy, Leiden University, Japan National Institute for Environmental Studies, and Institut für Interdisziplinäre Forschung und Fortbildung(IFF) Klagenfurt Austria (Adriaanse et al., 1997; Eurostat, 2001; Van der Voet et al., 2004; Bahn-Walkowiak et al., 2008), which are pushing nationwide MFAs for many economies in order to compare their TMRs and DMIs. These so-called national flow accounts are based on econometric data available from national statistics. They can be compiled without difficulties and thus have become popular for identifying trends of dematerialization.

On the level of substances, there are many SFAs appearing now on the market. Besides the academic community, it is also governments and administrations that have recently shown interest in SFA. The European Union is considering a new policy with regard to selected heavy metals such as lead, mercury, and cadmium. As an information base and in order to get an overview on kind and quantities of goods containing these metals, countries are advised to supply national flow and stock data. In the course of the new resource scarcity respectively availability debate, SFA becomes more popular, too.

An interesting example of the comprehensive use of MFA/SFA is found in Austria: first, a national standard for application of MFA and SFA defining terms and methodology has been established (ÖNORM S 2096–1) (Öster. Normungsinstitut.,

2005a, 2005b). Next, the software STAN (Cencic & Rechberger, 2008) has been developed that supports MFA/SFA on the basis of the Austrian Standard. Based on standard and software, agreements to apply MFA have been introduced in waste management: During the auditing process to become a certified waste management corporation, every company has to present a uniform, STAN-based MFA that observes the mass balance principle and includes anonymized, qualitative information about the recipient of the exported goods. This facilitates the certification process insofar as (1) data about imports and exports can be easily cross-checked for plausibility, (2) the building up of waste stocks in an enterprise becomes visible allowing a better assessment of risks and opportunities, not only for the certification process but also for insurance, and (3) the information allows verification whether exports leaving the company are complying with existing regulations. In a next step, it is planned to integrate an MFA module into the national waste database systems that has been established to follow the country-wide waste flows. The objective is to produce a tool that fulfills the requirement of the authorities to control waste flows as well as the need of entrepreneurs for an easier, more straightforward certification process for International Organization for Standardization (ISO), Eco-Management and Audit Scheme (EMAS), and others. It is expected that such an MFA-based nationwide governance regime improves waste management practice considerably.

An important development for the practical application is the development of software to support the establishment of MFA/SFA such as STAN and SIMBOX. The most important issue here is that engineers and practitioners have now tools for performing MFA/SFA that force the user to apply a certain methodology. This methodological corset can have a similar, though slower, effect than a standard or regulation: Applications and results might convene around the methods embedded in these software products, and a certain degree of uniformity of MFA/SFA studies can be expected. It remains to be seen if and which software tools are going to penetrate the market as anticipated.

Future Developments
MFA/SFA is on the brink of broader development and application. More and more research projects with MFA/SFA as a main objective are funded across the globe, and the number of projects that integrate MFA/SFA in their methodological part is increasing, too. International institutions such as the OECD and United Nations Environment Programme (UNEP) are getting engaged in the field (Graedel, 2010), and authorities realize the great chance for effective materials management by applying MFA/SFA in various parts of the economy, with the main emphasis still on environmental protection and resource conservation. This trend will become stronger because each new result will show the effectiveness of the methodology.

Linkage of MFA and GIS

The field remains of high interest to the research community because methodological development is still a challenge for many areas. The combination of geographical information systems with MFA/SFA methodology offers attractive new possibilities to link material systems with their flows and stocks to locations. This is of special interest if regional issues are addressed, when regional conditions determine the interaction of anthropogenic and geogenic flows and stocks. Another example is urban mining: Efficient recycling can only be achieved if geographical information about the anthropogenic material stock is available. Ultimately, the following information system is conceivable: For every anthropogenic process, there are inputs, stocks, outputs, and transfer coefficients as well as spatial coordinates available. This allows complete modeling of a region with all processes, resource requirements, and the necessary sinks in the environment.

Linkage of MFA and Economy

The link of MFA/SFA to economic science is still weak (Kytzia, 1997; Kytzia, Faist, & Baccini, 2004). Up to now, the fact that each process is not only a transformation, transport, or stocking of materials and energy but also a change in economic values has not been taken into account in a systematic way. This link is crucial when anthropogenic systems are to be analyzed, evaluated, and understood. It is the change in value by a process that drives the anthropogenic machine. Thus, future research should focus more on linking MFA/SFA to economic sciences. This is a great challenge particularly for SFA, because substances are—in contrast to goods— by definition not economic entities.

Data Acquisition

The current trend to collect vast amounts of data about goods (MFA, TMR, EF) and substances (SFA, LCA) is likely to continue. Most of recent data and the results produced thereby has not been screened for accuracy and compatibility. An exception is LCA: Society of Environmental Toxicology and Chemistry (SETAC) has taken the lead and created standards such as the ISO 14040 series providing guidance for application. Also, the ecoinvent database serving as an LCI with the numerical background necessary to create an LCA has become an unofficial standard. For SFA, regional, national, and global balances exist for many substances and from several different groups of authors. Some of the results do not correspond, and it takes great efforts to discern the reasons. It is necessary to coordinate the further development of MFA/SFA methodology, to create and share common databases, and to produce transparent results that can be compared and cross-checked with the results from the whole community. Finally, it will become necessary to standardize terms and methodology to increase compatibility and efficiency in the field.

A New Discipline Arises

Although more and more publications on MFA/SFA are appearing in a variety of journals, there is to date no journal dedicated to the investigation and assessment of anthropogenic systems. The new scientific discipline "metabolism of the anthroposphere" is not yet established and thus has neither specific journals nor organizations that support the growth of this new branch of science. To strengthen this discipline, it is necessary that new institutions be formed and that the institutions establish their platforms such as journals, conferences, and Web sites. It is noteworthy that the industrial ecology community represented by the ISIE (mentioned earlier) is hosting a chapter on MFA in close cooperation with ConAccount. The intention is to give the MFA community a platform for discussion. The informal platform has been fairly well used in the past, but it has not yet developed from a technical, application-oriented forum to a strategic body formulating scientific objectives, organizing means, and supporting the community to reach these goals within a feasible time frame.

The aim of industrial ecology is "to examine the impact of industry and technology and associated changes in society and the economy on the environment. It examines local, regional and global uses and flows of materials and energy in products, processes, industrial sectors and economies and focuses on the potential role of industry in reducing environmental burdens throughout the product life cycle" (International Society for Industrial Ecology, 2010). This goal comes close to the objectives of this book. However, the focus of the new discipline "metabolism of the anthroposphere" is neither on industry nor on the environment. The center of attention lies in understanding anthropogenic metabolism in order to design "better" anthropogenic systems. The power of MFA/SFA can be exploited best if it is applied to metabolic and anthropogenic systems. It is an excellent methodology not just for analysis and evaluation but also for design. To take full advantage of the MFA/SFA approach for design of the anthroposphere, it is necessary to combine it with other disciplines such as urban planning, architecture, and social sciences. A first and important step in this direction has been taken with the interdisciplinary research project SYNOIKOS (Baccini & Oswald, 1999).

4

Analyzing Regional Metabolism

Four activities, introduced in chapter 3, generate metabolic systems in human settlements. The region METALAND serves as an anthroposphere to illustrate the methodological instruments of MFA that are used to analyze and evaluate the essential characteristics of a regional metabolism. This chapter provides answers to the following questions:

• Why and how do we study regions to understand the metabolism of the anthroposphere?

• What types of material and metabolic systems do the activities generate in different steps of the cultural evolution?

• What are the metabolic essentials of a contemporary urban system at a developed economic state?

Even though humankind may be only a stage in an ongoing continuity, its advent represents a watershed. The two are not incompatible. Salamanders walk, fish don't; birds fly, reptiles don't. Yet a continuous chain of intermediates links the ones to the others. What distinguishes us radically from our primate cousins is our ability to understand the world and to manipulate it accordingly. Especially, it is the moral responsibility that goes with this ability.
—Christian de Duve, 2003

Regions as Anthropogenic Systems

About the Human Condition, the *Polis*, and Geographical Scales

Looking back at the manifold metabolic phenomena presented in chapter 2 and at the methodological approach in chapter 3, the following question is still open: How do we grasp the metabolic essentials of anthropogenic systems? Do we start from a concrete problem and, applying the MFA instrument, choose the appropriate metabolic system? This procedure would eventually deliver, in an arbitrary selection of problems, a large number of case studies on different scales, from the individual households, communities, states, and continents to a global anthroposphere. Another approach is the focus on medium-sized anthropogenic systems. Human beings have developed larger societal entities that give their members common goals and an obligation to care for a geographically defined territory. First of all, this territory is "earthly" with physical resources like water, soil, and so forth. Second, it is a region with its own cultural know-how and its historically formed identity, a polis. The resulting anthropogenic system is a combination of natural ecosystems, mostly colonized, and human settlements.

The word *landscape* emerged etymologically from the notion "to scape land," to make land (in German: *Land schaffen*). Landscape as a morphological property is an intrinsic aesthetic value of the anthropogenic system. It is, as the metabolic subsystem, a result of the four basic human activities introduced in chapter 3. The regional landscape, be it perceived as a garden (Harrison, 2008) or as a polis, forms a large-sized patch, which, together with its neighbors, gives the macroscopic picture of a settled continent. Regions serve as hinterland for other regions. In other words, the global anthroposphere is a patchwork of regional anthropogenic systems, all open, connected, and interrupted by giant oceans, mountains, wild forests, and ice. The goal of this chapter is to analyze the metabolic properties of such a patch.

Arguments for Choosing New Urban Systems

During the past 200 years, in the period of industrialization and de-industrialization, urban development concepts were based mainly on the paradigm of "core cities" and their peripheral agglomerates. Territorial management strategies in developed countries emphasized the necessity to strengthen the dense centers and to fight the urban sprawl. However, these strategies, implemented in territorial management laws and ordinances, could not prevent a transfer of the majority of the population from a rural type of life, and from the core city life, to a new urban lifestyle, which is mainly defined by the accessibility of urban commodities and services (see also chapter 1). From former physiologic studies (Baccini & Oswald, 1998), it became clear that such a development was only possible with a consequent and ubiquitous installation of the entire urban infrastructure independent of settlement densities and localities. It includes input infrastructure, namely the supply of energy, information, water, food, and construction materials, and output infrastructure such as effective waste management, as well as the buildings and connections between them that allow a high mobility of persons, commodities, and information. These studies also revealed the high degree to which settlements in developed countries depended on energy and materials from the "globally distributed hinterlands." It also became obvious that the key factor for urban development was the built-up area that already existed and not that yet to be built on open land.

In principle, there are two possibilities to gain an overview of the relevant properties of the metabolism of urban regions:

1. Select several regions as study objects, apply the same parameters for characterization, and derive generally applicable findings for the urban metabolism.

2. Design a fictitious region that contains the main properties of real regions and study the metabolic processes by changing the endogenous and exogenous factors.

It is evident that the two procedures are complementary. To follow the second procedure, data of real regions are required. In contrast, the first procedure is only successful if the workability of the findings can be tested in a model situation. A typical example for the first procedure is a comparison of cities in the United States (Herman et al., 1988). The 55 variables chosen comprehend demographic, territorial, economic, institutional, cultural, and infrastructural properties. A statistical analysis (factor analysis) helps to select the most sensitive variables (nine in total) to give each city a profile. These variables are as follows: average age of the population, perimeter, telephones per capita, hospital beds per 1000 inhabitants, expenditures per capita, flight passengers, vehicles licensed per capita, number of theaters, number of tall buildings (skyscrapers). The results are visualized in snowflake diagrams, which allow grasping similarities (e.g., Cincinnati and Seattle) and strong differences (e.g., between New York and Miami). The advantage of such an approach is its broad spectrum of variables from all areas of an urban life. The disadvantage lies in the fact that the method leads primarily to a descriptive pattern, based on statistical analysis, and does not permit a derivation of stringent cause–effect mechanisms.

In the following, the second approach is chosen. The model region is tailor-made for the investigation into metabolic processes by material flow analysis (MFA). The result is a regional metabolic system. The data are taken from separate investigations delivering information on production, consumption, substance concentration, transfer coefficients, and so forth. The advantage lies in the choice of a stringent, mass-balanced physical system that permits study of the metabolic processes (see chapter 3). The disadvantage is the fact that economic, political, and cultural variables are not included or are "hidden" in the physical parameters of the material system (e.g., preferences for a process due to its economic advantages or a substance not included due to a ban by environmental laws). In this chapter, an example will be presented in which material flows are combined with monetary flows.

Introducing the Model Region METALAND

The properties of the model region METALAND are based on the following criteria:

1. The region is an urban system as introduced in chapter 1. The population density is of the order 10^2 to 10^3 inhabitants per square kilometer.

2. The population size is of the order 10^6 to 10^7 inhabitants. It is a size that allows, by historical experience, a certain degree of autonomy with respect to political power (a nation, a state of a federation), a certain degree of economic self-subsistence (having a critical size of a home market), and a common cultural identity (the citizens feel solidarity with the sociopolitical project of their region).

3. The region belongs to a developed settlement area (i.e., its per capita income and its consumption of goods is high). There are relatively small percentages (<10%) of poor and rich people. Therefore, in a first approximation, average flows and average stocks per capita are valid for the majority of the inhabitants.

4. The use of the region's territory is intense (i.e., practically 100% of the area is cultural landscape, subdivided in agriculture, forestry, surface water, and settlement). There is no fallow land anymore, with the exception of the so-called natural parks. Forests are used fully as suppliers of biomass. Agricultural practice is oriented toward high yields of food production, within the frame of environmental protection laws. It follows that the region cannot count on a "reservoir of territory" for expansion. If the population grows, it has to increase its settlement density or it has to reduce the area for agriculture and/or forestry. If the material flow per capita is to increase, the region must either increase its regional exploitation or increase its imports.

5. Climate and topography are as follows:
 • Moderate precipitation in all four seasons (i.e., between 500 and 1500 mm per year).
 • Warm summers and cold winters with an annual average temperature of 10°C.
 • Relatively broad river valleys and moderate heights of hills (i.e., air streams and depositions are distributed in a relatively homogenous way over the whole region; local varieties exist but are not relevant).

6. The region is surrounded by similar regions. It means that METALAND cannot "profit" either from neighbors supplying "cleaner" air and water in the input or from downstream neighbors offering a high dilution potential for their waste export. In table 4.1, a list of real regions is given illustrating demographic territorial and economic properties in five continents. All seven examples are situated in moderate climate zones. The list illustrates the broad spectrum of population densities and land use, mainly due to the specific political and economic history of each region. At present, all regions belong economically to the developed countries, although the differences are still significant. Their ecological potentials with regard to reserve territories are quite different. Furthermore, the economic situations of their nearest neighbors differ strongly, not to speak of the variety of political stability.

It is evident that the model region METALAND cannot be a construct made of "average properties" of the listed examples given in table 4.1, or a statistically

Table 4.1

List of seven real regions (all provinces of their countries) having population sizes similar to that of METALAND

Region	Population (×10⁶ capita)	Population Density (capita/km²)	Area (×1000 km²)	Agriculture (% of area)	Forestry (% of area)	Settlement (% of area)	Surface Waters and Fallow Land (% of area)	GDP (thousands of dollars/capita)	Unemployment Rate (%)
Baden-Württemberg, Germany	11	300	36	46[a]	38[a]	15[a]	1[a]	50	5
New Jersey, United States	9	400	23	15[b]	35[b]	34[b]	16[b]	50[c]	9.5 (June 2010)
Tianjin, China	12	1000	12	41[d]	8[d]	25[d]	26[d]	10	4[e]
Shikoku Region, Japan	4	220	19	9[f]	73[f]	n.a.	n.a.	40 (2009)	5 (2002)[g]
Victoria, Australia	6	24	238	62[h]	32[h]	5[h]	1[h]	50	6
Buenos Aires, Argentina	14	45	307	n.a.	n.a.	n.a.	n.a.	8	n.a.
Gauteng, South Africa	10	520	17[i]	30[i]	66[i]	0.5[i]	0.5[i]	n.a.	23[k]
METALAND	10	400	25	55	30	12	3	40	—

n.a., information not available.

[a] http://www.statistik.baden-wuerttemberg.de/ (accessed September 25, 2010).

[b] USGS (2008).

[c] Wikipedia (data for 2009).

[d] http://tradeinservices.mofcom.gov.cn/en/local/2007-11-16/10781.shtml (accessed September 27, 2010).

[e] http://www.china.org.cn/english/features/ProvinceView/156218.htm (accessed September 27, 2010).

[f] http://tochi.mlit.go.jp/h13hakusho/setsu_1-2-2_eng.html (accessed September 27, 2010).

[g] http://www.mlit.go.jp/kokudokeikaku/monitoring/system/english/contents/04/4-1-4.pdf (accessed September 27, 2010).

[h] http://www.anra.gov.au/topics/land/landuse/vic/index.html (accessed September 25, 2010).

[i] State of the Environment South Africa—Themes—Land—Land use and productivity. Available at: http://soer.deat.gov.za/45.html (accessed September 25, 2010).

[k] http://www.southafrica-newyork.net/consulate/provinces/gauteng.htm (accessed September 27, 2010).

founded set of a hundred real regions, or a representative of best or optimal regions with respect to metabolic processes. What is presented in table 4.2 are properties that are needed (1) to calculate the material flows generated by the activities and (2) to design various scenarios for alternative metabolic systems (see chapter 5).

The topography of METALAND shows a broad U-valley with one main river that collects affluent from side valleys and that is interrupted by larger and smaller lakes. The water flow is, from a quantitative point of view, the largest material flow within the region. The form of the region is rectangular (figure 4.1) with side lengths of 180 km × 140 km. There are two big cities, each having 2 million inhabitants (40% of the total population). Fifty percent of the population lives in smaller units of between 20,000 and 50,000 inhabitants. Only 10% live in small settlements within a rural environment.

Water consumption of the order of magnitude 100 m^3/c.y leads to an annual demand of total 1×10^9 m^3 per region, or about 10% of the annual net input by precipitation. It follows from a quantitative point of view that in this region, freshwater is in abundance. The qualitative aspects will be discussed in a coming paragraph.

The endogenous agricultural production of 1.5 kg/c.d yields a gross energy flow of roughly 10 MJ/c.d, a quantity that would cover theoretically the need for nourishment of the regional population. However, the diet compromises a substantial quantity of meat. Therefore, ML has to import food. The supply of wood from forestry gives an energy flow of approximately 6 GJ/c.y, a quantity that contributes only a few percent of the total energy demand of 150 to 200 GJ/c.y.

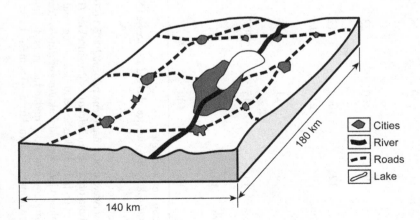

Figure 4.1
Topographic scheme of METALAND.

Table 4.2
Properties of the model region METALAND

Parameter	Data	Remarks
Area	25,000 km^2	
Mean height above sea level	300 m	
Population		
At time h	10,000,000	Population growth <1%/y
Density	400 capita/km^2	
Land use		
Agriculture	13,750 km^2 (55%)	
Forestry	7,500 km^2 (30%)	
Surface waters and fallow land	650 km^2 (3%)	
Settlements	3,200 km^2 (12%)	
Hydrology		
Precipitation	0.7 m^3 H$_2$O/m^2.y	A total of 1.8×10^{10} m^3/y
Evaporation	0.35 m^3 H$_2$O/m^2.y	
Inflow	1.0×10^{10} m^3 H$_2$O/y	300 m^3/s from neighboring regions
Lakes:		
Water volume	2.8×10^{10} m^3	Mean residence time: 2 y
Surface area	560 km^2	Mean depth: 60 m
Groundwater volume	2.0×10^{10} m^3	Assuming a constant volume
Outflow surface waters	1.9×10^{10} m^3/y	To neighboring regions
Biological production		
Agriculture	5.5×10^9 kg/y food	0.4 kg/m^2.y
Forestry	1.25×10^7 m^3/y wood	0.25 kg/m^2.y
Economy and employment		
Jobs	4,500,000	45% of population
Branch distribution		
Primary	7%	Agriculture/forestry/mining
Secondary	43%	Industrial sector (craftsmanship and industry)
Tertiary	50%	Service sector (trade, banking, transport, public services)

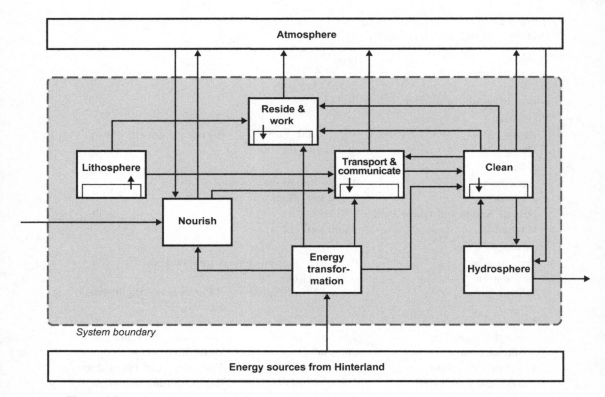

Figure 4.2
System analysis of METALAND. The four processes, referring to the four activities, stand for the sub-systems (see detailed systems later in the text) generated by the activities.

It follows that METALAND is, theoretically, "self-sufficient" with regard to water, but—based on current technology—not with regard to food and renewable energy on a biomass basis. METALAND needs hinterlands with which the region exchanges goods and tailor-made technological equipment to refine physical resources to satisfy its needs.

Metabolic Status of METALAND Described by Activities

A system analysis, based on MFA methodology, is given in figure 4.2. It consists of seven inner processes that are grouped according to

1. the four main activities (see chapter 3);
2. two environmental compartments supplying relevant primary goods;
3. one energy process (comprising all energy transforming processes).

Two outer processes comprise the third environmental compartment, the atmosphere, from which the region obtains the corresponding materials and to which it exports some of the regional outputs. The second outer process is the "global energy market" from which the region gets additional energy carriers.

In the following sections of this chapter, the four activity processes are analyzed in detail on the basis of the MFA method. At the end of the chapter, the system of figure 4.2 will be presented again, containing the quantified flows and stocks of METALAND (see figure 4.30 later in text).

TO NOURISH and TO CLEAN

The two activities are handled within one section because of their similarities with regard to the relatively short residence time of the involved materials, substances, and goods within the region. In other words, it will be shown that the metabolic dynamics of the two activities have—also because of the characteristic of the cultural evolution—a close relationship. For the activity TO NOURISH, the largest land area and the highest quantities of water and biomass are needed. The activity TO CLEAN is dominated by the highest water demands within the settlement areas.

TO NOURISH

An Outline of the Cultural Aspects of the Activity TO NOURISH
The flow of food from the environment to the human body is, trivially described, a "pull process." Human beings have to organize themselves to find, acquire, and prepare their daily diet. The environment is not, although often dreamed of, the medieval mythical country of Cockaigne, where roasted geese fly directly into one's mouth and cooked fish jump out of the water and land at one's feet. In the cultural evolution, the activity TO NOURISH shows a high diversity, as in the gathering and cultivation of plants and animals and as in the process of preparing the meals. At the beginning of the twenty-first century, it is obvious that every city offers a high variety of kitchens that have their roots in all continents of this globe. French eat Beijing Kaoya in Lyon, Argentinians eat pizza in Buenos Aires, Chinese eat French fries in Shanghai, Nigerians eat sushi in Lagos, and Canadians eat paella in Toronto. At first sight, this "urban sprawl of dishes" has the touch of worldwide marketing of successful brands. Looking at the origins and the significance of the dishes, one is impressed by the richness of the hidden technical inventions and of the rites in social gathering.

For a metabolic study, one has to be aware of the fact that the flow of food F depends on three main variables (after Baccini & Bader, 1996); namely,

$$F = f(N, R, T),$$

where N stands for the physiologic and cultural human need for food, R is the availability of resources for feeding, and T is the available technology (process engineering, economics and logistics in distribution) for transforming resources to individual daily meals.

N, the human need for food, is a function of

• The physiologic need: This need can be investigated by scientific methods. Dieticians give us data on the quality and quantity of food to be eaten to live a "healthy life." In principle, these data are based on chemical knowledge and medical experience.

• The cultural need: Eating is also social style, cultural rite, and communication. This need is investigated by methods offered by the humanities (philosophy, theology, history, sociology, psychology) and economics. Answers are given to the question: Why are humans eating what they are eating? Eating habits of individuals depend on their social environment and their personal idiosyncrasies developed within.

R, the availability of resources, is determined by

• The regional ecosystems (terrestrial and aquatic) given geogenically and transformed anthropogenically, considering the existing biodiversity.

• The regional climate and its hydrologic properties.

• The population of human beings.

• The disposal of exogenic resources such as energy carriers, water, fertilizer, primary food, and so forth.

The first two properties can be described with the methods of the natural sciences. Their development can be (although within certain limits) controlled by engineering processes. The third and fourth properties are objects of socioeconomic and anthropological investigations. If a region disposes of external resources, the population can grow without risk of encountering a famine. Otherwise, the society has to control population growth. There exist investigations of communities in remote alpine valleys (Viazzo, 1990) that are seasonally isolated from external resources for at least 4 months during wintertime. The inhabitants knew in late fall how much food in their stocks (for man and domestic animals) they had to depend on. Over centuries, they installed and practiced a relatively strict birth control, mainly supported by a social norm with regard to marriage licenses and, in secret, by abortion. The latter measure was not approved by the religious institutions but was socially tolerated. In addition, the daily food portion per person was strictly rationed. Even during periods of disadvantageous climate changes and low harvest yields, they could survive in their settlements.

T, the available technology, is determined by

• The agricultural engineering in seed and plant cultivation, animal breeding, and water management.

• The process "engineering" in preparing, conserving, and distributing the food.

• The cooking processes leading to the final consumption of meals, including all the instruments and methods involved in serving them.

It is evident that all these technical methods and instruments are based today on scientific knowledge. However, this knowledge can only be kept and developed because of an institutionally established education of professionals in each branch. This is a cultural effort. All these technologies are coupled with a tradition of professional careers such as farmers, gardeners, bakers, butchers, mechanical engineers, retailers, and many others. These professions also determine the diversity of lifestyles within a society.

The three main variables (N, R, T) are not independent from each other. In one of the classics in sociological literature, namely *The Process of Civilization* (Elias, 1939), the author dedicates a chapter to the topic "On Behavior at Table" in which he describes with concrete examples the interdependencies of the three variables. It follows that metabolic studies of the activity TO NOURISH have to consider the fact that the cultural paradigm in a certain region at a certain period of time determines the material flow systems.

The Material Systems Emerging from Cultural Periods

Three cultural periods are chosen (after Sieferle, 1993) to illustrate the forming of specific material systems: Paleolithic hunter-gatherer societies, agrarian societies based on solar energy, and industrial societies based on fossil and nuclear energy.

1. Paleolithic hunter-gatherer societies.

This period lasts between 100,000 and 10,000 BP with a standing global population of several millions of humans, mainly grouped in small tribes having 20 to 50 capita. Their material system for nourishment consisted only of two processes (figure 4.3a). Their sources were the geogenic terrestrial and aquatic ecosystems. The process *Hunting and Gathering* linked these sources with the process *Household*. Humans used the fire for cooking (and heating). Anthropologists claim (Groh, 1992) that this level of human life comprised already the ability to control consciously the ecological capacities, such as game population, and to adapt its own population (group size and growth rate) to the food limits given by the regional ecosystems. Furthermore, investigations support the hypothesis that some of these societies suffered drawbacks due to an overuse of their source.

2. Agrarian societies based on solar energy.

The emergence of agriculture began 10,000 years before the present. The swidden was the first globalized agricultural technique to gain productive soils from the tropical forests near the equator to the cold regions in Scandinavia. Within this period, the so-called Neolithic revolution started a population growth that reached

a)

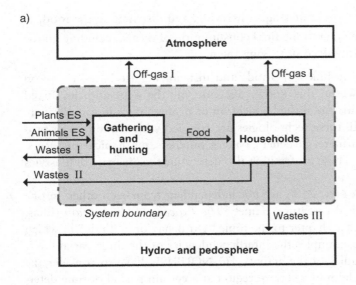

b)

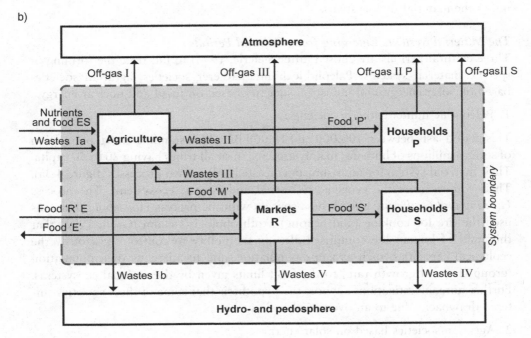

Figure 4.3
Material systems for the activity TO NOURISH in three cultural periods: (a) Paleolithic society; (b) agrarian society; (c) industrial society. ES, ecosystems; M+CI, mining and chemical industries; WM, waste management.

c)

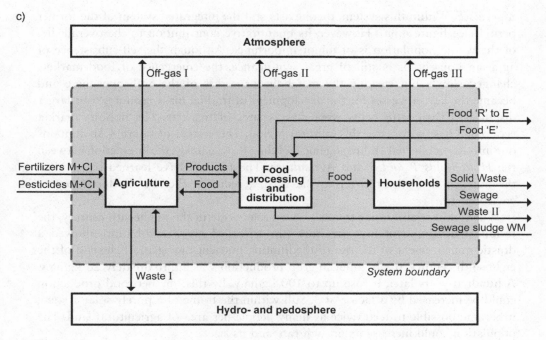

Figure 4.3
(continued)

a global size of 500 million in the eighteenth century. Urban settlements belong to the inventions of this period. New landscapes were formed that are still visible at present. Between 80% and 90% of the populations lived in a rural frame, worked for their self-subsistence, had to give tributes to their authorities, and could, to a small extent, sell their surpluses in marketplaces.

The energy system was purely solar. Depending on the climatic and territorial situation, hydraulic engineering became indispensable for a more or less reproducible annual yield to secure enough food. If the agricultural yields decreased because of oversalted or low-fertilized soils, people suffered from famine and had to move to fertile soils. Political hegemony and economic power was always based on a sufficiently large control over agricultural production.

In figure 4.3b, the corresponding material systems for this period show, compared with the former period, a replacement of the process *Hunting and Gathering* by *Agriculture*. The latter includes cultivation, breeding, harvesting, and processing of the raw products. These products are consumed in the households of the producers (*Households P*) or are sold on endogenous *Markets R* from which they reach the secondary *Households S* and other clients in the *Households P* and exogenous

Markets E. Within this system, there exists still the integrated system of the former period (see figure 4.3a). However, its quantitative contribution to the overall diet of the whole population is of minor importance. Although the self-subsistence of agrarian households is still of great significance, the emergence of food markets changed slowly but steadily the metabolic processes of the anthroposphere and became the key processes for the development of trading mass products over larger distances. Urban settlements were always sites of markets. The notion "market economy" has its roots in this cultural period. The system of figure 4.3b contains one important internal anthropogenic recycling flow; namely, the interaction between the *Households P and S* and *Agriculture*. The farmers have learned to improve the nutrient flows by recovering the wastes, reducing thereby the losses to the environment.

When industrialization extended on a broad scale in the nineteenth century, the technical development in agriculture (the so-called green revolution) allowed a drastic improvement of the use of the limiting nutrient nitrogen. At the end of the eighteenth century, the N input in crop production was approximately 20 kg/ha.y. A hundred years later, it rose up to 100 kg/ha.y. By this, the net food production could be increased by a factor of 3. Still within the frame of a purely solar system, it became possible to feed twice as many people per area of agricultural land. The population could increase its growth rate and its size.

3. Industrial societies based on fossil and nuclear energy.

Within the second half of the nineteenth century and during the first half of the twentieth century, humankind transformed its energy household from "recent solar energy" (biomass, water, wind) to fossil energy (coal, oil, gas). The agricultural food production went through an "industrialization process." Manual work in cultivation and harvesting was replaced by machines. Fertilizers with the limiting nutrient nitrogen and other essential elements such as phosphorus and potassium were synthesized in chemical factories. In addition, the discoveries of molecules that act as pesticides led to broad applications of "agrochemicals" preventing losses of crops. Having now abundant and relatively cheap energy sources, the "limits to growth" in food production were, within a hundred years, moved to far horizons. The preparation and distribution of food went through another industrialization process. New methods for storing and conserving food allowed packaging of consumer-ready food, and the low transport costs extended the distribution of mass products in the food branch from regional and continental to global scale. Nations that can afford economically the necessary supply of energy found on the world market and that invest in the new technological equipment for their food production can partially adapt their agriculture to the new economy. The world groups into developed countries, having abundant food, and into developing countries, suffering periodically from famine. The latter remain on the level of the material system of figure 4.3b. The "first world" is asked, for humanitarian reasons, to feed the poor. However, most developed countries have to cope with a socioeconomic change in agriculture of primordial significance.

The economic weight of agriculture, once more than half of a region's GDP (gross domestic product) and with more than 50% of all available workplaces, fell to a minor branch contributing only a few percent to the GDP and the labor market. Many nations had and still have to support this branch with subsidies (redistribution of tax money) and with higher tariffs for imported food to secure higher prices for domestic products. The intensive application of fertilizers and pesticides led to environmental damage, namely water pollution, degradation of soil qualities, and reduction of biodiversity. In the second half of the twentieth century, environmental laws were installed to prevent these negative effects (see also chapter 1). In the beginning of the twenty-first century, this paradigmatic situation of worldwide food production is still en vogue.

In figure 4.3c, the material system for this cultural period is sketched. Between the process *Agriculture* and the *Households* the process *Food Processing and Distribution* comprises the sum of all technologies and logistic procedures, such as all types of markets, serving the "ultimate consumer." The self-subsistent *Households P* of the system of figure 4.3b is practically nonexistent. In addition, the main sources of the system are *Mining and Chemical Industry* for the input of fertilizers and pesticides, exogenous *Food Markets*, and *Waste Management*. The latter will be treated separately in the activity TO CLEAN. Inner and outer sources and sinks are the processes *Atmosphere*, *Pedosphere*, and *Hydrosphere*. It is obvious that the system of figure 4.3c has become more complex. It offers not only more processes and a greater variety of goods but also more recycling paths.

From the foregoing arguments given, it is evident that the current worldwide situation of the metabolism, triggered by the activity TO NOURISH, still comprises all three material systems presented in figure 4.3. It can be postulated that the system of figure 4.3c is becoming more and more important with regard to the quantity of material flows. The system of figure 4.3b is about to transform to this state. In this context, the system of figure 4.3a is negligible. Therefore, the question arises whether the youngest and dominating system is apt to meet the criteria of sustainability. This chapter takes METALAND as a playground to quantify and qualify the key properties of the system of figure 4.3c. It will also include some observations on systems of the type of figure 4.3b.

TO NOURISH in METALAND

For the activity TO NOURISH, a first and very simple material system is given as part of the overall system presented in figure 4.2, comprising all four activities. For

Table 4.3
List of processes, goods, and indicator substances for the material system TO NOURISH (see also figure 4.4)

Process	Input Goods	Output Goods	Indicator Substances	Remarks
Chemical manufacturing	Energy, inorganic mining products, organic components from petroleum, air and water	Fertilizers, pesticides, wastes	N, P	Organic carbon is quantitatively negligible. Mining is not within METALAND.
Plant production	Fertilizers, pesticides, manure, sewage sludge, energy	Harvested products (fodder, cereals, vegetables, fruits, etc.), organic solid waste, emissions to air and water	C, N, P	See table 4.5 with production data. Fibers for textile fabrication see activity TO RESIDE&WORK.
Animal production	Fodder, air, and water	Animals (for slaughter), animal products (milk), manure, emissions to air and water	C, N, P	Other output goods for clothing (skins, furs, wool) see activity TO RESIDE&WORK
Industrial processing and marketing	Output goods of plant and animal production, air and water, energy	Food, solid wastes, emissions to air and water	C, N, P	Distribution over markets included. For food see diets in table 4.4.
Household	Food, air, and water, energy, appliances	Solid waste, urine, feces, wastewater, off-gases	C, N, P	Food consumption, includes household processing of food (storing, conserving, cooking) and all types of catering (restaurants, hotels, etc.)

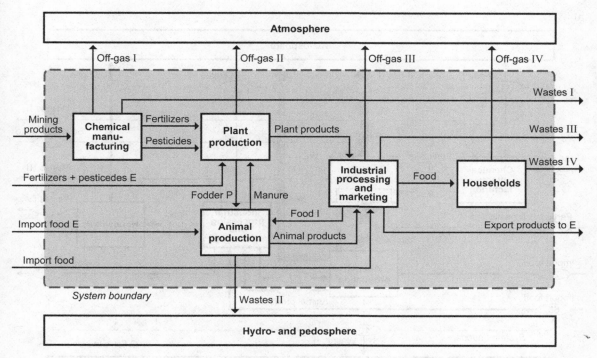

Figure 4.4
METALAND: Material system TO NOURISH (flows of water and air flows are not taken into account).
E, external markets.

a more detailed investigation, the system TO NOURISH is extended, comprising
five instead of only three processes (table 4.3 and figure 4.4). In all these processes,
the transport of the output goods to the target processes is included. In addition,
relevant source and target processes out of the system are given in figure 4.5. It is
an open system. Therefore, relevant goods are imported and exported. It is assumed
that METALAND does not dispose of endogenous mining sites. However, the
process *Waste Management* is within the region (but out of the system, belonging
to the activity TO CLEAN) recycling some of its products back to the system TO
NOURISH. In the system of figure 4.4, not all the goods listed in table 4.3 are
indicated. Water and air, for example, although of metabolic importance in each
process, are omitted in order to enhance the readability of the scheme. However,
they will be considered when balancing the indicator substances. In total, there are
four inputs and nine outputs. The inner flows amount to eight. Only two of them
are recycling flows: the *manure* from *Animal Production* and the *fodder* from
Industrial Processing.

a)

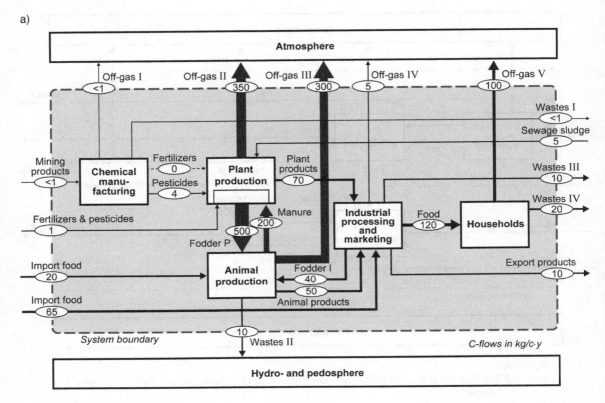

Figure 4.5
METALAND: The metabolic system of the activity TO NOURISH. (a) Carbon flow; (b) nitrogen flow; (c) phosphorus flow.

The Diet as a Cultural Phenomenon and as an Activity Designer

According to the statistics, the Japanese have the longest lifespan in the world. There is a great number of articles on the hypothesis claiming (some of them scientifically well supported, others pure speculation) that their diet is the main reason for it. True is that (1) there are different diets found around the world, and (2) eating habits are a decisive factor for the state of health of an individual. It is not the place here to enter a discussion on the relation between public health and eating habits. However, a culturally given diet of a whole region determines the properties of the material system presented in figure 4.4. In a market economy, *Industrial Processing* brings food to the market that *Households* ask for or are convinced to ask for. Agriculture grows plants and produces animal products that the food industry is inclined to refine for the market. In table 4.4, a comparison of four different diets is given to illustrate the qualitative and quantitative differences. The four countries chosen have, from an economic point of view, the same life standard. By purpose,

b)

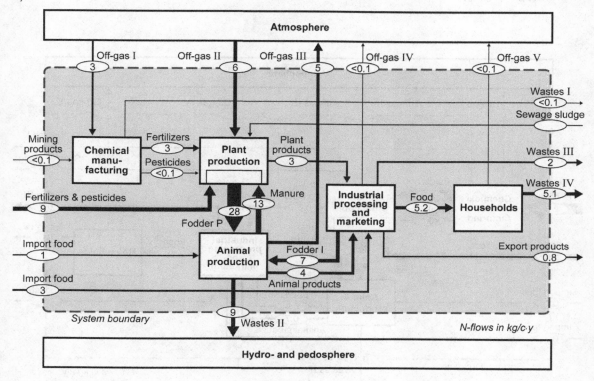

Figure 4.5
(continued)

Table 4.4
Comparison of four diets: Food consumption in kg/c.y (vegetables, fruits, oils and fats, and beverages are not included)

Goods	Japan[a]	United States[a]	France[a]	Switzerland[b]
Cereals	105	68	85	73
Potatoes and starch	37	32	76	46
Sugars	21	70	34	45
Total carbohydrates	163	170	195	164
Meats	38	117	108	59
Eggs	18	15	14	11
Fish	71	7	18	8
Dairy foods	82	267	357	159
Total animal foods (% of total)	212 (57)	401 (70)	498 (71)	237 (59)
Total	375	571	703	401
Top three making >70% of total	Cereals (rice), dairy foods, fish	Dairy foods, meats, sugars	Dairy foods, meats, cereals	Dairy foods, cereals, meats

[a]After Ridgewell (1993).
[b]After Faist (2000).

c)

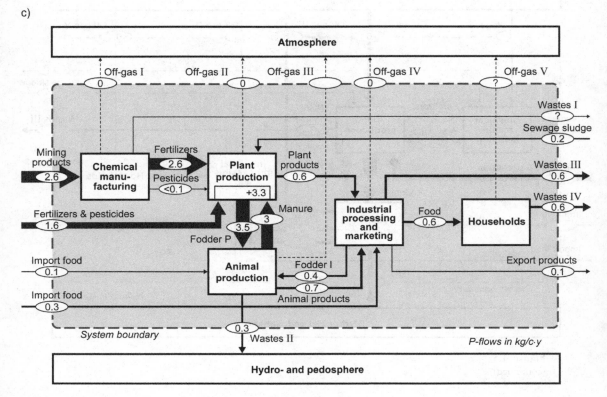

Figure 4.5
(continued)

not all food goods of a complete diet are included. The focus is placed on a comparison of the main carbohydrate donors and the main sources of animal food. (A complete picture of the METALAND diet is given later in table 4.6.)

The differences between the Asian country Japan and the European and the North American countries are significant. The staple of the carbohydrate donor is the cereal rice, which tops also the list of all food goods. Within the group of animal foods, fish in Japan is almost twice as important as meat. If fish stems mainly from the natural aquatic systems, meaning that no food sources from terrestrial plant cultivation are needed, the Japanese food consumption is also area efficient. Because the complete food mass lies between 700 and 900 kg/c.y (see table 4.7), the role of vegetables and fruits must be much higher in the Japanese diet than, for example, in the French or North American diet.

Switzerland is a country that during the past three decades has gone through a process of ecologically motivated adaption of the diet with less meat and more vegetables. Because of educational programs regarding eating habits, people have started to reduce fat-rich food. Therefore, the average Swiss diet differs from that of the Western neighbor France, although both have a long tradition as cheese producers. However, Switzerland has become a multicultural society with about one fifth of the population immigrating from Mediterranean countries and from Asia and bringing their diets into the daily life. It is one indicator of the dynamics of diets within two generations. Recent studies of the Chinese and the transition of their diet in the process of entering global interactions give an impressive picture of the dynamics on a very large scale:

During the first part of the major economic transformation in China (before 1985) cereal intake increased but decreased thereafter. There was also a long-term reduction of vegetable consumption that has now stabilized. Intake of animal foods increased slowly before 1979 and more quickly after the economic reforms occurred. While the total energy intake of residents has decreased, as has energy expenditure, large changes in the composition of energy have occurred. The overall proportion of energy from fat increased quickly. . . . Over a third of all Chinese adults and 60% of those in urban areas consumed over 30% of their energy from fat in 1997. Large shifts towards increased inactivity at work and leisure occurred. These changes are linked with rapid increases of overweight, obesity and diet-related non-communicable diseases (DRNCDs) as well as total mortality for urban residents. (Du et al., 2002)

Du and colleagues conclude: "The long-term trend is a shift towards a high-fat, high-energy-density and low-fiber diet. The Chinese have entered a new stage of the nutrition transition."

If a small society like Switzerland (with 7 million inhabitants) is changing its diet, the effects on a global scale are hardly observable. However, if countries such as China and India, comprising one third of the global population, move from an Asian to a European or North American diet type, the consequences with regard to necessary agricultural areas and yields are enormous.

A special aspect of the activity TO NOURISH is medical care by oral intake of various drugs. From a quantitative point of view, the flow of medicine is negligible (<1 kg/c.y) in comparison with the food flow. Most of the pharmaceutical products are synthesized along the process lines of the chemical industry and do not demand significant contributions from the agricultural branch. From a qualitative point of view, however, certain groups of medicines, such as analgesics and hormones, which are excreted again into the sewage water, can reduce the fertility of the aquatic fauna (e.g., estrogens in sewage water) and, if mixed with drinking water, can endanger human health. In environmental research, this relatively young topic is still loaded with many controversies (Burckhardt-Holm et al., 2005).

From the data in table 4.5, one could draw the conclusion that there is enough food worldwide to feed satisfactorily 6.8 billion people. However, the given diversity with regard to regional yields (depending on technology, climate stability, economic power, political stability, and governance quality) and to regional diets leads to

Table 4.5
Global production of the most important foods.[a] All data are mean values from the years 2005–2009 rounded to two digits

Food	Quantities 2009 (×10^9 kg/y)	kg/c.y (Population: 6.8 × 10^9)	Areas (×10^6 ha)	Yields (kg/ha)
Cereals	2500	370	710	3500
Vegetables and melons	920	140	540	1700
Roots and tubers	730	110	530	1400
Oilcrops	160	24	260	620
Sugar	160	24	280	570
Citrus fruits	120	18	90	1300
Milk	700	100		
Meat	280 (year 2007)	40		
Fish	140 (of which ca. 50 in fish farms)	20		
Eggs	70	1		
Total		850[b]	4900[c]	

[a]Data from FAOSTAT (Food and Agriculture Organization of the United Nations) and after Pretty (2008).
[b]The quantity of approximately 1000 kg/c.y of main foods produced corresponds with the same order of magnitude given in table 4.4 of 400–700 kg/c.y of net quantities (not including vegetables, fruits, and oilcrops).
[c]Total agricultural area at the beginning of the twenty-first century, giving a mean agricultural area per capita of 0.7 ha in the first decade of the twenty-first century (see also table 2.4, with 0.5 ha/c). In this total area of 4900 million ha or approximately 50 million km^2, roughly 50% is grassland for animal fodder (milk and meat production).

drastic differences. There is also a scientifically well-supported hypothesis that the limits of irrigation water, also due to climate changes, becomes again a decisive factor in global food production. An overview of the world food situation (von Braun, 2005) shows that at the beginning of the twenty-first century, approximately 0.9 billion people (or 12% of the global population) suffer from hunger and malnutrition, with more than 90% of them living in developing countries. On a global scale, agricultural production is still in the hands of small sized farming (85% with ≤2 ha per farm, and only <0.5% with >100 ha). Looking at the global food system on the market side, the top five retailers comprise only 14% of the global food market (von Braun, 2005). The global rules on trade policy (WTO, World Trade Organization), especially with respect to food products, are still objects of notorious controversial debates. However, the global tendency observable for hundreds of years is unbroken; namely, that growing economic welfare is coupled with a transition from a diet with high percentages of cereals and various vegetables to a consumer basket with more and more animal food. Therefore, it is postulated that the "food challenge" is most acute in the twenty-first century, and its handling is strongly dependent on people's consumption pattern (Pretty, 2008).

The Metabolism of the Activity TO NOURISH in METALAND

The region METALAND shows a food system that has the technological properties of a developed country with regard to its agricultural production and its industrial processing. Because of the diet of its inhabitants and its economic strategy (main effort on high-value products), its food supply depends on significant imports from the global food system and some of its food products on global markets. The metabolic status quo, shown in figure 4.5, is based on the two data sets given in tables 4.6 and 4.7.

Agriculture in METALAND on an area of roughly 1.4 million ha (see also table 4.2) or 0.14 ha/c produces approximately the same quantities of plant and animal food per capita (table 4.5). However, the animal products need roughly 70% of the total N and P input. The diet chosen in METALAND (see table 4.7) demands a dominant contribution, with respect to energy and the nutrients N and P, from animal products (meat, dairy products). It is assumed (see table 4.6) that at least 1 m^2 of agricultural land is needed to grow 1 kg of animal fodder (grass, cereals, roots, and tubers). The average transfer coefficient of the fodder into animal

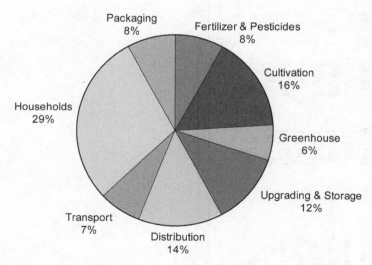

Figure 4.6
Primary energy requirements of the activity TO NOURISH (after Faist, Kytzia & Baccini, 2000).

Table 4.6

Nitrogen and phosphorous flows in agricultural production of METALAND (N and P concentrations refer to the overall educts, fertilizers, and fodder needed to produce the agricultural goods). The region depends on several hinterlands from which fertilizers, pesticides, and food are imported

Goods	Area, ×1000 ha (% of total)	Yield, kg/ha	Production, kg/c.y	Nitrogen Concentration, g/kg	N Flow, kg/c.y (% of total)	Phosphorus Concentration, g/kg	P Flow, kg/c.y (% of total)
Cereals	260 (19)	5,000	130	15	2 (11)	5	0.7 (15)
Vegetables	30 (2)	30,000	100	6	0.6 (3)	1.8	0.2 (4)
Roots and tubers	40 (3)	30,000	130	2	0.3 (2)	1.1	0.14 (3)
Fruits	30 (2)	30,000	80	4	0.3 (2)	1.7	0.14 (3)
Oilcrops	30 (2)	3,000	10	50	0.5 (3)	5	0.25 (5)
Sugarbeet	120 (9)	70,000	40	2	0.8 (4)	0.7	0.03 (1)
Milk	910 (65)	6,000	550	11	6 (33)	3	1.7 (36)
Meat	120 (9)	6,000	70	100	7 (39)	20	1.4 (30)
Eggs	3 (<1)	6,000	2	100	0.2 (1)	20	0.04 (1)
Total	1,400		1,100		18		4.6

Table 4.7

Food consumption and energy and nutrient flows in METALAND. Water for drinking and cooking is not included. Out-of-house eating and beverages are included, approximately 20% of total. Water and beverages amount to 80% of the total food mass but contribute only 5% of the total nutrient flow (N and P) (after Baccini & Brunner, 1991; Müller, Oehler & Baccini, 1995; Baccini & von Steiger, 1993; Faist, 2000)

Goods	Consumption (kg/c.y)[a]	Energy Content (MJ/kg)	Energy Flow[b] (MJ/c.y)	Water Content (g/kg)	Dry Matter[c] (kg/c.y)	C Conc.[d] (g/kg DM)	C Flow (kg/c.y)	N Conc.[d] (g/kg)	N Flow (kg/c.y)	P Conc. (g/kg)	P Flow (kg/c.)
Cereals and farinaceous food	70	15	1000	120	62	450	28	13	0.91	2	0.14
Vegetables	90	1.3	120	900	9	450	4.1	4	0.36	0.5	0.045
Roots and tubers	45	2.5	110	700	13	450	5.9	3	0.14	0.4	0.018
Fruits	100	2.0	100	850	15	450	6.8	1	0.10	0.15	0.015
Oilcrops and fats	15	37	550	<1	15	850	13	0	0	0	0
Sugar and sweets	40	17	670	10	39	450	18	7	0.1	0.7	0.01
Dairy products	160	4.2	670	800	32	450	14	9	1.0	1.4	0.15
Meat and fish	60	10	630	600	24	450	11	30	1.8	2	0.12
Eggs	10	7	70	750	2	450	1	20	0.2	2	0.02
Beverages	110	2.1	230	900	11	450	5	1	0.11	0.1	0.01

Table 4.7
(continued)

Goods	Consumption (kg/c.y)[a]	Energy Content (MJ/kg)	Energy Flow[b] (MJ/c.y)	Water Content (g/kg)	Dry Matter[c] (kg/c.y)	C Conc.[d] (g/kg DM)	C Flow (kg/c.y)	N Conc.[d] (g/kg)	N Flow (kg/c.y)	P Conc. (g/kg)	P Flow (kg/c.)
Total private households	700		4150		160		107		4.7		0.53
Total with "out-of-house eating" (+10%)	770		4600[c]		180		120		5.2		0.58

[a] Consumption means the flow of food from markets to households and restaurants/hotels and so forth where meals are prepared. It includes fresh food and convenience food. In the latter case, the meal preparation is made in the process *Food Processing and Distribution* (see figure 4.3c).

[b] The annual energy flow per capita of 4.6 GJ corresponds with an average daily energy flow per capita of approximately 12 MJ. Considering the residues in meal preparations, cooking, and eating, the average daily intake varies between 10 and 11 MJ per capita.

[c] DM, dry matter

[d] Conc., concentration.

products is 0.2. The industrial processing has a transfer coefficient of 0.8. It follows that the METALAND consumption of 2.2×10^9 kg of animal products (see table 4.7) asks for a primary production that is at least sixfold, namely 13×10^9 kg, and would need a total area of at least 13,000 km², which is practically the total agricultural area of METALAND (14,000 km²). It follows that the region is not self-sufficient with regard to the activity TO NOURISH, not even theoretically in the sense that, with regard to energy and nutrients, there could be a balanced exchange of one food type (e.g., cheese) with fruits and vegetables. METALAND has a surplus in milk production and can export some dairy products, but it has to import cereals, fruits, vegetables, and meat.

From the data in tables 4.6 and 4.7, the corresponding material flow systems for TO NOURISH are derived (see figure 4.5). First, the material system, introduced in figure 4.4, is quantified for the flow of carbon (figure 4.5a). The highest flows are found within the agricultural processes due to the carbon assimilation in photosynthesis (*Plant Production*) and the refining of plants into animal products (*Animal Production*). The carbon output of the regional agriculture to the first target process *Industrial Processing and Marketing* is only a seventh of its input (transfer coefficient 0.16). This process has a relatively high efficiency with only 8% of the output being transferred to wastes and off-gases. The region needs an additional carbon input from external markets (the goods *import food* and *import fodder*) to cover the needs of the ultimate target process *Households*. The carbon balance of "Households" reminds us of the experiments of Santorio and Lavoisier; namely, that the human metabolism oxidizes the organic carbon mainly into carbon dioxide that is emitted as *off-gas*. There is a small fraction emitted as methane, a highly potent greenhouse gas from the digestion products of cattle. For carbon, it is assumed that the activity TO NOURISH does not form any significant stocks within the region. It is also assumed that any "carbon mining" from the stocks in soils (e.g., leaching) is only seasonal and is balanced by recycling processes over a whole year. In summary, the carbon metabolism of the activity TO NOURISH is essentially an exchange of carbon dioxide between the region METALAND and the *Atmosphere*. Not included in this balance is the carbon connected to energy carriers needed to run the various processes (see also figure 4.6 and figure 4.30 with an energy comparison of all activities).

The material system for the nitrogen flow is given in figure 4.5b. As observed with carbon, the nitrogen shows its highest flows connected with *Plant Production* and *Animal Production*. The main source is molecular nitrogen stemming from the atmosphere. On one path it is technically transformed (Haber–Bosch process) into the inorganic fertilizer ammonium nitrate. On the other path, the leguminous plants assimilate the molecular nitrogen directly from the atmosphere and transform it to organic nitrogen within the plant. Although the manure is the relatively highest contribution of nitrogen in *Plant Production* (40% of the total input), the high yields depend strongly on the input of inorganic fertilizers. The overall nitrogen input into the METALAND agriculture amounts to 21 kg N/c.y. Its output to the environmental compartments, *Atmosphere* and *Hydrosphere*, is a total flow of 14 kg N/c.y (roughly 70% of the input). It follows that the efficiency of the N-household in METALAND agriculture is low. In addition, the off-gases and the wastes are oxidized forms of the element. The nitrous gases support the greenhouse effect and are potent smog-forming molecules. In the leachates from the soil, nitrogen flows eventually into the surface waters and into the groundwater as nitrate, lowering thereby the quality of drinking water and raising the trophic state of aquatic ecosystems. The following process *Industrial Processing and Marketing* has a much better transfer coefficient (0.8) for its three products (*food, export food, fodder*). The regional target process *Households* gets eventually only about a fourth of the total nitrogen imported into METALAND. However, its wastes, mainly transported in the sewage water (transfer coefficient 0.85), have a potential to replace fertilizers in agriculture (see also the case study on phosphorus in chapter 5). As with carbon, the dominant nitrogen species have chemical properties that do not allow the formation of significant stocks in agricultural processes. On the contrary, agriculture can, in the absence of fertilizer input, lower the nitrogen stocks of the soil, also called "nitrogen mining" (Pfister, 2005).

In addition to carbon and nitrogen, the material system of phosphorus flows and stocks is shown in figure 4.5c. In correspondence with carbon and nitrogen, phosphorus shows the highest flows in the two agricultural processes. The imported fertilizers stem from phosphate salts. Because of the chemical properties of P, a lithophilic element, agricultural applications in METALAND lead to a rise of phosphorus stocks in the soil, shown in the process *Plant Production*. The quantification of stocks and sinks is given in chapter 5 (figure 5.4). The export to the atmosphere is negligible. Runoffs from pastures and fields hit mainly surface waters and their sediments. Because P is a limiting nutrient for primary producers in aquatic systems, agriculture of the type shown in METALAND is a P sink with a high potential for eutrophication of surface waters (see also figure 2.24 in chapter 2 and see chapter 5). Of the total regional input of 4 kg P/c.y, only 0.6 (or 15%) units reach the regional target process *Households* (out-of-house eating included). As with nitrogen, the regional waste management can play a significant role in the P recycling process. This aspect will be discussed with respect to the activity TO CLEAN. The question regarding an optimal phosphorus management is discussed in chapter 5 (see the case study on phosphorus management).

Energy Needed to Run the Material System of TO NOURISH

All processes presented in the material system of the activity TO NOURISH are supported by manifold technical equipment (e.g., tractors, harvesters, milking machines, boilers, stoves) for which energy is needed. In a detailed study (Faist, Kytzia, & Baccini, 2000), it was shown (figure 4.6) that the total amount of primary energy needed (solar energy for photosynthesis excluded), summing up to about 30 GJ/c.y, is distributed more or less evenly between the agricultural processes (30%), the industrial and marketing processes (34%), and the households (29%). The transports are responsible for 7%. As listed in table 4.7, the energy flow of the food itself (i.e., its energy content) has a value of roughly 5 GJ/c.y. A study for a Swiss region (Faist, 2000) comes to the conclusion that the energy efficiency of the activity TO NOURISH, defined as the ratio between the energy content consumed in the process *Households* and the primary energy needed (solar energy for photosynthesis excluded) to support the whole food chain, is approximately 0.2. It means that for every "food Joule" consumed, 5 J are needed to "run its metabolic system." The higher this ratio is, the more energy efficient is the system. In comparison with the total energy demand of the region, the activity TO NOURISH is only responsible for approximately 20% (see figure 4.30).

Economic Properties of the Activity TO NOURISH

In most developed countries, agricultural processes are subsidized to keep the "primary producers," the farmers, on an income level that is comparable with that in other economic branches having higher net productivity. Therefore, a concrete economic analysis is somewhat distorted by money flows coming from the actors of a partially "planned economy." For the Swiss food chain (Faist, 2000), it was shown that the consumer expense (*Households*) is distributed evenly between the farmer, the industrial processor, and the retailer. In other words, the farmer gets only one third of the consumer expense. However, the margin varies very strongly from the type of food. From a cereal product, the farmers get only 10% of the consumer price, whereas with animal products (milk and meat), their share can climb to 50%. It is evident that the diet on one hand (the "pull role" of the consumer) and the agricultural policy on the other hand (the "push role" of the subsidizer) form a dipolar economic system that shows special feedback mechanisms. In various developed countries with diets rich in animal products, the agricultural policy is inclined to subsidize animal production.

It does it, for example, by promoting fodder production and by supporting (and protecting) raw product prices. After a period of adaptation, the subsidizer is confronted with an overproduction of meat and milk and has to install production limits or has to support the consecutive actors (industrial processors and retailers) to find new applications and new markets. Various studies support the conclusions (summarized in Faist, 2000) that along the food chain, the relative costs for labor diminish (from 35% in agricultural production to 17% in retailing) and those for the materials increase from about one fifth to two thirds. It must be underlined that the energy costs (at world market prices at the end of the twentieth century) take only 1% of the overall costs for the food chain.

Summary and Conclusions
On the basis of the model region METALAND, having properties of a developed country in a moderate climate, the material system for the activity TO NOURISH shows the following metabolic idiosyncrasies:

1. The highest mass flows of goods and selected substances take place in the agricultural processes (*Plant Production* and *Animal Production*) for which the largest share of the regional territory is needed (55%). However, the supporting technical energy is only between 5% and 7% of the total energy needed for all activities (see the summary of this chapter). From a quantitative point of view, the total mass flow of the order of magnitude 10^6 kg/c.y does not build any significant stocks within the region or generates losses. Exceptions are on the level of trace substances.

2. From an economic point of view, the consecutive processes in the food chain (*Industrial Processing and Marketing, Households*) are more important with regard to the share of the net product. Food chain economy in developed countries is dominated by material costs and not by labor or energy costs. The driving forces are the diets of the consumer and the subsidizers in (national) agricultural policies.

3. From an ecological point of view, the crucial processes take place in the *agricultural processes*, indicated by the two main nutrients nitrogen and phosphorus. The main challenge is a nutrient management that mitigates the losses of farming to environmental compartments and promotes more efficient production technologies and higher recycling rates of nutrients for the whole system. A case study is presented in chapter 5 to illustrate the essence of nutrient management.

TO CLEAN

An Outline of the Cultural Aspects of the Activity TO CLEAN
The activity TO CLEAN can be defined as the separation of goods: "Unwanted" goods (dirt, grease, sewage, etc.) are separated from "wanted" goods (shirt, metal, water). The motivation for this separation may be technical, economic, hygienic, aesthetic, or environmental. This definition implies that cleaning is considered as a cultural construct. In biological evolution, the hygienic aspect can be observed, as

animals "clean themselves" to prevent infections by parasites and "clean their nests" to prevent a potential predator from finding it because of its odor. In this context, "to clean" is an anthropomorphic term. Animal excreta serve also other purposes such as marking the territory against competitors of its own species.

At present, the activity TO CLEAN takes place at many levels (table 4.8). There are innumerable processes associated with this activity. This is especially the case for cleaning processes in industry. In fact, if the definition for cleaning given above is accepted, a large part of all industrial processes has to be included in the cleaning processes (e.g., the farmer who separates the potatoes from the soil, leaves, and stem, or the food-processing industry, which separates the edible from the inedible fractions, or the petroleum industry, which separates the valuable fractions of crude

Table 4.8
Examples of processes and goods for the activity TO CLEAN

Process	Separation Between		Input Goods	Output Goods
Individual Level				
Humans (toilet)	Feces, urine	Human body	H_2O, food	Feces, urine
Personal care	Dirt, sweat	Human body	H_2O, soap	Wastewater
Laundry	Dirt, sweat	Textiles	H_2O, detergents	Wastewater
Dishwashing	Dirt, food	Dishes	H_2O, detergents	Wastewater
Cleaning	Dirt, dust	Dwelling	Cleaning agents, including vacuum cleaner	Waste
Car wash	Dirt	Car	H_2O, detergents	Wastewater
"Waste," etc.	Rubbish	Household	Household goods	MSW
Industrial Level				
Laundries	Dirt, sweat	Textiles	H_2O, detergents	Wastewater
Plating industry	Grease, oil	Metal	Solvent	Waste solvent
	Metals	Water	Wastewater	Plating sludges, water
Refineries, etc.	Impurities	Sugar	Sugar-brine	Wastewater
Community Level				
Sewage treatment	Sludge, gas	Water	Sewage	Gas, sludge, and water
Waste treatment	Noxious sludge	Harmless sludge	MSW	Products of treatment
Public cleaning, etc.	Waste, dirt	Streets, parks	Dirt	Products of treatment

oil from the petroleum coke and other oily wastes). Because of the great number of processes and goods associated with the activity TO CLEAN, it is not possible to treat all these examples in an encyclopedic way. In the following, we shall concentrate on the most important cleaning processes in private households, on a few exemplary processes in industry, and on the sewage and waste treatment at the community level.

The Material Systems Emerging from Cultural Periods

In the evolution of human society, rules had to be developed to handle the separation of excreta. Written testimonies of nomadic tribes (approximately 3500 years BP) illustrate such rules (the Fifth Book of Moses, Deuteronomy, chapter 23, verses 10–14):

10. If there be among you any man, that is not clean by reason of uncleanness that chanceth him by night, then shall he go abroad out of the camp, he shall not come within the camp:

11. But it shall be, when evening cometh on, he shall wash himself with water: and when the sun is down, he shall come into the camp again.

12. Thou shalt have a place also without the camp, whither thou shalt go forth abroad:

13. And thou shalt have a paddle upon thy weapon; and it shall be, when thou wilt ease thyself abroad, thou shalt dig therewith, and shalt turn back and cover that which cometh from thee:

14. For the Lord thy God walketh in the midst of thy camp, to deliver thee, and to give up thine enemies before thee; therefore shall thy camp be holy: that he see no unclean thing in thee, and turn away from thee.

This rule shows the significance of cleanness within the nomadic society. The authority of God is needed, who visits only a clean camp to give his favors to beat the enemy.

The most simple material system for TO CLEAN is given in figure 4.7a. The *household* (or the camp as named in the Fifth Book of Moses) has the function of a cleaning process; namely, to separate the wanted from the unwanted. A second process is the *waste deposit* (or a landfill) in which biochemical reactions take place such as anaerobic decomposition, and from which leachates and off-gases are exported to the environmental compartments. Archeological investigations of Neolithic settlements (pile villages) come to the conclusion that the main waste deposits were just below the platforms between the stilts. When the deposits grew eventually near to the platform, the settlements were abandoned, and new ones were built. One can postulate that the oldest and quantitatively most important cleaning substance is water. Most religious institutions have established washing rites to symbolize, with a physical cleaning of the body, the inner attitude of a "cleaned soul" before entering a sacral room or starting a liturgical ceremony.

The forming of urban settlements within river valleys of irrigated agricultural territories asked for new techniques to clean continually very densely populated settlements. Alluvial canalization was invented and installed as sewerage infrastructure. Human excreta, organic wastes from kitchens, and washing waters all were transported from individual households into pipes that were fed with running water from imported water, the transport vehicle, and led eventually back into surface waters. Archeological findings support the hypothesis that cities built 5000 years BP in Mesopotamia, at the Nile, and at the Indus were already equipped with such a sewerage system. They had also drain compartments to collect feces for fertilizing agricultural land. Most likely, they did not use siphons to prevent fouling gases from entering the living areas. Therefore, in periods of water shortages and high temperatures, the cities must have been filled with very strong odors of human excreta.

From the handling of urban cleaning within an agrarian society, a certain diversity of values and metaphors emerged. On the negative side: Foul odors come from foul deeds. The devil comes with ammonia and hydrogen sulfide odors. It is the smell of death and hell. Those who have to handle wastes belong to the lowest class. In ancient Rome, slaves and prisoners had to clean the city.

On the positive side: Excreta of man and animal are precious fertilizers. In agrarian China, it was (or still is) expected from a polite guest, after a good meal, to leave his bowel contents in the toilet of the host. The fouling odor is an indicator of good digestion and of good health. In Western culture, the producer of excreta has to pay the waste collector. In Asia, the producer gets paid, because he sells a valued fertilizer. In the cities of antiquity, like in Rome, waste collectors of urine (to use the substance urea to make ammonia hydroxide for soap production) were taxed by the emperor. The anecdote goes that Titus (who criticized his father Vespasian for taking taxes from urine collectors) had to comment on the odor of the money his father put under his nose: "Non olet!" ("It does not smell").

The material system TO CLEAN of an urban settlement in an agrarian society is given in figure 4.7b. In comparison with TO NOURISH (see figure 4.3b), at the same cultural level TO CLEAN seems to be somewhat more complex. It has to do with the fact that the activity TO CLEAN in an urban settlement embodies the effort to collect mainly "unwanted" goods that are finely distributed throughout the individual households and trade shops all over the city. Some useful components are handled economically (e.g., fertilizers, textile fibers). Therefore, separate collections are organized. Others, for hygienic and aesthetic reasons, are exported technically in water canals or deposited in landfills. This material system is still the state of the art in the slum areas of very large cities. Waste collecting and recycling is done by scavengers. Canalization of sewage water is not connected to sewage

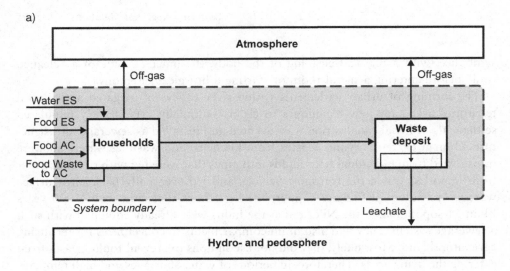

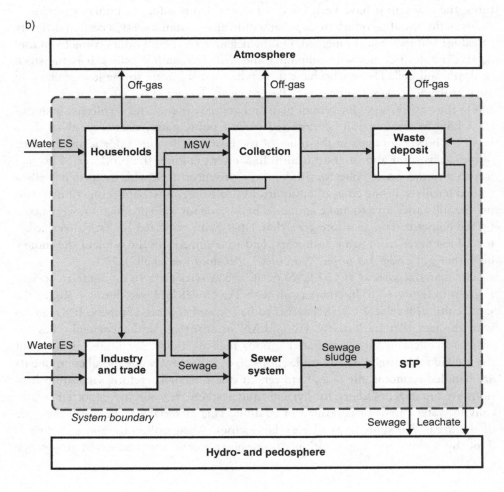

c)

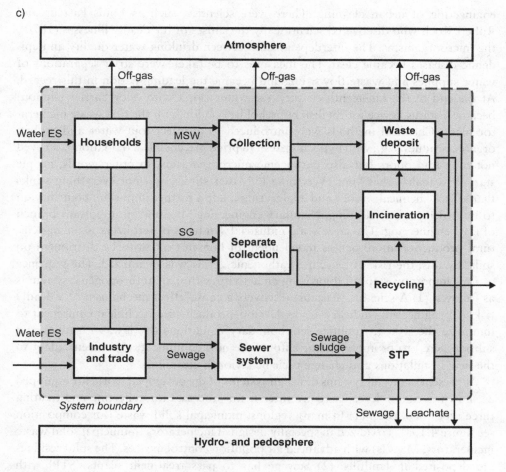

Figure 4.7
Material systems for the activity TO CLEAN in three cultural periods: (a) Neolithic and nomadic society; (b) urban society in agrarian systems; (c) industrial society. ES, ecosystem; AC, agriculture; STP, sewage treatment plant.

treatment plants. Waste deposits are also intermediate stocks of municipal solid wastes from the wealthy households. On and around these stocks, scavengers live and work as "urban miners." This cleaning system of urban settlements had and still has major shortcomings with regard to hygiene. Infectious epidemics such as cholera, typhus, and plague led to misery and death for large portions of a population. For hundreds of years, almost every generation had to cope with such epidemics.

One of the great achievements of technical development in the nineteenth century, a by-product of the epoch of industrialization, was a systematic improvement of the engineering of urban cleaning. There were scientists such as Louis Pasteur and Robert Koch who discovered scientifically the causes of the deadly illnesses (1870), the microorganisms. The interdependence between drinking water quality and epidemic diseases became clear. The measures to be taken were strict separations of water supplies and waste flows. England became the leading nation in this regard. At the end of the nineteenth century, the water closet with odor barrier (siphon) became a status device for English households. London built the first waste incinerator plant. Chemical methods were introduced to disinfect foul water and fouling organic residues (e.g., with hypochlorites). In principle, chemical oxidation destroyed not only foul odors but also pathogenic microorganisms. In other words, people started to realize that "not everything kills that stinks, and not everything stinks that kills." Chemical science and engineering made a further important contribution to improving urban cleaning. "Sanitary engineering" became an important branch of civil engineering. The large-scale industrial synthesis of fertilizers made agricultural production more or less independent of human excreta to be distributed on soils. By this, the risk of recycling pathogenic germs was minimized. The sequence of four important steps in improving an activity within an anthropogenic system is as follows: (1) A scientist (Pasteur) discovers a cause–effect mechanism of a deadly risk; (2) engineering–industry–trade develop–produce–offer technical equipment to minimize the risk; (3) politics sets boundary conditions (by law, with taxes and subsides, etc.) to promote and/or enforce the new equipment; (4) citizens adapt to the new installations and change their behavior in cleaning.

At present, material systems of urban systems of the developed world are equipped as shown in figure 4.7c. There are three main processes for collecting and treating three types of wastes. (1) In many regions, municipal solid waste (for composition see figure 4.10a) is oxidized in specially designed incinerators (municipal solid waste incinerators; MSWIs) with advanced air pollution control devices. The solid residues are deposited in landfills. (2) Sewage has to pass treatment plants (STP) with mechanical, biological, and chemical separation and degradation processes to purify the used water, which is eventually recycled back to surface waters. The main residue, sewage sludge, is deposited in landfills, or dried and incinerated, or used for biogas production, or used as fertilizer in agriculture. (3) The third main process, *Separate collection,* is a sum of collecting procedures that focus on (1) recyclable materials such as paper, glass, various metals (e.g., iron, aluminum, copper), and plastics to be reused directly as new raw materials; (2) materials that can be reused after further special treatments, such as kitchen and garden wastes in composting and biogas production; (3) technical equipment that has to be deconstructed first such as machines (e.g., automobiles, refrigerators, television sets, computers, mobile phones, etc.) to gain materials suitable for recycling according to path (1). Although, at first sight, this system (figure 4.7c) does not seem to differ strongly from the previous one (figure 4.7b), it consists of a large set of technical and logistic equipment (information and transport) that is far more sophisticated. In addition, the system is governed by an order of laws and ordinances to meet ecologically motivated quality goals that go beyond those for human health and hygiene.

However, it must be underlined that, at the beginning of the twenty-first century, a majority of the global urban population still cleans much more closely to the second system shown in figure 4.7b.

TO CLEAN in METALAND

METALAND's material system for the activity TO CLEAN corresponds with the setup shown in figure 4.7c. The cleaning technology is hidden in the six main processes. Therefore, the inner processes of each of these six main processes are presented in detail in the following.

TO CLEAN in Households

Private *Households* consist of a broad variety of cleaning procedures for which a large set of technical devices are in use. Five internal processes are chosen to describe the metabolic properties; namely, (1) toilet, (2) personal care, (3) laundering, (4) dishwashing, (5) equipment cleaning (figure 4.8).

In the process *Toilet*, the human body transfers its excreta (urine and feces) in a water closet to the urban sewage system (figure 4.8a). The quantitatively dominating flow is the flushing water. A fourth input good is the toilet paper. In some countries there is a standard additional cleansing device for the private parts, the bidet, equipped with a warm-water spray. Two important technical varieties of this worldwide standard in developed countries have to be mentioned. One is a combination of the water closet and the bidet in one apparatus ("spray toilet"), needing warm water and electricity, doing the cleansing with water, detergents, and with hot-air drying. This apparatus allows a paperless body cleaning. The other toilet type is an arrangement that separates urine and feces (no-mix toilet; Larsen, Rauch & Gujer, 2001). Feces drop into a compost box. Urine is collected, undiluted, in a separate container to keep a high nutrient concentration. In principle it is a remake, on a higher technical level, of an old setup, established in some agrarian societies to regain the nutrients for soil fertilizing. Because of the fact that this separation technique works "waterlessly" (no flushing) or with low water consumption and with a high nutrient recycling potential, it is considered to be "eco-friendly." In the frame of the whole material system TO CLEAN (see figure 4.13), the choice of the toilet type has severe consequences for the large-scale waste management of urban systems. The current type of water closet and its "upgraded" followers, the spray toilets, require centralized sewage treatment plants. An alternative, the no-mix toilet, would move an important part of the urban waste management toward decentralized cleaning systems.

a)

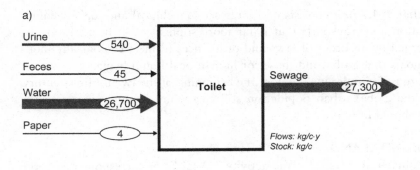

b)

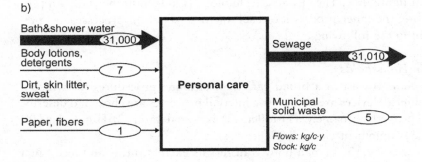

c)

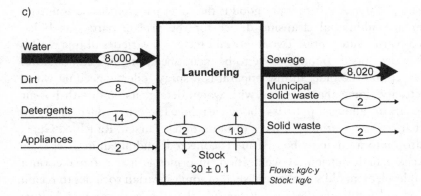

Figure 4.8
(a) TO CLEAN in private households (flows in kg/c.y): process *Toilet*. (b) TO CLEAN in private households (flows in kg/c.y): process *Personal Care*. (c) TO CLEAN in private households (flows in kg/c.y): process *Laundering*. (d) TO CLEAN in private households (flows in kg/c.y): process *Dishwashing*. (e) TO CLEAN in private households (flows in kg/c.y): process *Household Cleaning*.

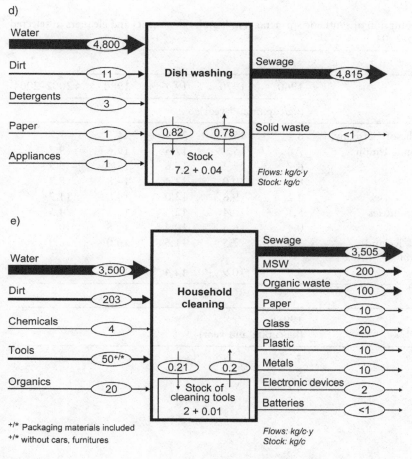

Figure 4.8
(continued)

The process *Personal Care* includes all procedures to keep the human body clean and healthy (figure 4.8b). It is mostly done by bathing and showering the whole body in and with water, typically daily. Hands are washed and teeth are brushed several times a day. Important additional goods in body washing are surfactants (see table 4.9), the main components of soaps, shampoos, toothpastes, and so forth. The great majority of surfactants are synthetic organic molecules that can, as some pesticides do, endanger the health of the aquatic flora and fauna. The results summarized in figure 4.8b are amazing: in order to clean the human body from about 20 g of dirt per day, a roughly 4000-fold amount of material (80 kg water/day) is used. With respect to material flow, the French custom of personal care, reported

Table 4.9
Per capita consumption of synthetic surfactants in laundry detergents and cleaners in selected countries, 1960–2004

Europe

Country	1960	1970	1976	1980	2002–2004[a]
	(kg/capita and year)				
Federal Republic of Germany	6.1	15.1	21.4	24.5	7.7
Benelux Economic Union	7.0	15.9	17.5	19.6	9.7
France	6.7	12.2	16.6	18.4	9.9
Great Britain	8.4	10.0	12.4	14.8	8.3
Italy	2.2	9.6	13.0	14.8	11.7
Scandinavian countries	5.1	10.4	12.3	15.1	4.5
Spain	0.4	6.5	10.4	16.8	10
Greece, Ireland, Iceland, Portugal, Austria	3.0	7.5	11.6	16.0	6.9
Western Europe	5.0	10.9	14.4	16.8	8.7

Asia

Country	1996b
	(kg/capita and year)
Japan	5
South Korea	5.2
Malaysia	5.5
Taiwan	6.5
China	2
Thailand	3
Philippines	3.5
Indonesia	2.8
World average	4

Data from Stache and Grossmann (1985); [a]Wind (2007); [b]Satsuki (1999).

from the sixteenth century, involved covering the body with all kinds of exclusive powders, which seems to have also had certain advantages; the mass of cosmetics needed to remove the sweat, skin litter, and sebum amounted to about one third of the mass of dirt on the skin.

In principle, all health care should be enclosed within this process (i.e., all medical therapy). Because health care has become a major economic and administrative branch, including all activities in developed societies, this aspect is excluded here, but it is included in all activities, due to their contribution as professional enterprises with all types of infrastructure and maintenance operations.

The process *Laundering* (figure 4.8c) consists of the washing and drying of textiles (clothes, linens, curtains, tablecloth, etc.) in private households by washing machines, tumblers, and by hand. The material flows associated with professional laundries are not included here (cf. laundry at the industrial level). The educts for the process *Laundering* are water, washing machine, washing powders and liquids, other chemicals (softeners, perfumes, etc.), TO CLEAN in private households (flows in kg/c.y), and dirt (including ca. 10^4 to 10^8 bacteria per gram textile). The products are wastewater (containing all chemicals and the dirt). The textiles or cloth are not included in the balance of the process, as these materials do not take part in the transformations and are assumed to enter and leave the process unchanged. In reality, there are small losses of textile fibers and other surface components (<1 kg/c.y).

As for the process *Personal Care*, the amount of water (20 kg/c.d) needed to remove the 20 g dirt from textiles is impressive. For the laundry in contrast, the mass of chemicals used to remove dirt exceeds the amount of dirt. The reason for this might be that the human skin, a complex living organism with a physiologically important microbial population, has to be treated with more care (and less concentrated products) than the dead fabric, which allows the use of much more concentrated chemicals for its purification and disinfection.

The goods applied in the process *Laundering* are constantly changing with time. The use of soap, which remains the most widely used surfactant worldwide, has been increasingly replaced by synthetic surfactants. Table 4.9 displays the development in the per capita consumption of synthetic surfactants in various parts of the world. Not only the amount but also the compositions of the surfactants have changed during this period (e.g., the replacement of the persistent, branched alkylbenzene sulfonates by the more biodegradable linear alkylbenzene sulfonates). The most important ingredients of today's household detergents are surfactants (e.g., alkylbenzene sulfonates), builders (chelating agents such as sodium tripolyphosphates and nitrilotriacetic acid; NTA), and bleaches (e.g., sodium perborates/percarbonates). In addition, numerous additives such as fabric softeners (quaternary ammonium salts, clay), anticorrosion materials (sodium silicate), anti-redeposition materials (cellulose ethers), enzymes (proteases, amylases), optical brighteners (e.g., stilbene derivatives), and fragrances/dyes and fillers (sodium sulfate) are contained in modern laundry detergents. In addition to the change in chemicals used for *Laundering*, the consumption of water and energy is also changing according to the actual economic situation and the corresponding improvement of washing machines (lower water consumption, lower washing temperatures). The stock (mass of washing machines in *Households*) in METALAND is roughly at 30 kg/c. The mean residence of these machines is about 15 years. It follows that the annual machine flow into the separate collection of machine devices for deconstruction and for shredding amounts to 2 kg/c.y.

The process of *Dishwashing* denotes the process of rinsing, cleaning, finishing, and drying dishes, glasses, and cutlery (figure 4.8d). It is either done by a dishwashing machine or by hand. The mass balance given assumes that about half of the dishwashing is done by hand and the other half by machines, a situation that is common in most developed urban societies. The goods involved in dishwashing are water, dishwasher, chemicals (see below), and dirt as educts and wastewater and scrap metals as products.

Chemicals used for dishwashing include nonionic surfactants for low foam formation, sodium metasilicate, pentasodium triphosphate or zeolite or NTA, fillers, ion exchange salts, anticaking agents, and chlorine-generating agents. As rinsing aids, mixtures of ethoxylated fatty alcohols, isopropanol, dehydrated citric acid, and deionized water are used. As for the other cleaning processes, the water consumption is large compared with the dirt to be removed. The relatively small amount of scrap metal produced by *Dishwashing* as opposed to *Laundering* is a result of the lower weight of the dishwasher and the need for a tumbler for laundering.

The process *Household Cleaning* (figure 4.8e) comprises the cleaning of all surfaces (e.g., windows, floors, tables, stoves, bathtubs, etc.), the cleaning of all mobile equipment (e.g., cars, bikes, shoes, electronic devices, etc.), and the separation of all solid waste (e.g., kitchen garbage, paper, glass, metal, plastic, etc.). The cleaning of all surfaces comprises a series of tools (e.g., vacuum cleaner, broom, wiper, cloth, paper, etc.) combined with water and various chemicals (detergents, organic solvents). The water flow amounts to 2000 kg/c.y. The flow of chemicals is roughly 3 kg/c.y.

Car washing is responsible for an amount of approximately 1500 kg water per capita and year, assuming 0.5 private cars per person and an average washing frequency of 4 weeks, with a need of 0.25 m^3 water per procedure, be it done in a carwash installation or by hand. The amount of detergents needed is less than 1 kg/c.y.

The separation of all solid waste from households depends on the waste collection system. For METALAND, it is assumed that separate collection is highly diversified into eight categories, namely municipal solid waste (MSW; for all mixed waste), organic waste (kitchen residues, gardening wastes), paper, glass, plastic, metal (mainly iron and aluminum), batteries, and electronic devices. All eight groups are treated separately (see following paragraphs). The handling of the biggest tool in mass, the private car, is treated in the activity TO TRANSPORT&COMMUNICATE.

The overview in table 4.10 emphasizes the following characteristics introduced by the six inner cleaning processes within the main process *Households* (figure 4.7c):

• The dominating cleaning good is water (approximately 75 m^3/c.y, or 75 Mg/c.y), exporting about half of the good "dirt and waste" in a ratio of about 200:1 to the sewage.

Table 4.10
Summary of the goods engaged in the processes for TO CLEAN in *Households* (in kg/c.y)

Good/Process	Toilet	Body Care	Laundering	Dishwashing	Household Cleaning	Total
Input						
Water	26,700	31,000	8,000	4,800	3,500	74,000[a]
Dirt and waste	585	7	8	11	203	814
Chemicals		7	14	3	4	28
Tools	4	1	2	2	50	59
Output						
Sewage	27,300	31,010	8,020	4,815	3,505	74,650
MSW		5	2		200	207
Separately collected waste			2	1	152	155
Total	27,300	31,015	8,024	4,816	3,957	75,012
Percentage of total output, %	36	41	11	6	5	100

[a]Corresponds with approximately 220 L water per capita and day.

• The other half of "dirt and waste" goes as solid waste mainly into two main fractions: a little more than half of it leaves as "municipal solid waste," and the rest goes to various separate collections.

• The main producers of material flows are the inner processes *Toilet* and *Personal Care*, responsible for about three fourths of the water flow. As already mentioned above, alternative toilet processes and measures for body care have the greatest effect on water saving.

• *Laundering* is responsible for half of the consumption of chemicals for cleaning. More than 90% of the chemicals applied in cleaning leave the households in sewage water.

TO CLEAN *in Industry and Trade*

As in *Households*, cleaning in *Industry and Trade* is traditionally based on water as a transport medium. In addition, the cleaning of surfaces by nonaqueous solvents has become increasingly important, and today the "chemical cleaning" of surfaces and goods is a very important process in modern manufacturing. In the following, cleaning in *Industry and Trade* is divided into cleaning processes using (1) water, (2) chemical solvents, and (3) other means to separate the "unwanted" from the "wanted."

Industrial Cleaning with Water

To obtain an overview of the cleaning processes in industry, the consumption of water by industry may be considered. The basis for this approach is the fact that more than 95% of water used in private households is used for cleaning purposes, and thus it may be assumed that likewise in industry an important fraction of water is used for cleaning purposes. Nevertheless, it has to be considered that very large amounts of water are used for cooling purposes and that a certain amount is also used for the production of goods.

Tables 4.11 and 4.12 demonstrate that—not considering agriculture—more than 60% of all water and about 50% of all surfactants are consumed for industrial purposes, with an increasing trend of surfactant use in the household sector. If the concentration of surfactants in industrial applications is of the same order of magnitude as in the household, the following water consumption for industrial cleaning may be assumed:

Water and Surfactants Used for Laundry, Dishwashing, and Cleaning in Households
Household use of surfactants: 1.1 kg/c.y (corresponds with ca. 20 kg/c.y of detergents), ratio chemicals to water 1:10,000–20,000, water consumption for cleaning (without toilet and body care) ca. 20,000 kg/c.y.

Table 4.11
The consumption of water by various consumers in Switzerland 1983 and 2000 (FAO, Food and Drug Organization o f the UN, 2010a). Note that use of water in agriculture depends largely on climate and crop

Consumer	Consumption per Capita		
	kg/capita and day	Mg/capita and year	Percentage, %
Consumption by Sector 1983 (Without Agriculture)			
Total trade and industry	420	152	59
Industry	330	120	46
Small trade	90	32	13
Municipal	294	108	41
Household	180	66	25
Losses transport and distribution	80	30	11
Public utilities	34	12	5
Total	714	260	100
Consumption by Sector 2000 (Including Agriculture)			
Industry	658	237	73
Municipal	215	77.5	25
Agriculture	19	7	2
Total water withdrawal	892	322	100

Table 4.12
Development of end-use markets for surfactants from 1982 to 2007[a]

End-Use Market	Consumption				
	1982 (kt/y)	1992 (kt/y)	1982–1992 (%)	2007 (kt/y)	2007 (%)
Industrial (industrial and institutional cleaning, process aids, and others)	1370	1880	54	1550	44
Household (laundry, dishwashing, cleaning)	760	920	30	1600	46
Personal care (toilet soap, shampoo, etc.)	400	500	16	360	10
Total	2530	3300	100	3510	100

[a]Data for 1977–1992 are from the United States, Western Europe, and Japan (Falbe, 1987), whereas data for 2007 applies only to the United States (Rust & Wildes, 2008).

Estimation of Water Used for Cleaning in Industry

Industrial use of surfactants: 2 kg/c.y surfactants $\rightarrow$ water consumption ca. 40,000 kg/c.y.

This very rough estimation does not take into account that in industrial applications, the concentration of chemicals may be different than that for household use, and that some surfactants are not used for applications in water (pesticides, dyes, etc.). Nevertheless, the amount of water calculated (40 Mg/c.y) compares favorably when compared with the total industrial water consumption (152 Mg/c.y) in table 4.11.

The value of 40 Mg/c.y can be further reviewed by investigating the use of cooling water and the difference between total industrial water consumption and cooling water consumption. In Switzerland, a survey in 1976 revealed that 56% of all water consumed by industry was used for energy purposes (cooling, heating, steam generation). Power plants were not included in the investigation. If this information is combined with the value for actual water consumption, 152 – 85 = 67 Mg/c.y, the value of 40 Mg/c.y compares favorably.

To investigate the processes and goods pertaining to aqueous cleaning in industry in detail, regional industrial activities have to be known. Regional differences may be very high due to the various branches of manufacturing. Chemical production, as well as pulp and paper industries, usually involves large amounts of water, which is the reason that chemical companies have traditionally been located in the vicinities of large rivers. Mechanical branches, in contrast, depend less on water for cleaning but use organic solvents.

Because no universal figures can be given that may be applied for any region, no further global values for goods associated with industrial cleaning are included here. In practice, such figures have to be considered in each particular region, including also the waters used by agricultural practice.

Industrial Cleaning without Water

So-called dry cleaning is an often used industrial process and is not only restricted to textiles but includes many other processes such as vapor degreasing, cold cleaning of metal surfaces, painting, stripping, and electronics manufacturing. In most of these processes, chlorinated organic solvents are the main chemical agent as well as the transport medium for the "dirt" (grease, wax, oil, etc.). In table 4.13, total European Union use and per capita use of selected processes involving chlorinated solvents are presented. Compared with traditional wet cleaning (laundry) processes, material turnover is highly reduced by dry cleaning. The other side of the coin are the negative effects of some chlorinated solvents on man and environment, a reason why for instance the use of trichloroethane has been phased out during the past 10 years.

Per capita values of table 4.13 may be used as average values of developed regions because they are collected from a large ensemble (the European Union with 350

Table 4.13
Estimated end use of selected chlorinated solvents for cleaning in the European Union

Process	TRI[a]	PCE[b]	DCM[c]	TRI[a]	PCE[b]	DCM[c]
	Mg/y			g/c.y		
Cleaning						
Vapor degreasing	63,000		4,800	180		14
Cold degreasing	25,000	16,000	6,800	71	46	19
Dry cleaning		26,000			74	
Cleaning agent	4,600		620	13		2
Paint stripping			29,000			83
Other						
Adhesives	6,900		15,100	20		43
Aerosol (foams)		5,200	9,500		15	27
Others and export	10,500	2,600	84,000	30	7	240
Total	110,000	148,000	150,000	314	423	429

[a]Trichloroethene (trichlorethylene); data for 1995 from Euro Chlor (1997).
[b]Tetrachloroethene (tetrachloroethylene, perchloroethylene); data for 2004 from Warwick (2005).
[c]Dichloromethane (methylene chloride); data for 1995 from Euro Chlor (1999).

million inhabitants). In practice, if regional studies are performed, individual processes have to be accounted for, and average values may be considered only for ubiquitous processes such as dry cleaning of clothes.

The chemicals used for industrial cleaning are not recycled in all cases. In the early 1990s, for example, it was estimated that for dry cleaning of textiles, 80% to 90% of the tetrachloroethene (PCE) had been emitted to the atmosphere. The remaining 10% to 20%, which contains the removed dirt, is handled as hazardous waste (incineration, etc.). The average total loss for this process was about 6–21 kg of solvent per 100 kg of textiles. Because of the improvement in dry cleaning technology, these losses have been greatly reduced.

For another application (foam blowing), the recovery rates are somewhat higher but also more difficult to determine. The range of emissions covers the wide area of 30% to 80% of dichloromethane (DCM) consumed for foaming. In any case, the actual processes applied regionally have to be investigated to establish reliable material fluxes. This is, of course, also true for other relevant industrial solvent applications.

Other Industrial Cleaning
As for *Households*, a subprocess *Cleaning Industry and Trade* also exists. The goods involved are the input goods in *Industry and Trade*, and as outputs the manufactured

goods as well as the solid wastes. Again, the kind and amount of goods associated with this subprocess are mainly determined by the manufacturing process. In general, the closer a manufacturing process is to primary industry (mining), the higher is the amount of waste materials generated. The largest amounts of wastes are produced by the exploitation and relining of ores in mines. This becomes obvious when the contents of metals and other raw materials in ores are examined (Baccini & Brunner, 1985). The "unwanted" material is in most cases much more abundant than the "wanted" material, which means that these wastes exceed any other wastes, including the end products of the consumption cycle, the MSW.

TO CLEAN in Waste Management

In urban systems, it has become established worldwide that large-scale centralized waste and wastewater management plants are built according to which various types of residues are collected. For the material system TO CLEAN in METALAND (figure 4.7c), we choose four types of plants as processes: *Sewage Treatment, Municipal Solid Waste Incinerator, Separate Collection*, and *Landfill*. In addition, the public spaces get cleaned such as buildings, streets, parks, beaches, and so forth producing additional sewage water, municipal solid waste, and material for recycling. For certain groups of industrial wastes (e.g., synthetic organic compounds, heavy metal salts) that are classified as hazardous wastes, special treatment plants are built, integrated in the process *Industry and Trade*. In the following, the major metabolic features of the four waste management processes are presented.

Sewage Treatment Plant

The material system for a sewage treatment plant (STP) is given in figure 4.9a. It is a generalized and simplified flow diagram to emphasize three consecutive main steps in an activated sludge-type STP. The *Mechanical Treatment* is equipped with screens and decantation chambers to separate large objects, sand/gravel, and fats and oils. The resulting homogenized suspension goes into an aerobic bioreactor (*Biological Treatment*) in which the organic substances are oxidized to smaller molecules and to carbon dioxide and water. The main nutrients nitrogen and phosphorus within the organics are transformed to inorganic nitrate and phosphate. There is broad variety of engineering techniques to run such bioreactors. An integrated denitrification step can reduce the oxidized nitrogen species to molecular nitrogen, which is exported to the *Atmosphere*. Another treatment with specified bacteria can accumulate phosphorus into the biomass to be separated. As an efficient alternative, the separation of phosphorus is realized in a chemical treatment; namely, to precipitate the inorganic phosphate as iron or aluminum salt. Depending on the threshold values for the STP effluents to the hydrosphere (given and controlled by the legal authorities), more or less engineering effort has to be invested to treat sewage water.

In the process *Sludge Treatment*, biological processes are installed to disinfect and to reduce the organic material by degradation, which can be realized aerobically or by anaerobic digestion, in both cases an exothermic process. In the latter case, methane gas is produced (biogas), which can be used as energy carrier for heating systems or electricity production.

The application of treated sludge as fertilizer (N and P) in agricultural soil is controversial, mainly due to its content of heavy metals (e.g., Cu, Cd, Pb) and of xenobiotic organic molecules with a toxic potential for soil organisms (see also comments to figure 4.9c). Dried sludge is also used as an energy source in cement production. It replaces partly, and free of charge, fossil fuel. The cement producer acts also as waste manager and is paid additionally for his service. For him it is economically very advantageous.

From an environmental point of view (see also chapter 2), it is evident that the ultimate goals of an STP are threefold:

1. Degrade as much organics to inorganic species as possible.
2. Mitigate nutrient flows into the hydrosphere.
3. Contribute as much as possible to the reuse of nutrients (N and P) and of the energy content.

Two example flows of substances passing an STP are illustrated in the following; namely, phosphorus and the non-ionic surfactant nonylphenolpolyethoxylate.

For phosphorus, the transfer coefficient of an STP for sludge is 0.4 without P sedimentation and 0.9 with an additional (chemical or biological) separation step (Chassot, 1995). For METALAND (figure 4.9b), it is assumed that the mix of about 200 STPs (or 1 STP per 50,000 inhabitants) has a mean transfer coefficient of 0.5 (i.e., only 20% of the regional STPs have an additional P separation). It follows that half of the P in the input is transferred to the hydrosphere in the purified sewage. One fourth is applied as fertilizer on the agricultural soil. The rest is either landfilled or incinerated (as energy source).

Whereas the amount of phosphorous entering an STP is equal to the amount leaving in purified water and sludge, for many organic chemicals an STP is a sink: Easily degradable organic substances are mineralized to CO_2 and water, and less degradable substances are partly transformed resulting in metabolites of various properties. Interesting examples are detergents because (1) they are the only chemicals that are made on purpose to enter the water path, and (2) they are not trace substances like PCBs (polychlorinated biphenyls) or heavy metals but are consumed and discarded into sewers in large amounts of several kg/c.y. In figure 4.9c, the fate of a specific non-ionic surfactant, nonylphenolpolyethoxylate (NPnEO), is presented during wastewater and subsequent sludge treatment. In the 1970s and 1980s, 0.3 mol/c.y of NPnEO was used in detergents for laundering, dishwashing, and other

a)

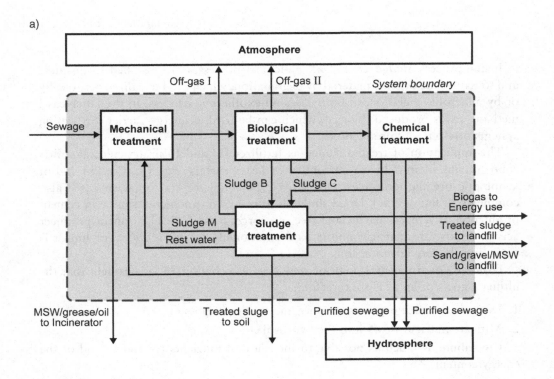

b)

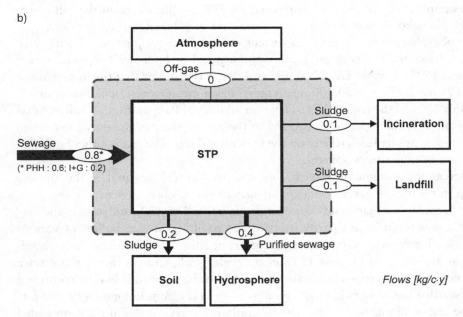

Figure 4.9

(a) Sewage treatment plant (STP): Material system. (b) STP: Phosphorus flow. (c) STP: Flow and transformation of nonylphenolpolyethoxylate (lower flow value) to nonylphenol (upper flow value) during sewage and sludge treatment.

c)

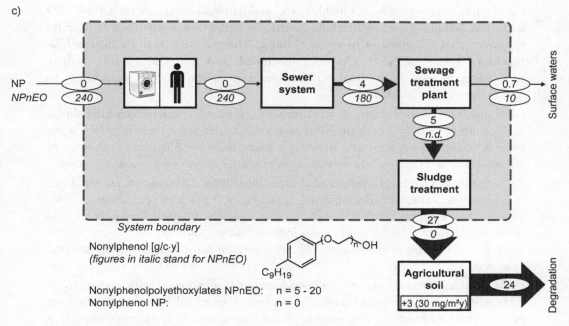

Figure 4.9
(continued)

applications. After sewage and sludge treatment, 40% (on a mole basis) of NPnEO was transformed into 0.12 mol of nonylphenol (NP) (Giger, Brunner & Schaffner, 1984; Soares et al. 2008). When sludge is applied in agriculture to make use of nutrients and humic substances, about 10% of NP present in sludge is not degraded and accumulates in the soil. Hence, due to the persistence, toxicity, and endocrine properties of NP, NPnEO has been phased out as a surfactant for most consumer and industrial products.

Lessons Learned from the Nonylphenol Case Study
• Substance flow analysis (SFA) proves to be a valuable tool to follow the path from source to sink not only of nondestructible substances such as metals but also of organic chemicals that are prone to degradation. This is especially the case when sophisticated chemical analytical methods are combined with SFA and uncertainty analysis. If balances are based on moles, new insights into systems behavior of chemicals can be observed: SFA reveals that in the sewer, about 25 mass % of NPnEO is "lost." Because 4 g/c.y of NP is found in the sewer, it is likely that the "loss" of NPnEO indicates first degradation of NPnEO and first formation of NP already in the sewer. Because NP is mainly formed under anaerobic conditions, this is also an indication about the state of the sewer system.

• It is interesting to note how a highly useful and harmless substance such as NPnEO with both lipophilic and hydrophilic moieties is being transformed into a harmful substance that accumulates in sewage sludge. The reason is that the hydrophilic functional group of NPnEO is being degraded, and the resulting intermediate metabolites as well as the final NP become more lipophilic, thus transferring to sludge (Brunner et al., 1988).

• Because NP in sludge is only slowly degraded (10 mass % accumulates after sludge application to soil), and because NP is an endocrine disrupter, there may be a long-term NP problem for soils and organisms living on soils. Also, small amounts of NP are emitted to the surface waters, with potential effects on aquatic organisms.

• Contaminations by surfactants and their metabolites are not trace contaminations in mg/kg (ppm) concentrations but are in the range 0.1–1 g/kg (per thousand) dry sludge. This is due to the fact that they are specifically designed for application in water and that they do not derive from indirect sources such as air transport or soil erosion. Such substances require special attention as potential environmental contaminants.

• The chemical fates of organic molecules have to be followed with respect to all metabolites and all products leaving the process. A focus solely on the degradation of the initial molecule in one output (treated wastewater) is inappropriate. It is necessary to include all outputs and to perform a full balance in order to evaluate STP as a treatment process. This is particularly the case for surfactants: It is not enough to test if they or their metabolites are still present in purified wastewater. The investigation has to include sewage sludge as well as off-gas from activated sludge process and other products.

• Organic molecules with metabolic properties such as NPnEO passing STPs either by treated wastewater, by sludge, or by off-gas should be banned from goods that are disposed of in sewage after use. This has actually been done for many applications of NPnEO in several countries.

MSWI Plant
Thermal treatments of wastes is performed by well-established technical equipment of regional waste management. The process oxidizes the organic contents of wastes and transforms the inputs to inorganic components. MSWI plants have the capacity to handle an input flow of up to 20 Mg per hour per furnace. By this they can serve a population of several hundred thousands to treat their solid waste. The composition of the waste is exemplified in figure 4.10a for a developed country. It is a mixture of vegetable and animal wastes, paper and paperboards, plastics, glass, ceramics, wood, textiles, and metals. The flow per capita and the percentage in weight of each component depend on several factors, such as consumer profiles (purchasing power, market preferences), collection schemes, and recycling offers.

a)

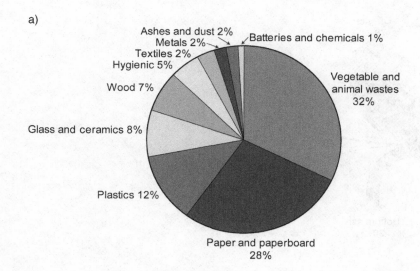

b)

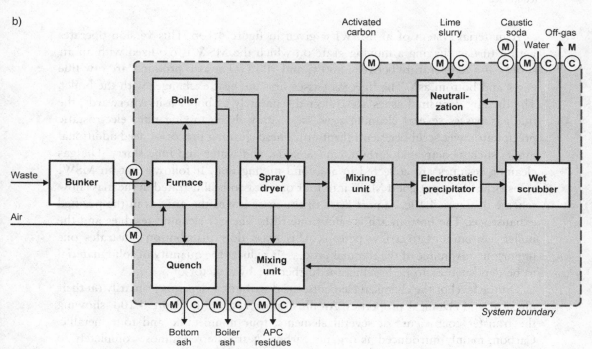

Figure 4.10
(a) Municipal solid waste incinerator (MSWI): Composition of MSW. (b) MSWI: Material system. M, mass flow analysis; C, sampling and analysis of substance concentration (c) MSWI: Mass flows. (d) MSWI: Flows of indicator substances (after Belevi, 1998). APC, gas cleaning.

c)

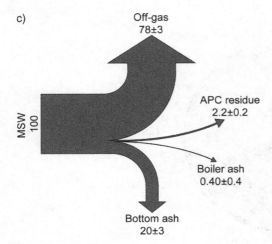

Figure 4.10
(continued)

A material system of an MSWI is given in figure 4.10b. This version operates with a furnace having a moving grate on which the MSW is oxidized with an air supply at a temperature between 800°C and 900°C. The two products are raw flue gases and bottom ash. The flue gas passes first the heat exchanger with the boiler. The thereby deposited ashes are collected separately as boiler ash. Afterwards, the flue gas passes several cleaning processes (spray dryer, mixing unit, electrostatic precipitator, wet scrubber, neutralization). These cleaning processes need additional inputs such as activated carbon, caustic soda, and water and lime slurry. The gas cleaning residues are collected in a second mixing unit. It follows that an MSWI transforms the input good MSW into four output goods, illustrated in the mass flow scheme of figure 4.10c. Almost 80% of the mass leaves the process in the purified exhaust gas. The bottom ash is about one fifth. The gas cleaning residues and the boiler ash amount to a few percent. This mass flow distribution illustrates one important advantage of the thermal process. It reduces the quantity of solid material to be deposited or to be handled for further use by a factor of 5.

On the level of the chemical elements, the distribution depends primarily on their physical and chemical properties. An illustration is given in figure 4.10d, showing the transfer coefficients of several elements, four nonmetallic and four metallic. Carbon, mainly introduced as organic carbon, is transferred almost completely to the exhaust gas as carbon dioxide. Phosphorus, a lithophilic element, is found at almost 90% in the bottom ash. Chlorine shows an atmophilic behavior and is found as hydrogen chloride in the raw gas. Because of the gas cleaning technique with a neutralization step, the chloride is withheld in a separation process as sodium or

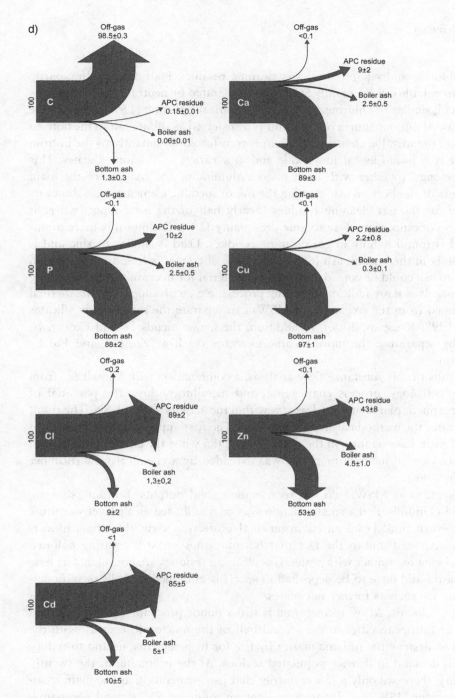

Figure 4.10
(continued)

calcium chloride and separated via gas cleaning residues. Half of the sulfur, partly oxidized to volatile oxides, mainly SO_3, can be separated by neutralizing the aqueous sulfuric acid, similar to chlorine. However, a significant portion (15%) leaves in the exhaust gas. About one fourth of the sulfur is transferred as sulfate salt to the bottom ash. Among the metallic elements, calcium is transferred dominantly to the bottom ash where it is found as calcium oxide and in a variety of calcium silicates. This element belongs, together with iron, silicon, aluminum, and oxygen, to the main components of the bottom ash. Among the minor metallic elements, zinc shows its preference for the gas cleaning residues (nearly half of the total input) where it comes to a concentration close to zinc ore quality (2%). Cadmium is more atmophilic and is found mainly in gas cleaning residues. Lead is similar to zinc and is found mainly in the bottom ash (60%). The bulk of copper is found in the bottom ash. Bottom ash could be considered as a raw material for iron and copper recycling and for pozzolans if an additional melting process (e.g., with support of the thermal energy gained from the oxidation of MSW) can separate the metals from silicates (Zeltner, 1998; Kruspan, 2000). In addition, the major metals Fe and Cu can be purified by separating the more volatile elements (such as Zn, Cd, and Pb) by evaporation.

The results of this substance flow analysis, a combination with knowledge from chemistry, petrology, process engineering, and metallurgy, show the potential of thermal treatment plants of mixed wastes within the activity TO CLEAN. The result also illustrates the methodical strength of MFA. The first application of this method goes back to an investigation in the years 1984–1985 when the classical mechanical engineering view of incinerator plants was extended by a chemical view (Brunner & Mönch, 1986).

At present, most MSWIs do not recycle their solid outputs. Bottom ashes are deposited in landfills, and a small fraction was or is still used as a gravel substitute in road construction. From an environmental protection view, this application is strongly questioned due to the fact that bottom ashes are still emitting polluted leachates, being in contact with water. Gas cleaning residues are considered as hazardous wastes and have to be deposited in specially equipped landfills or in underground storage such as former salt mines.

On a global scale, MSW incineration is still a minor procedure. It has its origin in the large European cities in the second half of the nineteenth century with the main aim to destroy the organic matter by fire for hygienic reasons and to reduce the landfill demand in densely populated regions. At the beginning of the twenty-first century, there are only a few countries that use incineration as the main treatment process for MSW (e.g., Japan, Scandinavian countries, Switzerland, Germany). For approximately 90% of the global population, landfilling is still the main process for waste management.

MSW Landfill

In principle, a landfill is a storage site in a topographically suited landscape that prevents a mechanical erosion of the deposited material into the environment. These landscapes are either natural, such as valleys (remote from human settlements), or man-made, such as former gravel pits or clay quarries. In global cultural history, "controlled landfills" (i.e., waste deposition sites having a strict input and output control with legally based quality standards) started in the second half of the twentieth century. The quality standards for controlled landfills are mainly based on environmental protection laws. They are managed technically to prevent hazardous emissions, to ensure long-term control, and to allow recultivation of the landfill surface after the end of deposition.

A new concept for landfills arose during the 1980s with the notion "final storage quality" (Baccini, 1988). The solid residues from the metabolism of the anthroposphere should have properties very similar to that of the earth's crust, such as natural sediments, stones, ores, and soils. It is a consequence of the principle of precaution and follows the guidelines of a sustainable development, namely not to leave the next generations any potentially hazardous sites. Based on this concept, two types of landfills were defined; namely, "reactor landfills" and "final storage landfills." Some countries began to revise or extend their environmental laws to set new quality goals for the material to be deposited. The potentials and consequences of the new concept for landfills will be discussed in a case study in chapter 5.

At the beginning of the twenty-first century, landfills of final storage quality are still very rare. On a global scale, the dominating controlled landfill is built and managed for MSW and is of the type reactor landfill. In the following, the metabolic idiosyncrasies of an MSW landfill are presented. The material system (figure 4.11a) is very simple. It is an open system (i.e., it is assumed that no complete enclosure is fabricated to isolate the landfill completely). There are the two input goods, precipitation water and MSW. The technical infrastructure (barriers, pipes for gases and leachates) of the landfill allows collecting and treating the landfill gas, mainly a mixture of methane and carbon dioxide, and catching all leaching water to be treated in an STP, either integrated in the landfill plant or in an external setup. The quantitatively most important "product" of the landfill is the residue.

The crucial question to be answered is as follows: How long does it take until an MSW reactor landfill has achieved a residue quality that needs, from an environmental protection point of view, no further treatment and control. Field studies come to the conclusion that the intensive reactor phase emitting landfill gases and large element flows comes to an end after ten to twenty years in deposition (figure 4.11b) (Baccini et al., 1987; Belevi & Baccini, 1989; Kjeldsen et al., 2002; Laner, Fellner & Brunner, 2010). However, the leachates continue to emit organic components for several centuries until a so-called final storage of the residue is achieved.

a)

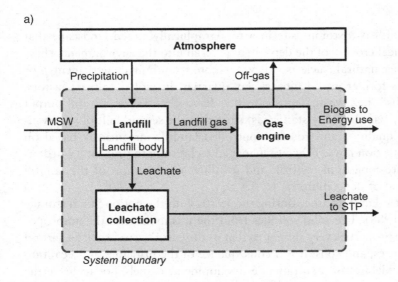

b)

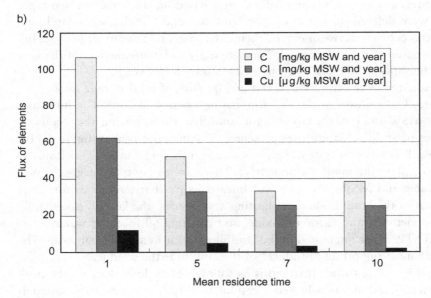

Figure 4.11
(a) MSW landfill: Material system. (b) MSW landfill: Element flows in the leachates of MSW landfills as a function of the mean residence time of the deposits (after Baccini et al., 1987).

With this perspective, it becomes clear that an ideally constructed barrier to keep the deposited material isolated postpones the principal problem of a reactive body to a time period when the barrier does not function anymore. In other words, an MSW landfill in the open system version has to be managed over several human generations, and in the closed version it must be ensured that the barrier maintains its function because the reactive potential of the deposited material persists "forever."

Sorting for Recycling
The fourth relevant process of the material system TO CLEAN in METALAND comprises all logistic and technical measures to separate "wanted" and "unwanted" goods" from *Households* and *Industry and Trade* for recycling; that is, by specific pathways to enterprises that are either equipped to upgrade the wastes to new raw materials (e.g., car shredder plants producing iron scrap for smelters) or by taking sorted material as it is, substituting primary sources (e.g., waste paper flowing to paper plants, aluminum metals flowing to aluminum smelters, glass bottles flowing to glass bottle manufacturers; see also chapter 3). At present, on a global scale, there is a high variety of collecting systems, controlled by different measures (economic incentives, laws, subsidies, waste management fees, etc.). The main aims are (1) mitigation of mixed wastes to be handled in consecutive treatments, (2) reduction of the input flows of primary resources, and (3) economic profit due to higher efficiency in resource management. Iron serves as one example of an established recycling procedure on an international scale to illustrate the metabolic characteristics. The national iron balance of Switzerland is shown in figure 4.12. Recycling has become, within the ideologically driven "Green Movement" in society, an almost dogmatic imperative. Who recycles is good, who wastes is bad. Therefore, it is not surprising that slogans such as "zero-emission production" and "no-waste communities" have evolved. All these movements have to be evaluated, from a metabolic point of view, under the aspect of the system boundary chosen. This aspect will be discussed in more detail in the case study on waste management in chapter 5.

Summary for TO CLEAN in METALAND
The waste flows quantified in table 4.14 give an overview of the material properties and the masses treated in the waste management of METALAND. The overview is supplemented with the material system in figure 4.13. The main mass flow comes from construction sites (see activities TO RESIDE&WORK and TO TRANSPORT&COMMUNICATE), a material that is inorganic. Only a small fraction of roughly 10%, listed separately as mixed construction waste, contains organic components. Sewage sludge is mainly an "aqueous flow." The separately collected wastes are of the same order of magnitude as that of the mixed wastes. The overall balance shows that roughly half of the material produced in the activity TO CLEAN

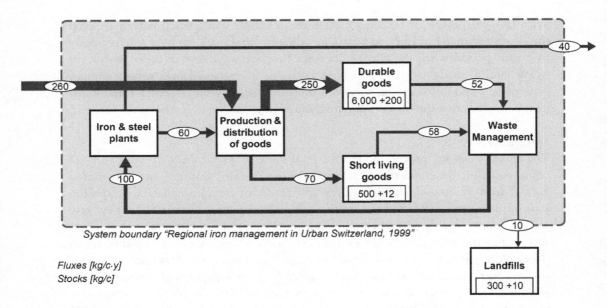

Figure 4.12
Sorting for recycling: Exemplifying the national iron balance of Switzerland (after Baccini, 2008).

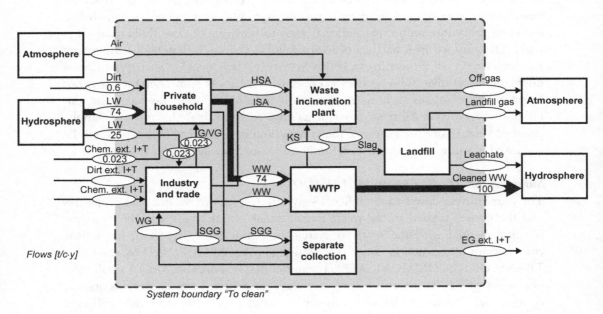

Figure 4.13
Material system for TO CLEAN in METALAND. Chem.ext., imported chemicals; EG, exported goods; HAS, municipal solid waste; KS, sewage sludge; ISA, industrial solid waste; LW, tapwater; SGG, separate collected goods; WG, recycling goods; WW, wastewater; WWTP, wastewater treatment plant.

Table 4.14
List of collected wastes in METALAND (flows ×100 kg/c.y). The total flow is subdivided into inorganic fraction and water fraction and three waste-management options

Type of Waste	Total Flow	Inorganic Flow	Water Flow	Waste-Management Option		
				Incineration	Landfilling	Recycling
Sewage sludge	4.5	0.1	4.2	2	0.5	2 (agriculture)
MSW	4	1.1	1.6	3	1	
Construction[a]	2.4	2	2	1.2	1.2	
Construction	20	20			10	10
Industrial[b]	0.4	0.2	<0.1	0.3	0.1	
Sorted[c]	3.6	0.4	0.3			3.1
Total	34.9	23.8	6.5	6.5	12.8	15.1

[a]Only mixed wastes from construction sites.
[b]Hazardous wastes included.
[c]In 100 kg/c.y: 1 iron; 1.5 wood and paper; 0.5 biomass; 0.3 various metals such as Al, Cu, Zn; 0.2 glass; 0.1 plastic.

is recycled, mainly due to the dominant mass flow of construction materials. Some landfills serve still as reactor-type landfills (storing wastes such as sewage sludge, MSW, mixed construction wastes, and industrial wastes). To deposit these components in landfills is still a controversial issue and will be discussed in chapter 5.

Because of the fact that cleaning processes serve as collectors (sewage systems with STP, waste collecting for incinerators and/or landfills) for population sizes of several hundred thousand inhabitants, the same processes can serve as monitoring and accounting sites to quantify urban metabolism (Baccini et al., 1993; Morf & Brunner, 1998; Kroiss et al., 2008). In combination with data on consumption, substance balances of large urban regions are feasible.

The energy demand for the activity TO CLEAN, in comparison with other activities, is relatively small, between 1% and 2% of the total in developed urban systems (Baccini, 2008). This is because a large aqueous portion is transported in sewer systems that function on the basis of gravitational force. Incineration plants transform chemical energy to thermal and electrical energy, resulting in a net energy output.

The cleaning costs per capita and year for the total wastes given in table 4.14 for METALAND and the other processes presented in this section range between US$100 and US$1000, depending on the specific transport methods and technical processes chosen. It is evident that the cleaning costs in an economically strong region are small in comparison with a mean available annual income per capita of roughly US$50,000. However, such a balance does not consider potential negative effects from hazardous sites or consequences from emissions distributed into the environment. This aspect will be discussed in chapter 5.

TO NOURISH and TO CLEAN: Metabolic Essentials

Both activities show key processes (i.e., processes that show the relatively highest material turnover and play the role as main sources or main actors). In TO NOURISH, it is the process *Agriculture*; in TO CLEAN, it is the *Households*. In both processes there is one key good; namely, the biomass and the water, respectively. In TO NOURISH, the available area of agricultural land with its soils grown over hundreds of years is the crucial resource capital. In TO CLEAN, the hydrologic system (climate, precipitation, drainage, natural and artificial water reservoirs) determines the availability and quality of water.

The two activities have three main properties in common:

1. The main materials involved (biomass and water, respectively) have short residence times within the anthroposphere (i.e., between hours and several months).

2. For affluent regions like METALAND, the growth rates (i.e., the growth of consumption of water and biomass per capita and year) are relatively small or tend to stay constant over several years. The growth rate per region is directly proportional to the population growth.

3. No relevant stocks of biomass and water are formed over years. If there are stocks (e.g., food or water), they are kept more or less constant.

The schemes for these similarities are given in figure 4.14. For the main substances carbon and water, the two material systems function as "flow-through reactors." There are two important exceptions. In the stock of the process *Agriculture*, the soil can accumulate phosphorus as described earlier (see figure 4.5c). The process *Landfill*, in which inorganic and relatively inert components are stored for long periods of time, is building up a stock. This flow, however, is two to three orders of magnitude smaller than the main material flow of this activity, namely water. It follows that the great majority of material systems of the two activities can be described mathematically with a quasi steady-state model (see chapter 3).

TO RESIDE&WORK and TO TRANSPORT&COMMUNICATE

An Outline of the Cultural Aspects of the Activity TO RESIDE&WORK

Sheltering Body, Social Group, and Technical Equipment

One of the essential idiosyncrasies of cultural evolution is the development of the sheltering technique, depending mainly on two factors: the climate and the available materials. Examples are caves or tents in moderate climates, igloos in the arctic zone,

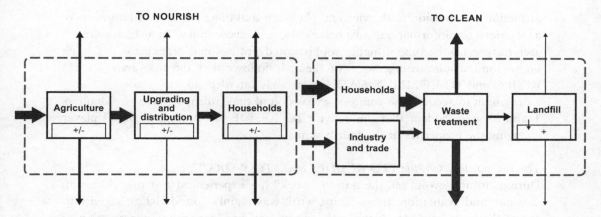

Figure 4.14
Metabolic schemes for the activities TO NOURISH and TO CLEAN. Both activities generate material systems that show very low stocks because the main goods (biomass and water) flow through the system. An exception is the process *Landfill* in TO CLEAN where a part of solid residues is stored. However, this flow, compared with water, is small in quantity.

or open wooden or clay huts in the tropical and subtropical areas. The constructive variety of shelters shows one common need of *Homo sapiens*: to have a protected site of its own to reside, namely to cook, to eat, to sleep, to raise children, and to construct their tools and gadgets. The notion "house," often synonymous with home, is found analogously in all cultures. The Indo-Germanic etymological root of the word *house* is the word *skeu*, meaning "to cover," "to wrap." It has become the name tag for a building where people reside but also a stem for any building extended by its function (e.g., schoolhouse, farmhouse).

Human beings owe their first skin to the biological evolution. The perception of the second skin, in its relevant functions, is based on scientific achievements. It is the atmosphere of planet Earth, a product of the biogeochemical processes during the past 4 billion years. The properties of the second skin, together with the given astrophysical boundary conditions and the terrestrial and aquatic ecosystems on the planet's surface, determine the climate. In the cultural evolution, one can distinguish two main steps of designing new "skins." The biological skin, a heritage of the primates, was expanded in a first step with clothing, the third skin, serving not only as protection against coldness, mechanical wounds, and sun radiation but also eventually as embellishment and status symbol in social communication. The second step was the building of houses. This fourth skin is a technical extension of the third, offering a better protection, moving toward an optimal inner climate independent of the changes outside of the house. Analogous to the clothing, the house has become a medium for the status of its residents. The house gives signals of attraction or repulsion to the viewers. The outer and inner surfaces of houses serve as carriers for information, advertisements, and decorations. In urban cultures, architecture has become a highly sophisticated professional branch for designing houses and whole arrangements of them. It follows that the two activities TO RESIDE and TO COMMUNICATE are strongly interdependent.

Taking into account the youngest expansion of the anthroposphere (see chapters 1 and 2), it is evident that the activity TO RESIDE has become a global player changing the properties of the fourth skin.

The Inseparable Couple "TO RESIDE" and "TO WORK"

During cultural evolution, the notion "work" has experienced a strong change in meaning and evaluation. In antiquity, work was mainly considered an activity to secure the basic needs for life, mainly agricultural production and construction of buildings and roads. This work was done by underlings and slaves. The workers had the lowest social status. An educated and well-off member of the ancient society did not work. Freedom and work were incompatible. The etymological root of the English word *work* is *uerg* or *erg*, meaning "to do" or "to make." In German, the word *Arbeit* ("work" in English) goes back to *arbejidiz*, meaning "tribulation" and "poverty." In the Christian societies, the role of work was gradually transformed from a negative to a positive image by introducing work or labor as human participation in God's creation of the world. Work was no longer a penalty but a service to God. The cities of the late Medieval period and the Renaissance showed in their housing structures the physical manifestation of this changed way of life. The tradesman, a respectable free citizen, did his craft in the basement and lived with his family on the upper floors. Houses of economically successful craftsmen, very often exclusively furnished, were part of the best quarters in town.

During industrialization, the new economic processes led to positioning labor as a basic element, together with capital, defining the evolving capitalistic system. More and more people moved from the rural world and the agricultural work to the urban, where there was a spatial distinction between the place to reside and the place to work. Within this system, "to have work" transformed to a human right. Humans without work (unemployed) have the lowest social status. Work has become a necessity for everyone, because it evolved to the crucial sense-giving activity. Society transformed to a new form of meritocracy. The better you work, the more you have, the better you are. Societies started to install compulsory insurance systems to prevent poverty of the unemployed. Communities and large companies entertain housing programs facilitating residents or employees with low income to rent or buy a decent home.

In societies that have already passed the period of industrialization, the dominant branch of the economy is the service industry. Here, the spatial distinction of the place to reside and the place to work dissolves. The new communication technology with mobile phones and Internet connections via personal computers allows people to do their business anywhere, anytime (i.e., not only at the company's offices but also on the road, at home, in a hotel, at the beach). The tight connection between residing and working of the Neolithic and agricultural societies across several hundred generations only loosened during roughly six generations within the period of industrialization, then became re-bound because of a technological innovation in communication. In other words, it is reasonable, taking into account the long period of cultural evolution, to handle the activities TO RESIDE and TO WORK as an inseparable couple.

The Material Systems Emerging from Cultural Periods

The material systems that emerged from this couple are given in figure 4.15. In Neolithic and nomadic societies, the system boundary is identical with the process *Household* (figure 4.15a). Within this process, there are two processes to define: one is *Manufacturing* (in the literal sense of the word, i.e., handmade) of a variety of products, such as building components, animal skins for clothes, hunting devices, agricultural tools, and pottery. This process represents, in combination with the process *Gathering and Hunting* (see figure 4.3a in TO NOURISH), one part of the activity TO WORK. The second process, named *Construction and Maintenance*, comprises all operations involved with building the house (or tent or cave) and maintaining it (including the fireplace) to function as the place to cook, to heat the room, to make ceramics, to store food and tools. The *Household* is the place TO RESIDE, because it gives shelter, social care, and a protected room to relax and to sleep. It is normally just one room. This activity leads to a formation of a material stock (i.e., the building with its equipment). In this cultural era, the residence time of buildings (or tents) is rather short (between one and two decades). From archeological findings, we know that Neolithic societies were already trading various tools. Therefore, import and export has to be considered.

Urban societies in agrarian systems show a more diversified material system (figure 4.15b). Specialized craftsmen have installed enterprises and manufacture and sell their products to the *Households* where not almost everything is produced as in the Neolithic era. The *Households* themselves have diversified their room structure. There is not just one "living room," but a space divided into several rooms with separate functions, such as living rooms, bedrooms, kitchen, working rooms, and storerooms (e.g., cellar). In addition, the rooms are equipped with furniture for various purposes such as tables, chairs, benches, cupboards, and beds. A town within an urban system shows an important innovation in the evolution of society, already realized in antiquity, namely the formation of a public space with public buildings. There is a clear distinction between the private space, where the intimacy of

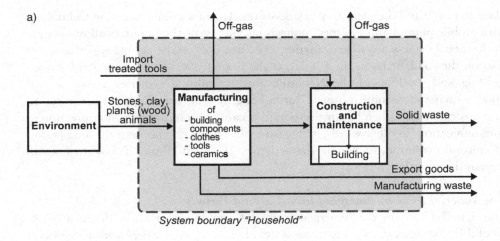

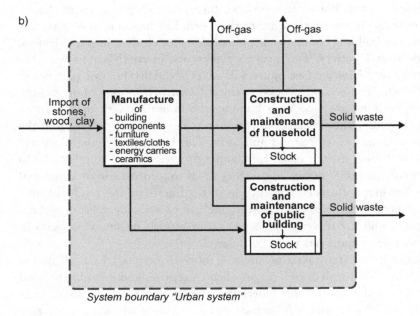

Figure 4.15
Material systems for the activity TO RESIDE&WORK in three cultural periods: (a) Neolithic and nomadic society; (b) urban society in agrarian systems; (c) industrial society.

c)

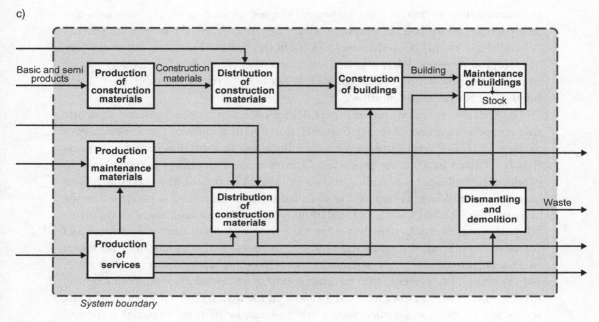

Figure 4.15
(continued)

Households is given, and the public space, open to everybody, where the economic transactions (marketplaces), political meetings (city hall), and the religious rites (temples, churches, mosques) are taking place. In between are the installations of the urban trade, open to the suppliers and the clients. Rural residing and working has still the character of Neolithic times, although technically more sophisticated and grouped in villages with a larger number of families residing and working together. Urban settlements, however, have created new institutions to regulate the activity TO RESIDE&WORK. The manufacturing is controlled by guilds. The private property of *Households* is secured by law. The public buildings are financed by taxes, feudal donations, and by partly enforced drudgery.

During industrialization, the urbanization started to break through the walls of the established cities (see also chapters 1 and 2). The places of urban residing and working sprawled out, and the people moved in. Because of technological and economic reasons, the mass production of goods such as textiles, tableware, and furniture took place in large-sized factories and replaced the small workshops of craftsmen. First, the specific "industrial cities" were to be found in England. Manchester, for example, was strongly criticized by contemporaries such as Charles Dickens, Friedrich Engels, and Alexis de Tocqueville (Lees, 1985) for the miserable living conditions of the laborers. The skylines of industrial cities showed brick chimneys whose permanent smoke promised good work and whose emissions and dust particles meant bad air. In spite of all the negative effects, the nineteenth century "changed the world" (Osterhammel, 2009). In the twentieth century, a new strategy of enterprise developed that was well summarized in Henry Ford's legendary phrase to justify the production of the Model T ("Tin Lizzy") on an assembly line: "I will build a car for the great multitude."

The resulting material system (figure 4.15c) reflects a complex process structure that comprises a higher diversity of production and distribution. The *Production of Services*, such as financing, planning, controlling, and so forth, has become a process that is involved in all other processes. *Construction of Buildings* is the process to produce the buildings for residing and/or working. The sum of physical connections between the buildings is part of the so-called infrastructure and is presented in the activity TO TRANSPORT&COMMUNICATE. In this system, there is only one relevant material stock, namely within the process *Maintenance of Buildings and Infrastructure*. From this stock the building, or parts of it, is eventually transferred to a de-constructional process named *Dismantling and Demolition* (in the former two systems also present, but integrated within construction and maintenance) producing an output either handled as waste in the activity TO CLEAN or as basic secondary resource in another production process, be it in the material system of TO RESIDE&WORK or of TO TRANSPORT&COMMUNICATE. Between production, construction, and maintenance, the distribution processes, representing the dynamics of a globalized market, have become indispensable transfer units within this material system. Therefore, almost every producer has an inner and an outer market to cope with.

Within this new framing, the activity TO RESIDE&WORK moved into a consumer market, guided by a permanent trade-off between shareholders of enterprises (the kings of capital) and union leaders (the lords of labor). The pursuit of happiness, an objective in the preamble of the Constitution of the United States of America (the first modern democratic society worldwide), stimulated a physical project with significant metabolic consequences. It was not the socialistic design of a classless society that took the metabolic lead. Within a liberal democratic society, a market economy helped to form a new "middle class," comprising roughly 80% of the population, with an own house in which to reside, equipped comfortably with automatic supplies of water and various forms of energy, with automotive cars to reach all urban services, such as markets, working sites, schools, health care, and various cultural institutions. The political concept was to limit the very rich and the poor, each at around 10%. The extremely wealthy were tolerated as very successful winners who accelerate the system. The poor are accepted as unlucky losers who slow down the system and have to be subsidized. "Life standard" became a politically relevant coin that had a promise for a "better life" on the front side and a "metabolic bill" on the back side. Better life meant permanently growing material flows and stocks to be handled by the socioeconomic system.

TO RESIDE&WORK in METALAND

The metabolic properties of the activity TO RESIDE&WORK are to be found in the life cycles of the buildings, their inner equipment included. For METALAND, we do not build up a new set of buildings "in the open countryside." We assume that there is a built urban system with a stock of buildings that is to be extended and transformed continually. Therefore, we have to choose:

1. Areas and volumes per inhabitant for residing and working within a given settlement space.

2. Material contents and composition of the built entities.

3. Residence times of buildings and their components.

4. Technical quality parameters of the building's facilities to determine the metabolic requirements for operating the buildings (e.g., energy demands).

5. Socioeconomic and political factors determining the growth rates and the life cycles of the buildings.

Looking at the above list, it becomes evident that because of the economic, technical, and cultural diversity given, the spectra of material systems possible and already created on a global scale is much greater than those presented for the two foregoing activities. There are national statistical data for some of the above five items but there is not yet a work study available to make a sound comparative evaluation on a global scale. In the area of statistics there exist some data to give an impression of the variation of the residing and working areas in different countries (table 4.15), expanded with the data for METALAND (see table 4.2). In agreement with the

Table 4.15
Comparison of TO RESIDE&WORK areas (mean values) in various countries and the choices for METALAND (in m^2/capita)

Country/Region	Residing	Working	Total	Reference
Germany	41			IFS (2006)
United States	68			
Japan	36			Janssen & Hertle (2007)
Switzerland	49	51	100	Baccini & Bader (1996)
World	4–9			Pacione (2005)
METALAND	50	40	90[a]	

[a] 900 km^2, stacked in the average on two floors, amounts to 450 km^2 settlement area, or 14% of the total settlement area.

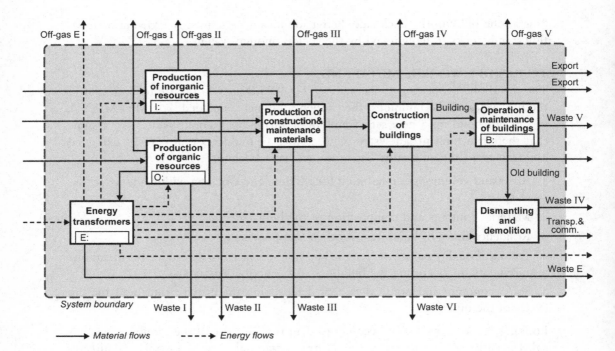

Figure 4.16
Material system for the activity TO RESIDE&WORK in METALAND. The indicated stocks are as follows: I, inorganic; O, organic; E, energy; B, buildings.

general character of the model region chosen, its gross areas per capita for the activity TO RESIDE&WORK correspond with a developed country.

METALAND's material system of the activity TO RESIDE&WORK is given in figure 4.16. It is a strongly simplified system for focusing on the buildings. All processes and goods involved in a "working activity" appear in the subsystems that emerge producing a product, a car, a washing machine, a mobile phone, a knife, a shirt, and so forth, described metabolically in the corresponding activity systems. Therefore, the system presented in figure 4.16 focuses on the "shelters," the buildings.

From a quantitative point of view, the furnishing of the building (apartments, offices, workshops) is of minor importance (Baccini & Brunner, 1991). META-LAND disposes of regional stocks of inorganic resources (I), such as gravel, limestone, metal ores, and so forth, and of organic resources (O), such as forests and coal. The corresponding processes (*Production of Inorganic Resources* and *Production of Organic Resources*) prepare the material for the next process, *Production of Construction and Maintenance Materials*, which comprises a broad palette of engineering technologies, such as cement manufacture, the timber industry, metallurgical procedures, and the plastics industry. The semi-products gained, such as cement, construction iron, wooden beams, and plastic pipes, are educts for the process *Construction of Buildings*. Its product is a building going into the process *Operation and Maintenance of Buildings* that has a stock of buildings B. Old buildings or parts of them are moved into a *Dismantling and Demolition* process. It produces secondary inorganic and organic resources that are either reused within the system by the process *Production of Construction and Maintenance Materials* or exported to other regional materials systems or to other regions. Operating the six processes means supplying them with energy delivered by the process *Energy Transformation*. This process has its own energy stock E (with biomass, mineral oil, coal, water reservoirs). In this activity (and the next one, To TRANSPORT&COMMUNICATE), the energy flow is crucial with regard to the overall energy consumption. Therefore, contrary to the two foregoing activities, the energy flows are integrated in the material system. All seven processes import and export materials and produce wastes (solid and aqueous) and off-gases. From a metabolic point of view, materials from the stocks I and O are transferred to the stock B. The spatial transport processes (referring also to the distribution processes mentioned in figure 4.15c) are not considered here. They are integrated in the material system of the activity TO TRANSPORT&COMMUNICATE.

Taking into account the diversity of buildings being constructed and added to a given stock, the changing technologies leading to new material compositions, and the socioeconomic factor stimulating replacement and demolition of buildings, the dynamics of the system has to be considered. In other words, the pattern of building functions and their material composition is time dependent. In figure 4.17, a schematic presentation of a time-dependent building stock of a region is given (Baccini & Pedraza, 2006). It shows a growth curve for a building stock (in mass units) over the time period of the twentieth century.

The chosen timescale is divided into four equally long time periods (I–IV). In some countries, there exist statistical data on the number of buildings, on the time of their construction, and on their functions; for example, "Detached Family House" (FH), Multiple Dwelling House or apartment house (MH), Office Building (OB), Trade and Production Building (PB). No statistical data are yet available to grasp the material composition of all buildings. Therefore, "real buildings" of each function in each of the four periods serve as data donators for a first approximation of the material contents of the total stock. It follows that each point on the building mass curve represents the sum of the products of the number of buildings times its material content. At a given time, the total stock consists of buildings of different ages. The "lifetime" of an individual building cannot be forecast. However, it is possible, on the basis of building statistics, to assign a mean value for the lifetime

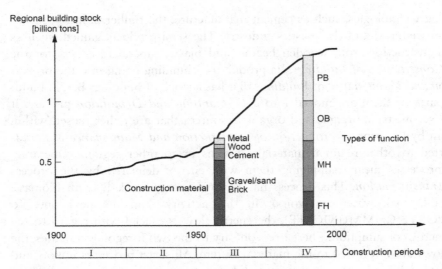

Figure 4.17
Schematic presentation of a time-dependent building stock of a region (example Switzerland; after Baccini & Pedraza, 2006).

of a building (Johnstone, 2001; Bader et al., 2006). It is evident that such a value is very specific for each region or even for each community.

Two basic properties of a building determine its material content; namely, the "construction system" and the "spatial structure" (figure 4.18). A first rough differentiation of construction systems leads to three main types (on the x-axis): wood, brick–stone, concrete. As a rule of thumb: The more concrete, the higher the material content (mass per building volume). The spatial structure is characterized with the ratio constructed volume (V_c) to total volume (V_t), shown on the y-axis. The higher this ratio (due to a higher subdivision of the inner space), the higher is the material content. These properties are illustrated in tables 4.16 and 4.17.

It is evident that a building, constructed mainly with concrete, has a higher material content (480 kg/m³) than that of a wood house (280 kg/m³). However, the construction system chosen for a "wood house" still includes essential elements made of concrete, resulting in 54% of the total material content. As a rule of thumb, to estimate the order of magnitude of a building stock, mainly based on concrete, one can choose a material content of 500 kg/m³.

In table 4.17, the buildings are grouped in the order of descending construction volume (V_c). It illustrates the fact that the reduction of the corresponding total material content is not strictly proportional. Although the general tendency is correct, namely the difference between FH and OB (see also figure 4.18, y-axis), variations in the choice of materials within a construction system can compensate

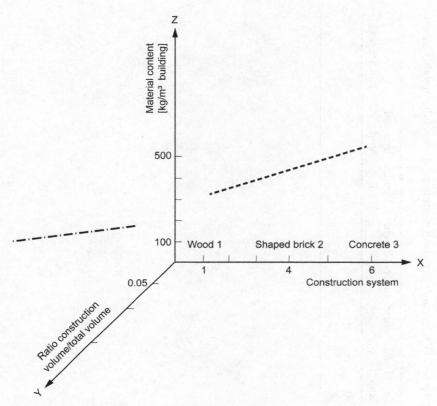

Figure 4.18
Schematic presentation of material content of building as a function of the construction system and the spatial structure (after Baccini & Pedraza, 2006).

this effect (see figure 4.18, x-axis). For the FH, more wood is chosen, reducing the gravel and sand content. By this measure, the total material content is significantly lower than that of the MH, in spite of the relatively highest construction volume V_c.

To sum up: The material quantity and quality of the stock B of an urban system (figure 4.16) can be determined if data are available on:

1. number of buildings with information on volume, construction period, and type of function;

2. material contents as a function of the construction system and the spatial structure; and

3. "life cycles" of buildings with regard to their function, their maintenance, and their "mortality."

Table 4.16

Material contents of buildings as a function of the construction system.[a] The content of the six materials is given as percentage of the total. The house investigated is a detached family house (FH) built in the fourth time period (see figure 4.17) with the same spatial structure but with two different construction systems. The individual material contents are given in percentage of the total material content. The six chosen materials do not sum up to 100% because other components (e.g., plastics, other metals, etc.) are part of the construction

Construction System	V_c (% of V_t)	Total Material Content (kg/m³)	Gravel and Sand (%)	Marl and Clay (%)	Cement (%)	Wood (%)	Iron and Steel (%)	Limestone (%)
Concrete	30	480	70	4	11	4	3	1
Wood	27	280	54	2	8	17		11

[a] After Baccini and Pedraza (2006).

Table 4.17
Material contents of buildings as a function of the spatial structure.[a] In this comparison, only the function type of the building is varied, and the construction system (concrete) is kept constant. The individual material contents are given in percentage of the total material content. Between 4% and 11% of the total mass consists of other materials

Function of Building	V_c (%)	Total Material Content (kg/m³)	Gravel and Sand (%)	Marl and Clay (%)	Cement (%)	Wood (%)
Family house (FH)	30	480	70	4	11	4
Multidwelling house (MH)	24	510	78	6	11	1
Production building (PB)	20	520	71	10	11	0.2
Office building (OB)	17	360	72	6	11	1

[a] After Baccini and Pedraza (2006).

With lack of sufficient statistical data, model approaches that are based on empirical and historical evidence can give first approximations to quantify the building stock (e.g., the "ARK-House-Method," after Lichtensteiger et al., 2006).

Cement Production, an Illustration of the Process Production of Construction and Maintenance Materials

Cement has become, on a global scale, the main binder in constructing buildings and infrastructure. In the first decade of the twenty-first century, cement production is of the order of magnitude 3×10^9 Mg per year (year 2008), corresponding with an average consumption of approximately 400 kg per capita and year. Asia has a two-thirds share in production (Cembureau, 2010). In cement consumption, the differences between continents and regions cover a range of three orders of magnitude. In some countries, the consumption is only between 10 to 20 kg/c.y, whereas in others, the peaks are around 2000 kg/c.y. Therefore, cement consumption is also taken as an indicator for economic growth because a nation's economic achievement, expressed in monetary units (GNP, gross national product), is mainly materialized in construction of buildings and infrastructure. This is mainly true for countries in development with strongly expanding urban systems. In highly developed countries, although in a modest economic growth process, the cement consumption per capita can stay constant or even decrease slightly over the past decades. This has to do with the fact that societies with a high standard of living and a small population growth tend to spend more money in areas like health, wellness, and security. Growth of new buildings and infrastructure is more and more replaced by renewing the existing stock.

On the metabolic level, cement production is a significant consumer of energy and an emitter of climate-relevant carbon dioxide. In figure 4.19, a simplified flow scheme gives the mass ratios of the main input goods in comparison with the output goods. The basic two educts are limestone (the calcium carrier) and sand (the silicate carrier).

Producing cement is an endothermic process forming a calcium silicate with pozzolanic properties. The applied energy carriers are mainly fossil fuels, which are burned with combustion air. The off-gas contains mainly the carbon dioxide from the quicklime and the oxidized organic carbon and, depending on gas cleaning installations, more or less dust. Per mass unit of cement, about the same quantity of carbon dioxide is emitted. Depending on the state of production technology, the energy needed per kilogram of cement is between 3 and 10 MJ. For METALAND a cement consumption of 300 kg/c.y is chosen with an energy demand of 5 MJ/kg. The corresponding energy demand amounts to 1.5 GJ/c.y. This energy flow has to be compared with the total demand of the activity system, namely 65 GJ/c.y (figure 4.20). The contribution of cement production lies within a few percent. However,

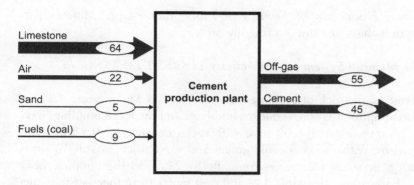

Figure 4.19
MFA of a cement production plant in Switzerland. Relative values in percent of total input mass, respectively output mass. (after Baccini & Bader, 1996).

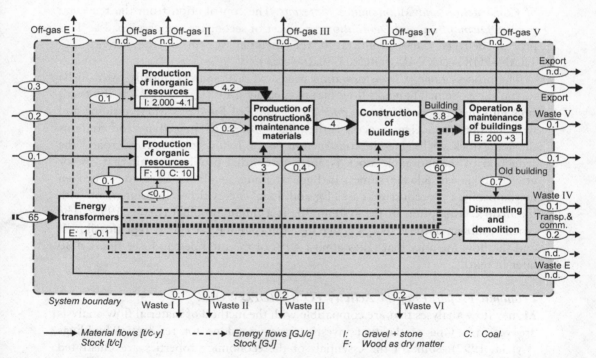

Figure 4.20
Quantified material system of the activity TO RESIDE&WORK in METALAND.

within the process *Production of Construction and Maintenance Materials*, its energy consumption share amounts to roughly 50%.

Quantifying the Material System for the Activity TO RESIDE&WORK in METALAND

The anthropogenic main stock of buildings amounts to 200 Mg/c (figure 4.20). It is based on the assumption that the residence buildings and the work buildings have each a volume of approximately 200 m³/c with mean densities of 0.5 t/m³. This stock, in comparison with the endogenic gravel and sand stock, is still 10 times smaller. The inorganic stock I decreases annually by 2%, and the building stock grows by 1.5%. Forests (see also table 4.2) and coal mines form together the inner regional organic stocks. Because the construction system is mainly concrete (see also tables 4.16 and 4.17), the main material flow stems from the inorganic stock The largest turnover of material (5 Mg/c.y) takes place in the process *Production of Construction and Maintenance Materials*. The contribution from the recycling process *Dismantling and Demolition*, producer of secondary resources, is 8%. This process delivers other products to the material system of the activity TO TRANSPORT&COMMUNICATE (0.2 Mg/c.y).

The process *Energy Transformation* needs external sources (mostly fossil fuels) to run the system. The main flow (approximately 60 GJ/c.y, or 90% of the total) runs into the process *Operation and Maintenance of Buildings* where residing and working take place. This process is the target process of the activity and contains the main stock. As long as the construction system follows the "concrete road," the reduction of the inorganic stock is not a primordial problem, as alternatives for gravel are given. Silicate stones, in huge amounts available, have to be broken mechanically, a procedure that asks for additional energy flows. In comparison with the main consumption of the system, this is a minor problem (Redle, 1999). The crucial problem of the activity's actual material system is the energy type (fossil fuel) and the flow quantity that takes almost 40% of the total demand (see figure 4.30 later in text).

Economic Properties of the Activity TO RESIDE&WORK

Money flow analyses that are compatible with the method of material flow analysis are yet rare. One of the first investigations was done for residential buildings (Kytzia, 1997). Some of the essentials of the economic properties are illustrated in figure 4.21. It is based on an investigation of concrete in a small city in Switzerland (Olten) and is strongly simplified. The central process is *Operation and Maintenance of Buildings* where the main stock of material is situated. It is limited to the residential buildings with about half of the total stock given for METALAND (figure 4.20). In principle, the money flows have the opposite direction to

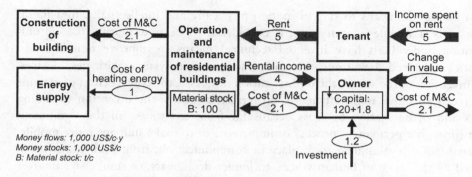

Figure 4.21
Monetary flows and stocks in the material system for residential buildings (after Kytzia, 1997; Baccini, Kytzia & Oswald, 2002).

the corresponding material flows. Two main protagonists are chosen: the owner and the tenant. For the owner, the material stock of the building has a value named as capital invested to build and maintain the building. The order of magnitude of US$120,000 capita for the value of the residential buildings (in the 1990s) has to be compared with the mean available income per capita and year, namely US$30,000. The capital invested in the residential building is equivalent to four annual incomes per capita (not per household). It illustrates the economic fact that buildings are not only very large material stocks but also belong to the largest investments within an economy. In other words, large parts of regional economic profits are eventually invested in buildings. The tenant's rent of 5 units includes here the energy cost of 1 unit. In this example, the main profit the owner of the residential building makes (depending on his obligations toward external investors) is the gain in value of the real estate and not the value gain of the building itself.

An Outline of the Cultural Aspects of the Activity TO TRANSPORT& COMMUNICATE

Striving for Mobility
In the literal heritage of all cultures, narratives of long adventurous journeys supply plenty of metaphors for the design and fate of a human life. To travel means to discover the new, to be challenged by unforeseen dangers, to find a treasure, be it jewels or knowledge, to return home rich and more mature and to be ready to settle. There was always a certain percentage of the travelers (people "on the road") who preferred to stay on the road all their lives or to choose a new land to settle in, far from their place of origin.

To be mobile means to stay in motion physically and intellectually. In a literal sense, only a mobile human being is considered to be free. During the cultural evolution, individuals have invented technical means to improve mobility, and societies improved their transport and communication systems with new devices. The invention of the wheel, the domestication of horses, the construction of wagons and roads to move more efficiently on terrestrial paths, the invention of sailing boats and of navigation to cross oceans far from coastlines, all this equipment went through a permanent process of improvement to make humans more mobile. An analogous development took place in communication, from oral information, limited by the reach of human voices, to louder drumbeats, to smoke signals over larger distances. The invention of script opened the door to the storage of accumulated knowledge to be accessible for future generations. The discovery of electromagnetic waves to become carriers of information was a quantum leap in information technology. The sound waves can be amplified transforming them to electromagnetic waves and transporting the information at a speed six orders of magnitude faster. At present, the "mobile phone" is the status symbol of a human being accessible anywhere, anytime. And the World Wide Web allows contact with people and libraries of all types around the world. The "network" has become the crucial notion for logistics in transportation and communication of a mobile society.

The Inseparable Couple TO TRANSPORT and TO COMMUNICATE
Within the vocabulary of metabolism of the anthroposphere, TO TRANSPORT& COMMUNICATE is defined as an activity that comprises *all processes that have been developed to transport persons, materials, and energy and to exchange information.* These processes range from road construction to education and administration. Some of the corresponding goods are streets, cars, railways, electrical wires, pipelines, ships, harbors, airplanes, airports, newspapers, books, telephones, and personal computers. It is not useful and hardly possible to make a clean separation between transporting and communicating, because the two were and still are interwoven in multiple ways. The messenger on horseback brings information from a sender to an addressee. It is an act of transport as an act of communication. A packaged kilo of coffee beans is transported in a truck, together with hundreds of other items, from a logistics center of a large distributor to a small grocery store. The client, who buys the coffee, finds information printed on the package, not only the name of the brand but also the exact weight, the origin, the price, and other items. Producers and traders communicate with the buyers on a much broader scale. The package is the last chain of an advertisement chain. The logos applied are periodically used in commercials in print and electronic media to attract the potential consumers. In cultural evolution, the networks built for transporting persons

and goods have become networks for communication, at least platforms to distribute information, not always necessary to exchange information. "Trade has shaped the world" (Bernstein, 2008), and the term *mobility* has become a notion that comprises the complexity of transporting and communicating within a large-scale urban system. Noteworthy is a young example with the label "New Mobility Culture in the Baltic Sea Area" (Kasprzyk, 2007). After the fall of the Iron Curtain and the integration of the Baltic states into the European Union, 18 cities and regions, from six different Baltic nations, formed a working platform "to promote sustainable modes of transport as part of a wider regional strategy." They focus primarily on the transport of persons because "most cities suffer from severe problems caused by transport: pollution, noise, space consumption, congestion." Recalling Le Corbusier's criticism of cities almost one-half century ago, the negative effects of urban mobility are still virulent. Contrary to Le Corbusier's concept to solve the problem by architectural means, the Baltic project sets its main effort on change of behavior of the inhabitants with regard to their mobility. It is, above all, a communication project on the diverse possibilities of movement in urban systems and on finding support within the population to restructure the transport system. The project leaders want to change "the mobility culture." A case study on mobility within the framework of METALAND is presented in chapter 5.

The Material Systems Emerging from Cultural Periods

1. *Neolithic and nomadic societies*   Within Neolithic societies, transport of materials was done simply by carrying the goods, moving on foot. On terrestrial paths, there were no technical vehicles yet, except in some advanced groups where domesticated animals had to pull plows and sleighs. Humans in subsistence economies did not have to overcome large distances. The oldest transport vehicle is probably the boat (made of wood, reed, fell), and the oldest transport networks are the natural waterways (rivers, lakes, seashores). Transport of persons and goods on boats was (and still is) energetically more efficient than that on terrestrial paths. Handicrafts produced containers (cloths, baskets, pottery) to carry babies and to gather food. Communication was mainly oral, supplemented by ornaments in clothing, painted skin, and hairstyle. The narrative heritage was forwarded with music, dancing, and wall paintings describing the fates of human existence. What we call a piece of art today in a sophisticated society is, within this terminology, part of human communication. It is evident that the activity TO TRANSPORT&COMMUNICATE had, already in Neolithic times, a strong impact on technical innovation. In figure 4.22a, the corresponding material system considers mainly the two processes *Production and Maintenance of Vehicles*, namely boats, containers, and pack animals, and *Production and Maintenance of Media* (music instruments, paints, decorations, jewels). The only vehicle that needed additional operational energy was the pack

a)

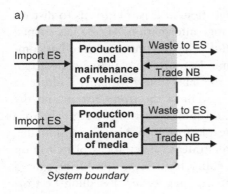

b)

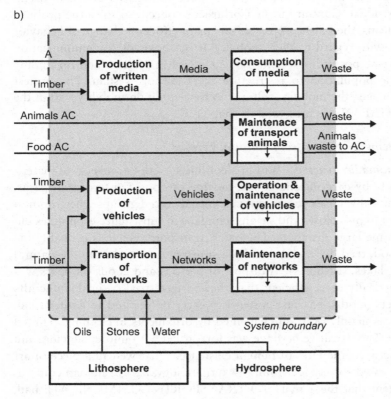

Figure 4.22
Material systems for the activity TO TRANSPORT&COMMUNICATE in three cultural periods: (a) Neolithic and nomadic society; (b) urban society in agrarian systems; (c) industrial society. A, from agriculture; AC, agriculture as source or target process; ES, from ecosystems; NB, neighbors.

c)

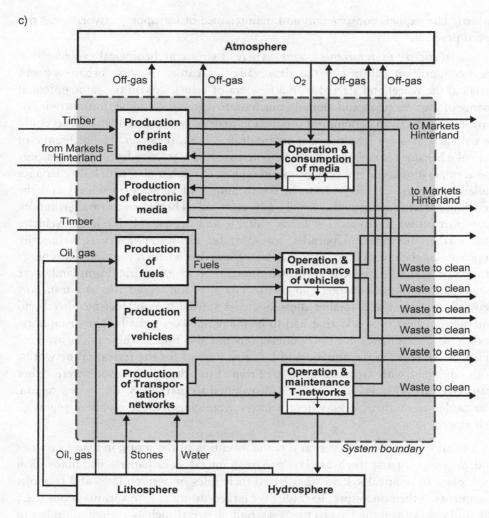

Figure 4.22
(continued)

animal. The explicit construction and maintenance of transport networks was not yet in practice.

2. *Urban society in agrarian systems*   There is strong archeological evidence that the first agrarian cultures in the river valleys, creating the first urban systems, invented the wheel and started with a new era of vehicles. With the same potential energy of muscles (man and animal), much more material could be transported. The terrestrial paths from agricultural production to densely populated settlements could be enlarged to serve freight transport by draft and draft horses. The technique of the yoke became refined and allowed an increase of the tractive force. The transport speed of persons could not only be raised on horseback but also with light carriages pulled by horses. The interaction of the technique of vehicles on wheels and the breeding of transport animals started a new economical branch in cultural evolution. Transport on wheels asked for roads, suitable and dependable for larger vehicles over farther distances. In parallel, transport on the waterways was technically improved by the invention of sails that transformed wind energy into kinetic energy of the vessels. Large rivers, in their main function to irrigate agricultural land, were extended to a system of larger and smaller channels and served also as a transport network for larger and smaller ships. Societies started to construct terrestrial and aquatic transport networks that had to be maintained continually. The metaphoric notion *network* implies not only connections but also nodes. These nodes are provided by harbors for the aquatic and by *caravanserais* for the terrestrial networks, both equipped with large warehouses. From there, the stored goods were either transferred to other vehicles or finely distributed to markets of the node's region. The nodes were also places where vehicles were constructed, repaired, replaced, and stored.

A great leap in communication was the invention of the script, in many varieties and on various pads (such as clay, papyrus, animal skin, paper). The innovation took place in political systems that based their rules on written laws and religious testaments of their ancestors. To trade over larger distances with various goods (e.g., silk and spices from the East to the West, bulk material such as grain and timber in the other direction) stimulated the invention of a suitable currency (at the beginning mainly silver). Business became money based and substituted for the barter trade. The flows and stocks of goods together with centrally managed agricultural production asked for an accounting based on arithmetic rules. The abacus is one of the oldest known calculation aids to keep track of the masses, numbers, prices, and so forth. Written documents became the status of legally binding witnesses of agreements between partners in business and politics. A new branch of art emerged, the art of written works such as poetry, novels, and written music. The leading people knew how to read and write or had always a staff around them with this ability. A new class was formed, the literates, for thousands of years a small percentage of the total population. Within this period, the technique of producing written documents evolved from handwriting to printing, allowing eventually a mass production for an increasingly literate population. It was the beginning of the end of the sovereignty of interpretation by small oligarchic elites. The distribution of knowledge changed society. It accelerated the process of globalization as much as the technical improvement in transport systems, although the corresponding material and energy flows were very small in comparison with the bulk. In figure 4.22b, the expanded material system is given. In comparison with the foregoing system (figure 4.22a), the processes *Production of Transport Networks* (terrestrial and aquatic) and *Production of Written Media* (at the end of this historical period also printed media) are added, followed by the corresponding processes for maintenance. The main source for the animals in transport is agriculture, a process mainly oriented toward the activity TO NOURISH. The strong expansion of animal breeding for transportation increased the demand of fodder production (i.e., the limiting factor for "biological vehicles" was the production of fodder). In case of low harvest yields, the demand for fodder came into conflict with people's demand for their nutriment. In this material system, the speed of transport and communication over larger distances (>5 km) was between 3 to 20 km per hour (i.e., until the beginning of the nineteenth century, nobody saw anybody or anything travel faster than a horse).

Before the next technical leap would change transportation and communication in a substantial way, the past centuries with "solar-driven anthropospheres" developed the global trade system as follows (Bernstein, 2008):

• During the sixteenth century, the exchange of crop species such as corn, wheat, coffee, tea, and sugar between continents revolutionized the world's agricultural and labor markets.

• By the early seventeenth century, Spanish and Dutch mariners had decoded the last great secrets of the planetary wind machine, allowing them to cross the vast expanses of the world's oceans with relative ease. By 1650, goods of all kinds and people of all nations ranged over most of the globe.

• The discovery of huge silver depots in Peru and Mexico produced a new global monetary system. The most common piece of currency, the Spanish eight-real coin, was as ubiquitous as one of the worldwide valid credit cards today.

• The seventeenth century saw the rise of a completely new trading order—the publicly held joint-stock corporation. These organizations had considerable advantages over what had preceded them: individual peddlers, their families, and royal monopolies. Large corporations soon came to dominate global commerce.

3. *Industrial society*   The technical inventions such as the steam engine (1769), the aircraft (1849), Bessemer's steel (1855), the combustion engine (1860), the dynamo (1866), and the steam turbine (1883) were some of the essential elements for

innovations such as railroads, steamships, automobiles, airplanes, and cities lighted by electricity. The global and regional trade was ready to introduce new transport networks with vehicles that (1) were much faster, eventually by one or even two orders of magnitude, (2) offered much higher transport capacities in weight and distance, and (3) were cheaper than the previous transport systems, measured in dollars per ton-kilometer. A decisive innovation was the exploitation of a new energy source, the stocks of fossil fuels (coal and oil) in the upper layers of the earth's crust. It allowed a decoupling from the limiting factor energy in biomass, produced in agriculture and forestry, and other "solar sources" such as energy from water and wind.

The material system of the activity expanded in complexity (figure 4.22c), mainly due to the introduction of the main energy carrier fossil fuels and the propagation of electronic media. The main transport routes, terrestrial and aquatic, were maintained, extended by aviation. The production of transport networks remained a regional matter, whereas vehicle and media production became a global business. Three important material stocks were formed and continually increased, the media stock (libraries and archives, infrastructure for them included), the vehicle stock, and the transport network stock. What changed within two centuries (nineteenth and twentieth centuries) were the quality, density, and diversity of vehicles. Some substantial replacements took place. In this context, detailed studies were made with data from the United States (Nakicenovic, 1986). On the waterways, the propulsion technique changed from sails to steam to combustion motors (figure 4.23a). In the construction of merchant vessels, metal replaced wood (figure 4.23b).

On terrestrial ways, automobiles with combustion engines replaced horses (figure 4.24). In intercity passenger transport, the car peak was reached in 1960, the railways and buses are the loser, and the winner is the airways (figure 4.25). For the case of the United States, Nakicenovic concluded that the replacement of old by new technologies in the transport and energy system lasted about 80 years (Nakicenovic, 1986). The technological changes within road vehicles, however, were more rapid. They lasted only a few decades. It is postulated that network expansions like railroads, motorways, and air transportation need considerably more time than that needed for technological substitution of vehicles within one type of network.

In densely populated regions, the public transport systems (railways, buses, and aircraft) have built big "hubs" for average daily passenger flows between 10^4 to 10^5 capita (e.g., airports Atlanta Hartsfield with 200,000, London Heathrow with 170,000, Tokyo Haneda with 150,000, Seoul Kimpo with 90,000, Hong Kong with 80,000). The underground railway system of Tokyo with approximately 200 stations has a daily flow of 8 million passengers; some of the larger stations have flows of several hundred thousand passengers per day. The commuter lifestyle in urban systems has led to a symbiosis within the big nodes of public transport; namely, to

a)

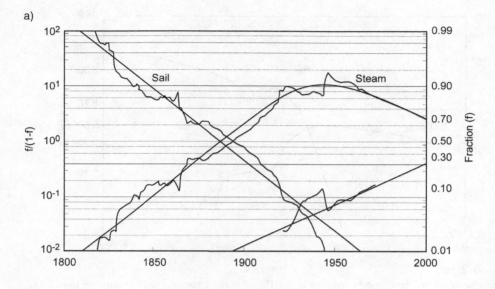

b)

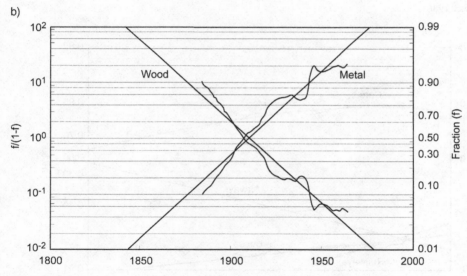

Figure 4.23
Substitution of vehicles due to technological innovations between 1800 and 2000: (a) Substitution in merchant vessels by propulsion system; (b) substitution in merchant vessels by structural material (reprinted with permission from Nakicenovic, 1986).

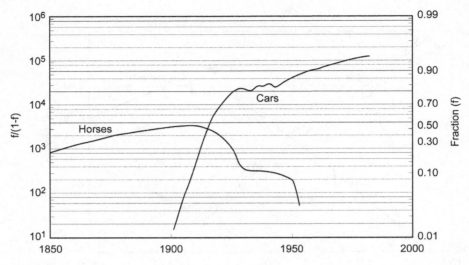

Figure 4.24
Substitution of non-farm horses and cars in the United States (reprinted with permission from Naki-cenovic, 1986).

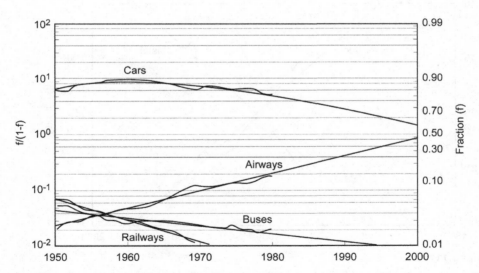

Figure 4.25
Intercity passenger traffic substitution in the United States (reprinted with permission from Naki-cenovic, 1986).

combine places for vehicle change with places for shopping. The daily transport of passengers in private cars reaches similar daily flows on certain segments of motorways. The supply service stations ("gas stations"), however, are linearly distributed and do not have the character of big nodes.

Changes in communication technology were triggered with the installation of electrical power grids. Inventions like the telegraph (1844), the telephone (1876), the radio (1896), television (1925), and the computer (1941) belong to the technical equipment that revolutionized human communication, first for business, political administration, and military purposes, but more and more as daily instruments for private social purposes. To talk around the world synchronously seems to be the logical continuation of socializing on the ancient agora, now an electronically connected virtual square of the global village.

The resulting material system is, in comparison with the former period, more diversified (figure 4.22c). The printed media still exists beside the electronic devices. There is still a demand even for handwritten messages delivered by a postal service. There are still sailing boats and horseback riders, mainly for recreational and sporting activities, besides the armadas of motor vessels and cars. For business purposes, however, the transportation and communication system chosen is based mainly on economic and, to a lesser extent, on ecological criteria. These aspects are to be qualified and quantified in the next chapter with METALAND.

TO TRANSPORT&COMMUNICATE in METALAND

Transport Networks

The assessment of material flows generated by the construction of transport networks is based on three basic types of information:

1. the actual "stocks" of transport networks;
2. their mean "lifetime";
3. their growth rates.

Excluded from the assessment are the buildings connected with these networks (e.g., railway stations, airports, storehouses, etc.). They are included in the rough assessments for the activity "to reside and work" (see figure 4.20).

Seven regions with affluent economies were chosen to compare the orders of magnitude for different types of networks. The data are drawn from the United Nations Transport Statistics. In table 4.18, the transport lengths of four types of networks, namely road, railway, waterways, and pipeline, are compared.

With respect to the length of man-made networks, roads form the most important transport infrastructure, followed by railways. In European countries, the railway network amounts to about one tenth of the total road length. In the United States

Table 4.18
Comparison of transport networks from United Nations transport statistics for year 2003. Length in meters per capita, obtained by dividing the total length (year) by the population (UN, 2008)

Country	Road	Railway[a]	Waterways[b]	Pipelines
Federal Republic of Germany	2.8	0.4	0.02	0.03
France	16.7	0.5	0.06	0.1
Netherlands	8.1	0.2	0.22	0.02
Sweden	15.6	1.2	0.06[c]	n.d.[d]
Switzerland	9.7	0.7	0.003[c]	0.02
United Kingdom	7.1	0.3	0.003	0.07
United States	22.1	0.5	0.003[c]	0.89
METALAND	12	0.7	0.01	0.04

[a]Tracks, sidings included.
[b]Canals only.
[c]Data from 1985.
[d]No data available.

the railway network is even less important. The lengths of waterways and pipelines are relatively small. Only the Netherlands has an exceptionally extensive waterway network. With respect to pipelines, their network in the United States is as long as that for railways. For the following rough assessment of the material stock in transport networks, waterways and pipelines are neglected.

Although there are significant differences in the specific network lengths (i.e., the length per capita) between the listed countries, the order of magnitude is, with a few exceptions, the same. It is most likely that the specific length decreases with increasing population density. In a country with a low population density, a longer network is needed to achieve the same exchange rate of passengers and goods. Countries like Germany (250 inhabitants/km^2), the Netherlands (350 inhabitants/km^2), and the United Kingdom (230 inhabitants/km^2) have lower specific lengths than those of countries like the United States (25 inhabitants/km^2) and Sweden (19 inhabitants/km^2).

For METALAND, the material stocks in roads and railroad tracks are estimated as follows: The lengths per capita are 12 m and 1.5 m, respectively. For roads, a mean width of 9 m, for railways 7 m, including sidings, were chosen. The resulting traffic area amounts to approximately 120 m^2 per capita. A mean depth of 0.8 m for roads and railroad tracks was chosen, consisting mainly of gravel and sand, having a density of about 2 Mg/m^3. The top layer of roads of about 10 to 20 cm consists of bituminous gravel (bitumen content 5%). The resulting material stock would thus amount to about 190 Mg per capita from the traffic network. This anthropogenic material stock has about the same mass as the stock in buildings (about 200 t per capita).

For railroad tracks, the iron rails and the sleepers are also considered to estimate a further part of the material stock:

• 120 kg iron in rails per meter track;
• a sleeper per 0.6 m track, assuming 30% in steel (74 kg per piece), 50% in concrete (280 kg), and 20% in wood (70 kg).

The resulting material stock per capita, assuming a mean length of 1.5 m, amounts to 0.24 Mg iron, 0.35 Mg concrete, and 0.035 Mg wood.

The contribution to the main stock of concrete in buildings (about 160 Mg/c) and wood (about 4 Mg/c) is less than 1% and therefore negligible in a first approximation. However, the iron stock of about 0.2 t increases the stocks in the activity "to reside and work" (about 2 t) by 10% and must be taken into account.

To assess the order of magnitude of annual material flows for the construction and maintenance of networks, a mean annual growth rate and a mean lifetime of goods and materials in the stock have to be assumed (table 4.19).

During the past 50 years, the growth rate of the road network was about 10 times higher than that in railway construction. The resulting flows for maintenance work for roads are currently about the same as those for construction. However, for the uppermost layer, the M/C (maintenance/construction) ratio is already about

Table 4.19
Assessment of the material flows in METALAND for the process *Operation and Maintenance of Transport Networks*[a]

Material	Stock (Mg/c)	Annual Growth Rate (% of stock)	Mean Lifetime (y)	Maintenance/ Construction[b]	Total Flow (Mg/c.y)
Gravel and sand in roads	145	1	100	1	3
Bituminous layer in roads	35	1	20	4	2[c]
Gravel in railways	20	0.1	30	6	0.1[d]
Iron in railway tracks	0.2	0.1	20	80	0.01

[a]Data after Redle (1999).
[b]Ratio maintenance flow/new construction flow.
[c]20% of bituminous gravel recycled.
[d]Upper 20 cm.

4 and 6, respectively. For iron in railroad tracks, maintenance work is responsible for the main flux (M/C = 80).

Vehicles

Transport vehicles are grouped into passenger vehicles and freight vehicles for four different transport ways: road, railway, waterway, and airway. A first set of data provides a comparison of the stock of vehicles, given in numbers per capita.

The specific numbers (number of vehicles per capita) of different types of passenger vehicles differ by orders of magnitude (table 4.20). In all countries, the passenger cars are in the hundreds per 1000 inhabitants, followed by motorcycles. Among the freight vehicles, trucks form the largest group by number (table 4.20), about 10 times less than passenger cars. Comparing the numbers of the different countries, one observes the special position of the United States with the highest specific numbers for passenger cars, trucks, buses, and tractors.

On railways, the mobile equipment is grouped into locomotives, passenger carriages, and freight wagons (table 4.21). Passenger transport on roads is one to two orders of magnitude higher than that on railways. Again, the United States has an extreme value in passenger transport on railways, this time on the low side. The load capacities of freight wagons are of the same order of magnitude as those of freight vehicles on the road. Only the Netherlands and the United Kingdom show freight capacities that are about 10 times lower.

For the Netherlands, this apparent lack in load capacity for freight is compensated by the corresponding capacity on ships. Analogous data for the United Kingdom and Sweden are not available in the consulted statistics. Only inland ships are considered. Freight transport on waterways is certainly an important part of the overall transport on a global scale. For a regional metabolic study, it is not included. The contribution of aircraft is relatively small; for example, for Switzerland 0.48 aircraft per 1000 inhabitants and 0.02 seats per capita amount to minor passenger and freight capacities with respect to other vehicles. Therefore, they will be neglected in the assessment of material stocks. However, transport by aircraft will be considered in the process *Operation and Maintenance of Vehicles* (see figure 4.22c and the later tables 4.25 and 4.26).

The material stock in vehicles for METALAND is assessed as follows (table 4.22):

• Based on the data in tables 4.20 and 4.21, a mean number of vehicles per capita (within the range documented by the data from the arbitrarily selected countries) is chosen.

• For each type of vehicle, a mean mass per unit is chosen.

• Ships and aircraft stocks, based on the comments for tables 4.20 and 4.21, are not considered.

Table 4.20
Vehicle stocks on roads for passengers and for freight per 1000 inhabitants in 2003 (UN, 1986, 2008)

Passenger Vehicles per 1000 inhabitants

Country	Passenger Cars[a,b]		Motorcoaches (Buses and Trolleys)		Motorcycles (and Mopeds)		Bicycles
	1985	2003	1985	2003	1985	2003	1985
Federal Republic of Germany	424	546	1.14	1.05	62	45	i.d.
France	380	493	0.87	1.23	91	41	i.d.
Netherlands	340	425	0.82	0.70	76	32	i.d.
Sweden	377	454	1.64	1.53	i.d.	24	i.d.
Switzerland	405	510	1.77	5.92	143	104	370
United Kingdom	313	453	1.36	3.02	33	22	i.d.
United States	552	626	2.48	2.16	23	15	i.d.

Freight Vehicles

Country	Trucks		Tractors		Trailers		Total Load Capacity (kg/capita)	
	1985	2003	1985	2003	1985	2003	1985	2003
Federal Republic of Germany	24	31	4.57	0.22	5.7	4.88	165	i.d.
France	54	90	2.48	0.37	2.5	0.59	123	250
Netherlands	28	58	1.82	0.40	6.2	i.d.	205	i.d.
Sweden	31	47	0.41	0.07	41.3	6.52	185	274
Switzerland	31	38	0.72	0.13	10.9	2.42	123	161
United Kingdom	35	52	1.71	0.50	3.4	i.d.	i.d.	i.d.
United States	159	27	4.81	0.64	14.3	9.64	i.d.	i.d.

i.d., incomplete data.
[a]vehicles per 1000 inhabitants with four seats per vehicle.
[b]Including taxis.

Table 4.21
Vehicle stocks on railways and waterways. Vehicle number per 1000 inhabitants in 2003 (UN, 2008)

Country	Locomotives	Passenger Carriers		Freight Wagons		Ships	
	Number	Number	(seats per 1000 capita)	Number	Load Capacity (kg/capita)	Number	Load Capacity (kg/capita)
Federal Republic of Germany	0.07	0.25	n.a.	3.1	183[a]	0.056[a]	54
France	0.08	0.26	21	1.7	85	0.026	42
Netherlands	n.a.	0.17	15	n.a.	n.a.	0.54[a]	380
Sweden	0.07	0.21	14	1.9	80	n.a.	n.a.
Switzerland[a]	0.27	0.71	44	4.5	148	0.054	90
United Kingdom[a]	0.13	0.26	0.017	1.0	32	n.d.	n.d.
United States	0.07	0.01	n.a.	n.a.	452[a]	0.109	144

n.a., no data available.
[a]Data for 1985.

Table 4.22
Material stocks in vehicles for METALAND

Vehicle	Vehicle (V/c)	Mass (kg)	Stock (kg/c)	Percentage (%)	Lifetime (y)	Flow (kg/c.y)	Percentage (%)
Car	0.4	1,000	400	57	10	40.0	65
Bus	0.002	10,000	20	3	10	2.0	3
Truck	0.03	5,000	150	21	10	15.0	25
Motorcycle	0.1	50	5	<1	10	0.5	<1
Bike	0.37	10	4	<1	15	0.3	<1
Tractor	0.004	2,000	8	1	15	0.5	<1
Locomotive	0.0003	100,000	30	4	40	0.8	1
Coach	0.0007	35,000	25	4	40	0.6	1
Freight wagon	0.004	15,000	60	8	40	1.5	3
Total			700	100		60	100

V/c, vehicle per capita

The sum of all stocks amounts to about 700 kg/c, of which the passenger cars (57%), the trucks (21%), and the railway vehicles (16%) add up to more than 90% of the total. Roughly 80% of the vehicle mass is transported on the road, where 60% of the material stock is needed to transport passengers and 40% to move freight.

For the assessment of annual vehicle flows through METALAND, only the above-mentioned four most important vehicle types are considered in a first approximation. For this stock, an annual growth rate of 0.5% is chosen. Road vehicles show the fastest turnover (mean lifetime about 10 years) in comparison with that of railway and waterway vehicles (40 years). The consequence is a material flow of about 60 kg/c.y. Almost 70% stems from passenger cars. Thus, it is necessary to know the chemical composition of passenger cars in order to assess the material flows by vehicles passing through the anthroposphere. Passenger cars were introduced at the beginning of the twentieth century. Their material composition has changed since then, and the mean residence time in the anthroposphere has shortened. There is no appropriate data yet available to evaluate this process over a longer time period. First assessments, on the basis of producer data, give an overview of the contents of some goods and indicator materials in passenger cars entering shredder plants in the mid-1980s (Wutz, 1982; Burg & Benzinger, 1984; Franke, 1987; Palfi et al., 2006). It is a mixture of cars having an average age of about 8 to 10 years with a range between 1 and 20 years. For this mixture, an incomplete "goods/material matrix" is presented in table 4.23. It must be emphasized that the new cars entering traffic at the end of the 1980s showed already lower content of steel, higher

Table 4.23
Chemical composition of passenger cars[a]

Goods/Material Matrix on the Basis of Construction Goods[b]

Construction Good	Mass	C (g/kg)	Al (g/kg)	Fe (g/kg)	Cu (g/kg)	Zn (g/kg)	Cd (mg/kg)	Pb (mg/kg)
Steel and cast iron	690			690				
Non-iron metals	45		25		7	7	500	1000
Rubber	52	36						
Plastic	45	36						
Glass	39							
Various	129	13		20				
Total	1000	85	25	710	7	7	500	1000

Goods/Material Matrix on the Basis of Shredder Goods[c]

Shredder Good	Mass	C (g/kg)	Al (g/kg)	Fe (g/kg)	Cu (g/kg)	Zn (g/kg)	Cd (mg/kg)	Pb (mg/kg)
Scrap iron	770			750				
Hand-sorted non-iron metals	30			5	5			
Sieve residues	20		20					
Sieve fraction	180			10	2	2	10	700
Total	1000	70		760	7	7	10	700

[a]After Baccini and Brunner (1991).
[b]Data from Burg and Benzinger (1984), Franke (1987).
[c]Data from Baccini and Zimmerli (1985).

content of aluminum and plastics, lower content of cadmium, and other changes in chemical composition. Material flow analysis in a car shredder plant (Baccini & Zimmerli, 1985) gives a comparable goods/material matrix for which complete data are also lacking (table 4.23). It must also be noted that in car shredder plants, often appliances such as washing machines and refrigerators (up to 30% of the total input) are treated, too.

Nevertheless, the following conclusions can be drawn:

1. The shredder residues with a mass portion of about 20% are organic materials (organic carbon content 40%, mostly from plastics) with relatively high concentrations of copper, zinc, cadmium, and lead. They are products of waste management that should not be deposited in landfills per se but treated properly in subsequent plants (e.g., further metal separation and incineration).

2. According to the rough material balance, it is probable that an important fraction of the nonmetals (e.g., cadmium) is transferred with the scrap iron into the steel furnaces in recycling of iron. Many steel factories working with scrap as educt use about one third of the shredder material in the input. Therefore, it is indispensable to know the transfer functions of steel factories, which are important processes in the material flow connected with the activity TO TRANSPORT&COMMUNICATE.

For the process *Operation and Maintenance of Vehicles* (figure 4.22c), two main aspects for assessing material flows have to be taken into account:

1. energy consumption;
2. maintenance of the vehicle.

For an assessment of the energy consumption, a data set for average annual vehicle distances is given in table 4.24. Only vehicles with engines are considered. Tractors are neglected. The different regions show transport distances of the same order of magnitude.

A rough estimation of the fossil fuel demand serves as a cross-check to test the energy consumption values presented in table 4.25 (extended by METALAND and updated): About 0.5 passenger cars per capita (see table 4.22) travel approximately 14 000 km/y (table 4.26). For cars let us assume a gasoline consumption of 0.09 L/km. The energy content of 1 L gasoline is 0.034 GJ. The resulting energy consumption for passenger cars equals

$$0.5 \text{ [car/capita]} \times 14{,}000 \text{ [km/car and year]} \times 0.09 \text{ [L/km]} \times 0.034 \text{ GJ/L}$$
$$= 21.4 \text{ GJ/c.y.}$$

Table 4.24
Average annual transport distances for vehicles (×1000 km)[a]

Country	Passenger Cars	Buses	Motorcycle	Truck	Locomotive
Germany	10	48	1.9	26	62
Japan	7				
France	13	41	n.d.	4	64
Netherlands	10	48	2.2	24	45
Sweden	12	n.d.	n.d.	n.d.	64
Switzerland	12	15	3.7	28	82
United Kingdom	12	41	3.1	23	55
United States	23	13	2.7	21	26
METALAND	13				

n.d., no data available.
[a]After Baccini and Brunner (1991); for passenger cars, 2006 update from OECD (2007b).

Table 4.25
Annual energy consumption by transport mode (UNECE, 2008)

Country	Total (GJ/c)	Contribution of Transport Mode (%)			
		Road[a]	Railway	Waterway	Airway
Germany	32	85	3.0	0.4	12
France	36	84	2.0	0.6	13
Netherlands	39	74	1.2	1.9	23
Switzerland	40	77	3.8	0.2	19
United Kingdom	38	74	1.9	2.2	22
United States	91	83	1.7	0.5	13

Table 4.26
Emissions to air from transport vehicles for 2005 (BAFU, 2010)

Vehicle	$\times 10^3$ km/vehicle and year	Vehicles per capita	CO	NOx	HC	CO	NO_x[a]	HC[b]
			g/km			kg/capita and year		
Passenger cars	14	0.52	2.5	0.5	0.4	17	3	3
Trucks and buses <3.5 t	16	0.05	3.3	4.2	0.4	3	3	0
Total						20	7	3

[a]NO_x: nitrogen oxides
[b]HC: hydrocarbons

A comparison of the annual energy consumption by transport mode (table 4.25) illustrates that the traffic on roads comprises 75% to 85% of the total for the activity "to transport." Transport by aircraft has become an important energy consumer, too. This is due to the high energy demand per kilometer vehicle transport. Its contribution to overall passenger and freight transport is between 10 and 25%. The significantly higher energy consumption in the United States in comparison with European countries (by a factor of about 2.5) can be explained by the higher vehicle stock (factor 1.5), higher energy consumption per kilometer (factor 1.5), and longer transport distances (factor 1.1). In European countries, 80% of the energy is consumed on the roads of which again about 80% is used for passenger cars. The resulting 64% of the European Union average total of 37 GJ/c in table 4.25 amounts to 23 GJ/c.y, which is in agreement with the first assessment calculated above.

For a first approximation, the regional material flows generated by the process *Operation and Maintenance of Vehicles* can be assessed by calculating the fossil fuel consumption by road vehicles. The average 22 GJ/c.y corresponds with about

650 L gasoline/c.y or 440 kg/capita and year. Adding up to 100%, assuming fossil fuel as primary source for practically all transport activities, a material flow of about 0.5–0.6 Mg/c.y is used for transporting passengers and goods. This flow is one order of magnitude higher than the flux resulting from the metabolism of vehicles (ca. 0.06 Mg/c.y) and one order of magnitude lower than the metabolism of the network (3–4 Mg/capita and year).

From this it becomes evident that the chemical composition of the fossil fuel and the transfer functions of the combustion engines in vehicles are decisive factors for the quality of the resulting emissions. An estimation of the resulting air pollutants, namely, CO, NO, and hydrocarbons (HC), for METALAND is given in table 4.26. A second type of emission to be considered is used motor oil with a flow between 4 to 6 kg/capita and year (Brunner & Zimmerli, 1986; Koehn, 1987). Although this flow is small compared with the overall flow of about 500 kg for fossil fuels, it must be treated separately in waste management (e.g., incineration) to prevent water or soil pollution.

Operation and Consumption of Media

A rough assessment of the metabolism of printed and electronic appliances is given for METALAND in table 4.27. The stock of approximately 40 kg/capita is about one order of magnitude lower than the vehicle stock of 700 kg/capita (see table 4.22). The material flow is dominated by printed matter (newspapers, books, etc.); that is, paper is still the most important information carrier with respect to the material quantity needed. In spite of the exponentially growing information flow with electronic devices, the paper as information carrier has kept its dominant role from a metabolic point of view. However, paper has a short mean residence time (<1 year), can be recycled, and is based on the renewable resource wood. The paper flow of about 80 kg/capita (for printed media) is of the same order of magnitude as the vehicle flow (see table 4.22). The energy flows needed to produce and print papers and to run the electronic appliances are approximately 2–4 GJ/c.y (Baccini & Bader, 1996) or between 1% and 2% of the total energy demand for META-LAND. Nothing is mentioned here with regard to the complexity of the chemical composition of all the electronic devices. In many countries, there are legally based sorting and specialized recycling procedures to prevent flows of these appliances into the MSW and to recuperate mainly various metals. It has to be co-financed, as the market prices for these metals do not cover the expenses for the whole "cleaning procedure," with a so-called prepaid disposal charge. The user is obliged to return his "old" device to the appliance dealer, who has to take it back free of charge and forward it to a certified recycler.

In summary, the process *Operation and Consumption of Media* does not strongly increase the flow of goods illustrated in table 4.22, apart from the paper flow.

Table 4.27
Assessment of stocks and flows of printed and electronic appliances for METALAND

Appliance	Number per Capita	Mass (kg) per Gadget	Stock (kg/c)	Residence Time (y)	Growth Rate (% of stock/y)	Flow (kg/c.y)
Telephones[a]	1.5	0.3	0.5	7	1	0.08
Entertainment electronics (total)			32			3.4
Television	1	30	30	10	1	
Radio	1	2	2	10	1	
Others[b]	0.5	0.3	0.15	5	2	
Computers (total)			20			
Workstations	0.1	100	10	10	2	
Personal computers	0.5	20	10	7	2	
Printed matter (newspapers, books, etc., as papers)		100	100	<1		80
Copiers (for paper copies)	0.5	20	10	7	2	1.7
Total			160			85
Network cables[c]			300	30	1	10

[a]Stationary and mobile (for "residing and working").
[b]Calculators, iPods, calendars, and so forth.
[c]Not only used for media purposes, therefore separately listed, see also case study on copper in chapter 5.

The synthesis of the three contributors (transport networks, vehicles, media) to overview the overall activity is given in figure 4.26. The dominant material stock is in the transport network. Two orders of magnitude lower are the stocks in the other two processes (vehicles and media). Because the residence times differ (transport networks several decades to a century; vehicles and media days to decade), the corresponding flows differ only by one order of magnitude at the most; that is, the flows of the network are only several Mg/c.y and the vehicle flows, mainly due to the combustion of fossil fuels, sum up to the same order (off-gases). The media flow

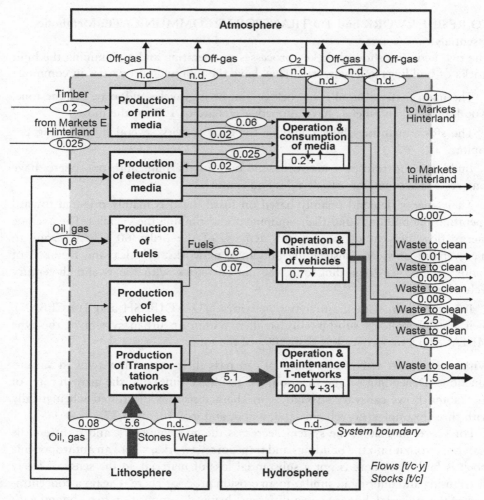

Figure 4.26
Quantified material system of the activity TO TRANSPORT&COMMUNICATE.

is dominated by the printed matter on paper. In other words, the activity TO TRANSPORT&COMMUNICATE in contemporary affluent societies induces material flows in the following decreasing order of quantity: passengers > freight > information. That is, mobility in contemporary urban systems of developed countries means a metabolic effort that has its emphasis on the transport of persons. Although the electronic information flow is still increasing exponentially, it did not yet lead to a decrease in the physical flow of persons. On the contrary, the per capita stocks and flows of vehicles, their networks, and of the media appliances are still growing.

TO RESIDE&WORK and TO TRANSPORT&COMMUNICATE: Metabolic Essentials
The two activities show their key processes in operating and maintaining the built stocks of buildings and networks. They have the following properties in common:

1. The main material used is of rock origin, be it gravel, sand, clay, or limestone. Wood, plastics, and metals have altogether a share of 10% at the most.
2. The stock quantities built comprise for each activity several hundred Mg per capita.
3. Buildings and transport networks, the bulk of the built anthroposphere, have long residence times, namely decades and centuries.
4. Their energy demand (mainly based on fossil fuels) is mainly oriented toward operating the buildings and their equipment and running the vehicles. The average energy flow per capita is, for each activity, of the order 60 GJ/c.y. From an energy point of view, constructing buildings, networks, vehicles, and media is of minor importance (regarding only the current stock with the twentieth-century technology).
5. In comparison with the first two activities (TO NOURISH and TO CLEAN), their area demand is substantially smaller. Within an urban system of the type METALAND, they cover less than 20% of the total.

Within these two activities, a material can pass their material systems in various functions. Depending on the function, the residence times and the growth rates of stocks and flows can vary strongly. This characteristic is illustrated schematically with three examples, gravel and sand, wood, and iron (figure 4.27).

For gravel and sand, the system decreases the geogenic stock and transfers the materials from it into the buildings and transport networks, where an anthropogenic stock is built up. There is not a substantial loss of material, in the sense of finely distributing and diluting it, similar to an erosion process. There is again a concentration of this material, however, in a different chemical composition (e.g., bound with cement in concrete). Because of the fact that the system is not in a steady state, the recycling flow available for reuse is relatively small.

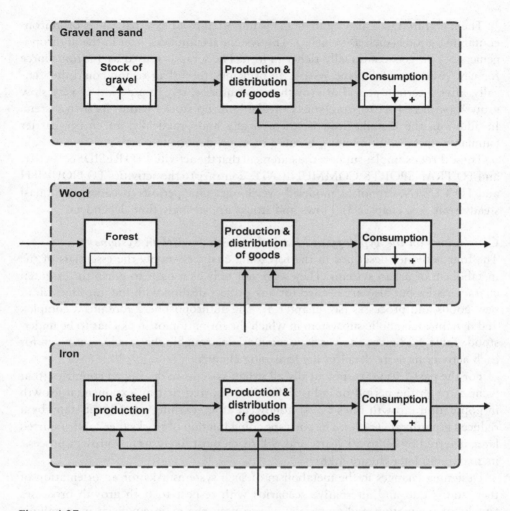

Figure 4.27
Schemes for the metabolic characteristics of gravel and sand, wood, and iron (reprinted with permission from Baccini, 2008).

For wood (after Müller, 1998), the source process *Forestry* is kept in a steady state. The main flow goes, similar to gravel and sand, into the anthropogenic stock (construction wood). Part of it has a short residence time (days to weeks); namely, the cellulose fibers (extracted from wood) in paper, which is recycled. Eventually all of it, the old construction wood, the old paper, and the firewood are used as energy sources in heating and in the incineration of MSWs. In other words, the assimilated carbon in photosynthetic activity of the forests is returned to the atmosphere in the same chemical form (carbon dioxide).

The metal iron (see also figure 4.12) is imported from external mines or in iron-containing goods (such as vehicles). The system accumulates iron in the anthropogenic stock. It gets continually richer in iron. The scrap is reused as an iron source for new products within the system. In principle, the urban region could theoretically, after a certain period of growth over centuries, rely in a period of very slow growth (with respect to iron) solely on the built up stocks within its own system. In addition, the metallic iron is, theoretically and practically, an energy carrier (similar to metallic aluminum), because of its oxidation state zero.

These three examples support the statement that the activities TO RESIDE&WORK and TO TRANSPORT&COMMUNICATE, contrary to the activities TO NOURISH and TO CLEAN, cannot be modeled over longer time periods (decades) in a quasi steady state (see chapter 3). Flows and stocks are strongly time dependent.

Conclusions on the Typology of Material Systems for Urban Systems
The four activities described in the foregoing chapters supply the essentials of the metabolism of urban systems. They serve not only as a basis to grasp the essential characteristics but also are needed for any project dealing with the introduction of new goods and processes (see chapter 5). The anthroposphere contains a complex and dynamic metabolic subsystem in which the formation of stocks has to be understood. Table 4.28 gives an overview, summarizing roughly the conclusions given for each activity in more detail in the foregoing chapters.

For the past 150 years, practically all urban systems in developed countries (that went through the period of industrialization) showed first an exponential growth in population (growth rates >2%), followed, after reaching an affluent state, by a reduced growth rate (<1%). The corresponding buildup of stocks (per capita) started later, deferred by 30 to 50 years, and will reach most likely, as any growth process, its maximum later (figure 4.28).

Designing changes in the metabolism of such systems asks for an orientation of their status quo and alternative scenarios with respect to both growth processes (stocks of population and materials). Up to now, the main emphasis in metabolic changes was put on flows (especially in environmental protection and in economic concepts) and not on stocks. Contemporary systems are oriented toward growing systems (figure 4.29a) where the main material flows come from primary sources and increase the stocks. Internal stocks as secondary sources are of minor importance. An urban system that reaches nearly a steady state (from a metabolic point of view, not of an economic one) could depend strongly on the endogenic stocks built up during the growth phase (figure 4.29b).

A second important aspect is the grasp of the complementary energy flows. In figure 4.30, an overview is given, showing synoptically the sizes of material and energy flows for each activity.

Table 4.28
Synopsis and characteristics of anthropogenic stocks

Activity	Key Process	Key Material	Residence Time	Formation of Stocks	Energy Demand	Type of Model
TO NOURISH	Agriculture	Biomass	Short (days to weeks)	Small	Small	Quasi steady state
TO CLEAN	Private household	Water	Short (hours to days)[a]	Small	Very small	Quasi steady state
TO RESIDE&WORK	Operation of buildings	Gravel and sand	Long (decades to centuries)	Large (buildings)	Large	Dynamic
TO TRANSPORT& COMMUNICATE	Operation of vehicles for persons	Gravel and sand	Long (decades to centuries)	Large (networks)	Large	Dynamic

[a]Exceptions in residence times for water are watersheds in transport systems.

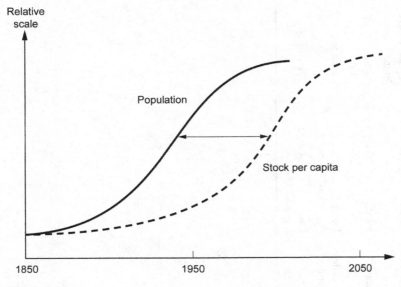

Figure 4.28
Scheme for logistic growth in population and stocks per capita (reprinted with permission from Baccini, 2008).

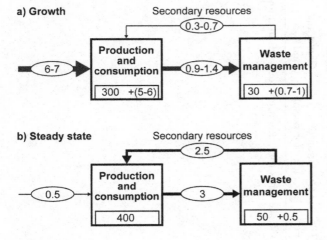

Figure 4.29
Flow characteristics of an urban system (a) in the state of growth and (b) 20 years later in a quasi steady state (flows in Mg/c.y; stocks in Mg/c) (reprinted with permission from Baccini, 2008).

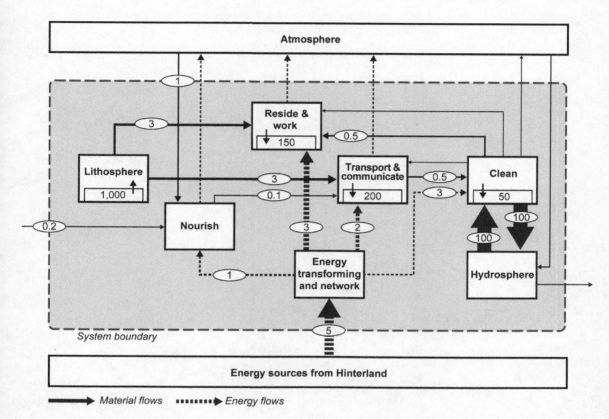

Figure 4.30
Simplified metabolic system for the two activities TO RESIDE&WORK and TO TRANSPORT&
COMMUNICATE (Baccini, 2008). Material flows in Mg/c.y, material stocks in Mg/c, energy flows
×1000 W/c. Gray energy embedded in goods imported from the hinterland is not included (reprinted
with permission from Baccini, 2008).

This figure, together with the synopsis in table 4.28, illustrates the differences in
the "daily perception" of metabolic processes within urban systems. We experience
the large flows of waters while cleaning, the transport flows of persons and goods
on the roads and railways, and the flows needed to get food from stores to private
households. What is hardly anticipated is the change of stocks. Only after decades,
when comparative photographs illustrate the changes in familiar landscapes, is the
importance of stock dynamics realized. It is the inertia of the stocks within the
anthroposphere that hides the metabolic risks in the long-term perspective. In contrast, the stocks host the metabolic potentials for a metabolic improvement of the
urban systems. Within this field of tension, any metabolic design has to test its
quality.

5

Designing Metabolic Systems

Designing is, to begin with, an intellectual effort to create sketches for any human activity. Designers of metabolic systems produce blueprints for the physiologic setup of a human society. Metabolic designers are partners in a transdisciplinary enterprise called "designing the anthroposphere." This chapter provides answers to the following questions:

• How do we position the metabolism in the context of an anthropogenic system?

• How do we set the objectives in metabolic design in the context of early recognition for a sustainable development?

• How can we design scenarios for optimal metabolic systems?

Immortal life with Kalypso or in Elysium or in the garden of the sun has its distinct appeal, to be sure, yet human beings hold nothing more dear than what they bring into being, or maintain in being, through their own cultivating effort.
—Robert Pogue Harrison, 2008

Preliminaries to the Design of Anthropospheres

Designing new anthropospheres is a literary method that has a long tradition. Thomas More wrote *Utopia* in the year 1515 while he was on a diplomatic mission to Flanders for King Henry VIII of England. A contemporary of Erasmus of Rotterdam, Martin Luther, Leonardo da Vinci, Michelangelo Buonarroti, Niccolò Machiavelli, Albrecht Dürer, and Hans Holbein, More in his book presented a new commonwealth on a fictitious island. The Utopian society served as a mirror to reflect the political situation in Europe in the sixteenth century. More criticized the failures of the upper class: living like parasites on the labor of others; going to war for false reasons; creating unemployment instead of providing work. In More's Utopia, people are peaceful, live by communistic principles, refrain from privacy, and dedicate all their services to the welfare of the community. In the book, More lets a mariner, Raphael Hythloday—a fictional character who had traveled around the world and had visited Utopia—relate his impressions of a "New World." The reporter's name is a play on Greek words (Marius, 1994): "Hythloday" means "expert in nonsense" or "good place" and indicates the humor and the ambiguity of the literary design of a new society.

There are very few examples in history that permit us to follow the designing process of real settlements based on detailed documentation. In the urban planning of the twentieth century, three realized concepts for new large-scale settlements illustrate the architecturally dominated design of new societies, such as Chandigarh in India (Albert Mayer, Matthew Nowicki, and Le Corbusier, 1950–1961), Canberra in Australia (W.B. Griffin and M.M. Griffin, 1913), and Brasilia in Brazil (Oscar Niemeyer, 1956–1965). All three cities, planned from scratch, have the function of new capitals. Form follows function. In each project, there is a hidden or outspoken vision of societal qualities to be attained by the future inhabitants. For postcolonial India, the new city became a strong icon of its move to modern times. For the young nation Australia, it was a political compromise to detour the rivalry of Sidney and Melbourne and to set a new node in the "bush" signaling the urban move to the west. Brasilia combines the arguments of the former two to support the development of a pioneer society.

In the traditional planning process of new settlements, the design of its metabolic processes is of secondary importance. Thomas More did not waste any sentence on the metabolism of his Utopia. At the beginning of the sixteenth century (500 BP), there were no doubts about the given quality of the physical system underlying the functioning of a society. In the twentieth century, Le Corbusier wanted "to free the cities from the tyranny of the streets" (Le Corbusier, 1970). He paid special attention to separating the roads for pedestrians, bikers, and automobiles because he considered traffic as "the deadly enemy of the children." For Canberra, the garden city movement, based on Ebenezer Howard's concept from the beginning of the twentieth century (Howard, 1902), had some guiding function.

With the exception of the garden city concept, which included agricultural production within the urban territory to ensure a certain degree of self-sufficiency in nourishment, the modern urban centers were not meant to have a clearly defined hinterland for their metabolism. As a node within a global network of trade, any city should be able to get its supply on a free market. In the second half of the twentieth century, the ecological movement stimulated a renaissance of the "garden city" and initiated also a new scientific branch of "urban ecology" (Richter & Wieland, 2010).

At present, there is a manifold literature on the guiding principles for future urban systems. There is not (and probably never will be) a generally accepted "theory of the city" that can help us distinguish the "good" premises and methods from the "bad" ones—good in the sense of being more useful for analysis and synthesis (Baccini & Oswald, 2007). A subjective manner of perceiving human settlements determines the examination method chosen. The different premises result in a subjective selection of questions relevant to the city. The observers' premises and objectives also influence the tools they deploy. The methods used, not always with a scientific foundation, define the settlement type, which is thus a construct of the observer. Architecture, history, sociology, political science, anthropology, the natural sciences, engineering, or any other discipline is not predestined to play the leading role in urban development. In all probability, the best method for defining the "good city" is based on participatory, transdisciplinary work. We can only learn what a "good city" could be in the relevant social context. The fact that all designers bring in their culturally emerged values asks for a political discourse on the responsibility of humankind to decide on what should be done and what should be omitted.

It follows that "a design of the metabolism of the anthroposphere" has to lay open its premises, its methods, and its limits, within the context mentioned above. The scheme presented in figure 5.1 illustrates the essentials of the context underlying the four case studies presented in this chapter.

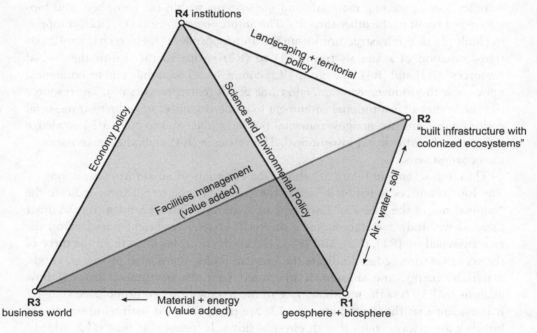

Figure 5.1
Scheme of resources correlations within a regional urban system.

We distinguish four types of resources, (R1–R4). Resources are cultural constructs. A short lexical definition describes the term *resource* as "an economic or productive factor required to accomplish an activity, or as means to undertake an enterprise and achieve desired outcome. Three most basic resources are land, labor, and capital; other resources include energy, entrepreneurship, information, know how, management, and time." There are various proposals to structure resources (Ohlsson, 1999). The natural resources are "first-order resources," whereas social resources are called "second-order resources." Ohlsson emphasizes the social consequences of (first order) natural resource scarcity. It is "the perceived inability to mobilize an appropriate amount of social efforts to accomplish the often large structural change required for adapting to natural resource scarcities. It is thus a particular kind of scarcity, a scarcity of natural resources, entailed by the imperative to manage scarcities of natural resources."

In the context of metabolic studies, we define natural resources, (R1), as the sum of materials and organisms that have evolved within the pre-human geosphere and biosphere and that still evolve beside the cultural evolution of the anthroposphere. Their "existence as resources" is purely anthropomorphic and depends on the reflected and documented discovery of the earth and the universe by human beings. In other words, naming materials and life systems within the geosphere and biosphere is a result of the other three Rs. The anthropogenic resources, (R2), comprise the built physical infrastructure within the anthroposphere (see also chapter 4). The transformation of materials from (R1) to (R2) is due to the use of the "social resources" (R3) and (R4). The group (R3) comprises all economic and technological know-how to produce, maintain, and trade goods (entrepreneurship). The resource (R4) embodies all institutional equipment to guide, organize, and administrate social entities (governments, nongovernmental organizations) and to cultivate knowledge (educational and cultural institutions). The resource (R4) embodies governance in the broadest sense.

Thus, the scheme in figure 5.1 shows the essentials of an anthropogenic system. The four resources individually set at the corners of the tetrahedron indicate the "equivalence of the four Rs." Each type of resource is interdependent on the other three. If we study a metabolic (or a material) system, we orient ourselves on the face stretched by (R1), (R2), and (R3). The connections between them (or three of the six tetrahedron edges) indicate the formation of systems with processes, goods, materials, energy, and substances structured with the instrument material flow analysis (MFA). It is the metabolic face of the tetrahedron "anthropogenic system." It is evident that the connections to (R3) are partly coupled with monetary flows, but they are always coupled with physical flows. It means that the "value added," formed by transformation of matter and energy within the metabolic face, is always based on physical properties.

The three other faces are stretched with (R4). They have two "ruling edges" and one "metabolic edge." Historical documents from various phases of cultural evolution contain reports on the rules that were to be followed for use of land, water, biomass, and so forth. Metabolic boundary conditions trigger political measures to control. Political measures stimulate inventions and innovations to overcome metabolic bottleneck situations (e.g., insufficient transport capacities) and resource scarcities (e.g., in short supply of energy). The face stretched by (R2), (R3), and (R4) is the "economical triangle" in which enterprises make their profits. The face stretched by (R1), (R3), and (R4) is the "engineering triangle" where technological innovation takes place. The fourth face, stretched by (R1), (R2), and (R4), is the "ecological triangle" where all scientific activities (natural sciences and humanities) are resident, where all our knowledge (*logos*) about our house (*eco*, synonym here for "houses" on all scales) is developed and stored for coming generations.

Where are the four activities in this picture? The four basic activities chosen are the driving forces (needs and deeds) that make the static scaffolding of the tetrahedron a dynamic anthropogenic system, a system that is too complex to be handled just with one method. The target "sustainable development" asks for a design or redesign of the whole system (see chapters 1 and 2). It is evident that a metabolic design is restricted to the metabolic face in figure 5.1. However, such a design hangs on the policy strings that are moved by the values of a society and its individuals. These values are partly written down in constitutions, laws, charters of human rights, and religious testaments. Therefore, one has to distinguish between the roles of the designers and the makers. The makers in policy and enterprises decide, within their field of competence and sovereignty (a field that has not always clearly defined boundaries), on what to do by whom. The designers of all branches (in architecture, engineering, economy, policy) offer the decision makers a variety of "routes to reach a goal." These offers are usually oriented toward a concrete target given by a potential client and are accompanied by a rationale of the designer. The rationale of metabolic designers is based on the analysis and evaluation of a metabolic problem encountered. The design of a setup of solutions is restricted to the "metabolic face of the resource tetrahedron." However, it gives the relevant links to the three other faces that are relevant for any maker.

Each of the four case studies that follow in this chapter starts with a relevant resource problem for which a sustainable solution can only be found by reforming the anthropogenic system as a whole. Therefore, each case study lays open, as well as possible, the properties of the policy strings to which the "metabolic triangle" is connected. How should we use phosphorus on a global scale? How should we handle the copper stocks in the constructed anthroposphere? How should we choose our strategies for waste management and for mobility infrastructure? Some of the answers to the "How Questions," but very clearly not all of them, can be found in metabolic systems. These "How Questions" are always accompanied by "Why Questions." Answers to the second group, in the framing of a metabolic design, will be given as assumptions of value sets that base the policies within a given context. There are also "Who Questions" to be answered, a task that, in our opinion, cannot be answered competently by metabolic designers, who operate mainly with the tools of natural sciences and engineering. Unfortunately, this position is often not shared by those metabolic designers who do not hesitate to mix their evaluations with dilettantish recommendations for decision processes in any societal field.

Case Studies for the Design of Metabolic Systems

Phosphorus Management

The Essential Substance Phosphorus
The substance phosphorus (P) was first discovered by the alchemist Henning Brand in 1669 while concentrating urine and was identified to be a chemical element by Antoine Lavoisier in 1777 (cf. chapter 2). It is a multivalent nonmetal with the atomic number 15. In the periodic table, it borders on the essential elements carbon, nitrogen, oxygen, silicon, and sulfur. It exists in two major forms—white P and red P, but because of its high reactivity (e.g., with oxygen), it is never found as a free element in nature. Phosphorus is stable in the 5^+ oxidation state, which is illustrated by the wide range of phosphate minerals available on the planet. Phosphorus is fundamental for the biosphere and living cells: it is a constituent of all genetic material such as DNA, RNA, ATP; cell membranes are formed by phospholipids; and calcium phosphate (apatite) is instrumental for bone structure and strength. An adult person consumes and excretes about 1–2 g of phosphorus per day and has a body content of about 500–650 g of phosphorus, with 85% to 90% in bones and teeth. A wide range of organophosphorus compounds has been designed as pesticides; many of these substances are hazardous for higher organisms. Fluorinated phosphate esters are among the most potent neurotoxins known.

The amount of P on the planet is large: the earth's crust contains about 5×10^{14} to 10×10^{14} Mg of P; the amount in the oceans (8×10^{10} Mg) is four orders of magnitude smaller. In 2009, global "reserves" were estimated at 1.6×10^{10} Mg (figure 5.2); "reserves" stands here for the amount of P rocks that can be economically extracted and used for production of P with today's available technology. Not included in those figures are significant new phosphate resources that have been identified on the continental shelves and on seamounts of both the Atlantic Ocean and Pacific Ocean. Phosphorus is a limited, so-called nonrenewable resource and can only be formed by nuclear reactions, such as by fusing two oxygen atoms together, a reaction taking place above 1000 megakelvin in large stars with masses greater than 3 solar masses. Before the twentieth century, the major source

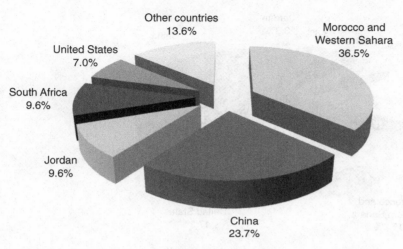

Figure 5.2
Global P reserves by countries 2009 (in percent). Total reserves as of 2009 amount to 1.6×10^{10} Mg (USGS, 2010a).

of phosphorus was bone ash and guano (bird droppings that accumulated over thousands of years on islands off the western coast of South America). Phosphate rock containing calcium phosphate was first used as a raw material in 1850, and, after the introduction of the electric arc furnace in 1890, became the key source of phosphorus. Today, total world production of P from phosphate rock amounts roughly to 1.6×10^{8} Mg per year (figure 5.3).

The first commercial use of P was developed in 1680 by Robert Boyle who applied P to ignite sulfur-tipped wooden splints, thereby inventing matches. Global demand for fertilizer during the second half of the twentieth century led to a large increase in phosphate (PO_4^{3-}) production. Today, the main P products are phosphorus, phosphates, and phosphoric acid, and the most important commercial use of phosphorus-based chemicals is the production of fertilizers for agriculture and horticulture. Some further applications of P include use for plasticizers, flame retardants, pesticides, detergents, water treatment chemicals, baking powder, food additives, explosives, and as a constituent of steel, bronze, special glasses, fireworks, semiconductors, and toothpaste.

Rationale for Selecting P for a Case Study on Design
Phosphorus is one of the key elements for sustaining life on planet Earth and an essential nutrient for the biosphere: Together with carbon (C), hydrogen (H), oxygen (O), nitrogen (N), and sulfur (S), P forms the chemical fundament of every biological system. It is not possible to substitute other substances for P. In ecosystems, P is often a limiting factor, thus governing the growth rate of organisms (cf. also

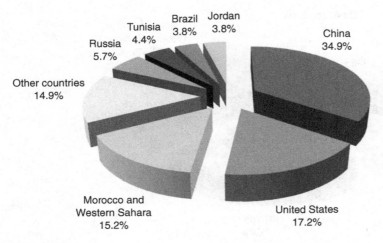

Figure 5.3
Global production of P by countries (in percent). Total world-wide production amounts to 1.6×10^8 Mg (USGS, 2010a).

"Managing Oligotrophic Ecosystems within a Eutrophic Anthroposphere" in chapter 2). For human beings, P is a key substance, too: A lack of P in the food intake results in retarded growth of infants and in deficits in the development of teeth and bones (rachitis). Too much P in the diet disturbs the metabolism of other minerals and leads to calcium deficiency.

From a substance flow analysis (SFA) perspective, P serves as a metabolic indicator for a substance and for area (cf. table 5.8 later in text). Both short-lived phenomena (food) as well as long-lived stocks (soils and landfills) are representative for P. As a nutrient, P is of primordial importance for the activity "to nourish" and also, for some countries, of relevance for the activity "to clean." Phosphorus is regulated by the three normative criteria constitution and law, economy, and technology. Phosphorus is well suited to metabolic studies because there is abundant data available along the chain resource extraction, agricultural and industrial production, consumption, waste management, and final sink.

Phosphorus is at the center of public debate: Topics such as "peak phosphorus" are creating a new awareness about the potential scarcity of P (Elser & White, 2010, White & Cordell, 2008)). Algal blooms and poor surface water qualities are often presented as pictures of the negative impact P can have on aquatic ecosystems. And the recent discussion about how to manage P-containing wastes such as sewage sludge has expanded from an expert debate to the public level, raising again appreciation for the nonrenewable nutrient P (Schick et al., 2008; Lederer & Rechberger, 2010).

Contemporary P Metabolism

Phosphorus flows and stocks of different regions vary considerably according to the economic structure. Urban countries such as Singapore or Dubai rely heavily on imported food, whereas other countries such as the United States are net suppliers of food and require large amounts of fertilizer. Also, personal income as well as dietary habits differ from one region to another, particularly in the north–south direction. Thus, many different P metabolisms exist on planet Earth. In the following, the P metabolism of a particular region, namely METALAND introduced in chapter 4, is used for discussing the design of P-management systems. The reason is that METALAND has been defined as a representative advanced urban society with agricultural production that contributes substantially—but not sufficiently—to inland food supply. This situation is typical for many urban regions in affluent countries such as the United States, those of Europe, Japan, and others. It is likely that many other regions will encounter a similar P regime in the future.

The contemporary P flows and stocks of METALAND are presented in detail in figure 5.4 and are summarized in aggregated and normalized form in figure 5.5

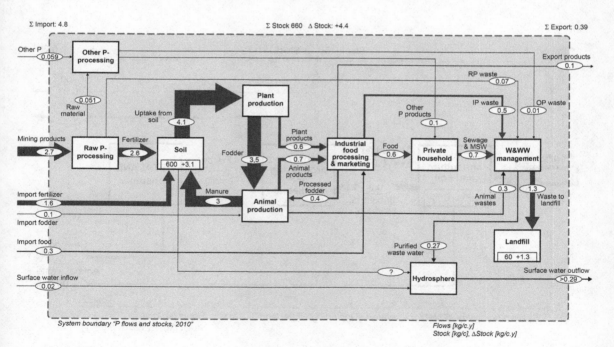

Figure 5.4

Flows and stocks of P in METALAND. IP waste, Industrial food processing & marketing waste; P, phosphorous; RP waste, waste from raw phosphorous processing; OP waste, waste from other P-processing ; W&WW, waste and waste water.

a)

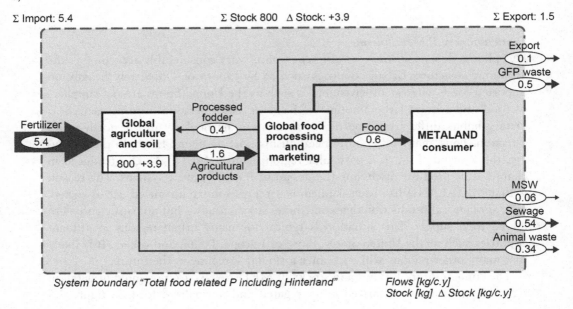

b)

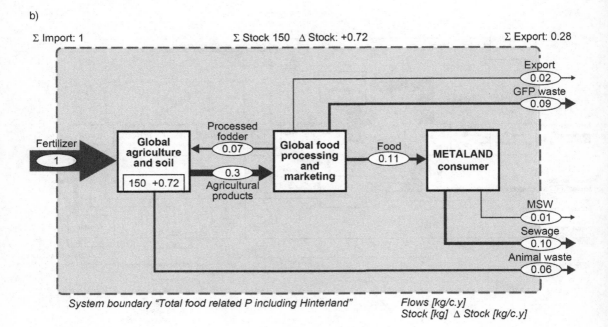

Figure 5.5
(a) Aggregated and (b) normalized flows and stocks of P in METALAND. GFP waste, waste from global food processing and marketing.

(see later for comments on calculations of figure 5.5; see also chapter 4, "TO NOURISH and TO CLEAN" and figure 4.5c).

To supply nourishment for the consumer, the following processes are instrumental: production of raw products such as fertilizer and educts for other P-containing products (process *Raw P-processing*), production of food in agriculture (processes *Soil*, *Plant production*, and *Animal production*), and *Industrial food processing*, *distribution and marketing* of food for men and animals. Because in METALAND detergents do not contain any phosphates, agriculture accounts for more than 90% of all applications of P; other processes and flows (e.g., for water treatment or pesticides) are of little relevance (cf. process *Other P-processing*). On the disposal side, the two key processes are sewage treatment and waste management, summarized in the process *Waste and wastewater (W&WW) management*. The main receivers of P are the processes *Agricultural soil* and *Landfills*. When comparing figures 4.5c, 5.4, and 5.5, it becomes obvious that for design purposes, an extended and more sophisticated P system is required. In the description of METALAND (see figure 4.5c), a first analysis and evaluation of P flows and stocks is given. The additional information presented in figures 5.4 and 5.5 that includes the hinterland allows designing a new P-management system.

There are two important differences between figure 5.4 and figure 5.5: (1) the first figure shows all P flows and stocks in METALAND but focuses just on META-LAND. The latter figure concentrates only on food-related P flows and stocks but includes the hinterland, too. In addition, in order to visualize the most important yields and efficiencies, the results are normalized with respect to P import. The reasons for establishing figure 5.5. are as follows: (1) Because the activity TO NOURISH is the most important activity regarding P metabolism, comprising more than 90% of all P flows, it is essential to take a closer look at those processes that are associated with flows and stocks of food. (2) It is important to include the hinterland because about one fourth of all food and fodder is imported from the hinterland into METALAND, hereby inducing a large fertilizer flow in the hinterland. Thus, to understand and quantify the whole food chain from fertilizer to human consumption, all processes, not only within METALAND but also in the hinterland, have to be investigated.

For establishing figure 5.5, fertilizer use has been recalculated as follows: Figure 5.4 includes 0.3 kg P/c·y as food import for human consumption and 0.1 kg P/c·y as animal fodder import. To produce this imported food and fodder, fertilizer is used outside of the system boundary in the so-called hinterland (sometimes, P used outside of the boundaries is called "virtual P," in analogy to "virtual water" in the case of production water used outside of the product system boundary). This amount of fertilizer is not contained in the SFA of figure 5.4, as this figure includes only P crossing the METALAND border. However, in figure 5.5, all P necessary to produce the 0.6 kg P/c·y that is required for human nutrition is taken into account.

The reason is to be able to calculate the total P flow necessary to supply 1 unit of P to the consumer.

The SFA displayed in figure 5.4 has been established using the literature and statistical databases about agricultural practice and household consumption. The resulting quantitative metabolic picture calculated by SFA for today's most advanced economies is astonishing:

1. The main imports into METALAND are mining products (phosphates), imported fertilizer, and imported food for humans and animals. All other imports as well as P imported by surface water inflow are marginal.

2. There are few P exports from the region, amounting to less than 10% of imported P. The main export path is surface waters: three fourths of all P leaving the region is by means of river water, and the remaining export consists of commercial export products.

3. There are two large stocks in METALAND: The content of P in agricultural soil amounts to about 600 kg P/c and grows by 0.5% per year, resulting in a doubling time of roughly 200 years. The amount of P in landfills (60 kg/c) is about 10 times less, but grows faster with 2.3% per year, equaling a doubling time of around 40 years. In total, more than 90% of the P imported remains and accumulates in the region: 65% in the agricultural soil and 29% in landfills. Thus, there is a large global redistribution of P taking place: P is extracted as phosphate mineral from the lithosphere in the global hinterland and subsequently accumulates in the pedosphere of METALAND.

4. By food intake, each person in METALAND consumes 0.6 kg P/c·y. To produce this amount of food by today's agricultural practice, in total (including the hinterland) a ninefold amount of P in fertilizer (5.4 kg P/c·y, see figure 5.5a) is required. In other words, 10 units of P input result in only 1 unit of P utility (food). At first sight, this result might differ widely from country to country, depending on the economic structure: For highly urbanized countries with little agricultural production and thus large import of food, the total P import over the SFA system boundary might be much closer to the actual P consumption figure. However, a rigid SFA reveals the following: If the amount of fertilizer is taken into account, too, to produce the imported food outside of the system boundary (hinterland), the ratio 1 unit of P consumed versus 9 units spent for production remains roughly the same.

5. By far, the most important processes for P management are *Soil*, *Plant production*, and *Animal production*. These three processes are linked as a cycle that is supplied by domestic as well as imported fertilizer. Fodder import for animal production is of less importance. About 3–4 kg P/c·y is recycled in this agricultural subsystem as a result of cultivation of soils, crop production, animal production, and manure application. For the overall flow of P, the fraction of animal production

versus crop production is important, with animal production increasing the P demand in agriculture significantly.

6. Thirty-three percent (1.6 kg/c·y) of P imported into METALAND ends up in liquid and solid wastes. The main fractions originate from *Animal production*, *Food processing and consumption*, with minor amounts from *Other processing* operations. Wastes from *Industrial food processing and marketing* and *PHH* are of equal importance, contributing together about 1.2 kg P/c·y of waste, which corresponds with 25% of the total P import.

7. Sewage from *PHH* is by far more important as a carrier of P than MSW: nearly all P in wastes of private households derives from urine and feces collected by the sewer system, and only little more than 10% of P is amassed by municipal solid waste (MSW) collection. Assuming that other wastes treated in the process *Waste and wastewater management* have a similar distribution between liquid and solid fractions, and taking into account that treatment plants in METALAND are equipped with sophisticated P elimination technologies (transfer coefficient TK_P sludge = 0.8), a remaining flow to surface waters of 0.2–0.3 kg P/c·y results.

8. Although the general pattern of P use is similar for most advanced regions, there are specific regional differences that are relevant for the design of P management: The dilution potential for P in surface waters depends highly on the regional hydrologic conditions. Topography and soil properties are—together with agricultural practice—responsible for erosion and surface runoff. Thus, P management systems must be tailor-made according to regional characteristics.

Challenges of Contemporary P Management

Phosphorus Is Essential for the Biosphere
The key issue of P is that it is an essential element that is required for all living systems. Unlike other substances such as copper or polyethylene, it cannot be replaced by substitutes that can fulfill the same functions. Thus, it must be ensured that enough P is available to supply the food required to feed 7 billion persons today and probably 9 billion to 10 billion persons in 2050.

Present P Use Is Not Efficient and Leads to Large Stocks in Soils
Figure 5.5b shows the inefficiency of the present-day system to feed affluent populations: To produce 1 unit of P suited for human nutrition, a ninefold amount of P in fertilizer is required. The losses along the food production chain are very large. In particular, three fourths of P is accumulated in the soil, with—above a certain limit—little additional value for plant production. Today's agricultural practice is overloading soils with P ("luxury fixation of P by soils").

Agricultural Policy Determines P Management
It is interesting to compare the differences in the metabolic problems of the United States in the 1930s with those of today: During the beginning of the twentieth century, the concentration of P in agricultural soils was continuously depleted by contemporary farming practice. In a "Message to Congress on Phosphates for Soil Fertility," U.S. President Franklin D. Roosevelt complained in 1938 that "the phosphorus content of our land, following generations of cultivation, has greatly diminished (Roosevelt, 1938). It needs replenishing. The necessity for wider use of phosphates and the conservation of our supplies of phosphates for future generations is, therefore, a matter of great public concern. We cannot place our agriculture upon a permanent basis unless we give it heed." Thus, whereas in the 1930s the soil stock was too little, today's soils are overstocked with P. It is noteworthy that 80 years ago, P was a topic of presidential concern, and the question arises, if P will again reach presidential agendas, will it be for issues of scarcity, for environmental protection, or for agricultural practice?

The Dilemma of P Mobility in Soils
The main reason for today's large stocks of P in soils has to do with the properties of P in soil environments: Soil P is found in different chemical speciations, such as organic P and mineral P, with 20% to 80% of P in the organic form that is not available for plants. Soluble P is of value only if it is located inside the rhizosphere of plants. Because the rate of diffusion of P in the soil is slow, high plant uptake rates deplete P around the root zone. Soil microbes are important for the mobility/immobility of P in soils, as they can mobilize immobile forms of P and vice versa. The generally low overall availability of P in the bulk soil limits plant production. To support high growth rates of crop plants, modern agriculture uses large amounts of readily soluble P. Unfortunately, during a growing season, crop plants recover only a small fraction of the P applied. More than 80% of P is immobilized and becomes unavailable for plants because of adsorption, precipitation, or conversion to organic forms (Holford, 1997). Thus, intensive agriculture with high-yield crops and strong fertilizer inputs increases stocks in soils.

Phosphorus Stocks in Landfills Are Growing Fast
In addition to stocks in soils, there is also a stock in landfills that is about an order of magnitude smaller than the stock in soils. This stock consists of residues from sewage treatment plants (STPs) (sewage sludge and sewage sludge ash) and from MSW treatment (bottom ash from MSW incineration). In METALAND, sewage is treated in STPs with an average efficiency of P removal of 80%, and MSW is incinerated in state-of-the-art waste to energy plants ($TK_{P\ bottom\ ash} = 1$).

Recycling of P Is Hampered by Pollutants

Sewage sludge is not applied to agricultural land because of heavy metals and organic substances. From the 1960s to 1990s, heavy metals such as lead, cadmium, and zinc were of most concern for sludge application on land. However, these loads have been decreased by input-oriented measures, generally by enforcing pollution prevention programs in the metal production and application sector, and specifically, for example, by limiting lead as a constituent of gasoline or by introducing compulsory mercury separation in dental offices. Thus, concentrations of most metals in sewage sludge have been reduced since the 1980s, with the exception of copper and possibly zinc, which are subject to weathering and corrosion from still growing (inner and outer) surfaces of buildings and infrastructure (Brunner, Müller & Rebernig, 2006) (see also the copper case study later in this chapter). Actually, the rapid and significant decrease of many metals in sewage sludge in the past 30 years confirms the statements of chapter 2 that due to environmental protection measures on the production side, today's production emissions are of less importance than end-user–induced emissions.

Balancing of Interests—Environmental Protection versus Resource Conservation

In METALAND, the main reason that P contained in sewage sludge is not applied on agricultural soils is persistent and hazardous organic substances. In particular, pharmaceuticals and endocrine substances have been detected in critical amounts in sewage sludge (Giger et al., 2003; Hale, 2003). Observing the precautionary principle, authorities in METALAND have decided not to release these substances into the environment and thus are either co-firing sludge in cement kilns or incinerating sludge and disposing of the resulting ash in sanitary landfills. In both cases, P is lost, either as a constituent of cement or as a component of ash in landfills. Thus, decision makers of METALAND have concluded that at present, the risk of polluting the environment by persistent organic substances and heavy metals is greater than the risk of running out of P.

Building up a Long-Term Eutrophication Risk in the Soil

Regional studies such as Henseler, Scheidegger, and Brunner (1992) show that under conditions of METALAND, up to 17% of the annual P input to soils can be transferred to surface waters. Because the flow of P to soils is large, and because most P is accumulated in the soil, this source of P for surface waters becomes even more important than P in purified wastewaters from urban and industrial wastewater treatment plants. If current P management prevails for the next 200 years, the concentration of P in the soil will increase by a factor of 2. Although the increase of P is no problem per se for the soil, it contributes to the amount of P transported to the surface waters by leaching, erosion, and weathering of soil. Assuming that 10% of accumulated P reaches the surface water, this will double the amount of P in the surface waters. Thus, the stock in soils and the resulting erosion and surface runoff determines the future eutrophication of surface waters in METALAND. For future water quality protection, the focus on long-term accumulation of P in soils is now equally as important as was the rapid expansion of sewage treatment plants in the beginning of water pollution control.

Controlling P Flows to Surface Waters Requires MFA/SFA
Based on figure 5.4 alone, it is not possible to assess if the load of P causes problems for surface water quality. In addition to P loadings, the volume of surface water flow and the volume and residence time of water in lakes within META-LAND as given in table 4.2 are required. With a lake volume of 2.8×10^{10} m^3 of water, a residence time of 2 years, and an annual flow of P from sewage treatment plants of 2.7×10^6 kg P (0.27 kg/person and year times the total population of 1×10^7 persons), an average concentration of 0.2 mg P/L can be calculated assuming that all purified sewage of METALAND is directed to the lake. Under these assumptions, the lake will become eutrophic, and water quality deteriorates. If only one tenth of the purified sewage reaches the lake water, the additional load will allow the lake to remain mesotrophic. If the same estimation is applied to flows of P from agricultural soil to surface waters, a similar picture evolves: assuming that 10% of P added annually to the soil stock is transported by erosion and runoff to the lake results in a lake water concentration of 0.2 mg P/L, causing severe eutrophication. This small example shows the sensitivity of the surface water regime of METALAND with regard to P loads. Assessment and design of measures regarding P and water quality issues in any region requires (1) basic knowledge of the hydraulic regime of the region (regional MFA of water), (2) a comprehensive mass balance of P (regional SFA of P), and (3) fairly detailed information about the location of the main P sources and receiving waters such as lakes, other surface waters, and groundwater [adding geographical positions to the information of (1) and (2)]. Soil monitoring is clearly not sufficient for P monitoring because it means "late recognition": accumulation (as well as depletion) of P and other soil constituents happens slowly. Because of the uncertainty of soil sampling and measurements, values about the concentrations of soil constituents are associated with comparatively high variances. Thus, it requires decades until a significant change in soil constituents can be detected by direct soil monitoring. In contrast, if today's concentrations and annual loads are known, MFA enables calculation of future soil concentrations in advance and in a timely way, acting as a true "early warning system."

Lack of Resource Conservation

If today's metabolic pattern of P utilization persists, and if no major technological breakthrough in prospecting and exploiting of P resources is achieved, P availability will eventually become limited. Estimates stretch from 80 to 400 years, revealing large uncertainties. The "true" and unknown range of P resources is dependent on (1) the rate of P consumption, (2) the development of new technologies for more efficient P extraction and production, and (3) the discovery of new deposits. As of 2009, global reserves amount to 1.6×10^{13} kg. With a world population of 7 billion persons, which equals about 2000 kg of P per person; taking into account a food requirement of 0.6 kg P per person and year and an efficiency of the P supply system of about 0.1, the current reserves last for about 400 years. If efficiency of the entire food production process doubles, this will solve the problem for another 400 years. This very rough calculation, which does not take into account population growth or changes in dietary habits, points out that at some point in history, "business as usual" cannot be a solution anymore for the nearing problem of P scarcity. How to react to this situation, which seems far away for some experts (Becker-Boost & Fiala, 2000) and frighteningly close for others (GPRI, 2010), is discussed in the next section. Because P is absolutely essential for the biosphere, it can be anticipated that with increasing concerns about scarcity, there will be a fast-growing future drive for new ways to use and produce P. Another technological quantum leap such as the one Justus Liebig initiated during the nineteenth century appears to be necessary at some future point in history. As inherent to all such developments, it is not perceivable what this development will be. As it was an inconceivable idea for a person living during the Middle Ages that one day we would heat up stones in an electric arc furnace and produce a chemical that enhances the growth of plants, it is beyond our horizon today to think of future mechanisms to use and produce the always needed P. A promising first step toward enlarging the reserve base of P appears to be the utilization of residues from iron production that are rich in phosphorus due to the P content of iron ore (Matsubae et al., 2010; Yamasue et al., 2010).

Designing a P-Management Strategy

The design of a metabolic system asks first for the set-up of political guidelines (see "Preliminaries to the Design of Anthropospheres" above and figure 5.1). If a society commits itself, in a constitutional amendment, to develop its territory in a sustainable way, it has to design and operate its metabolic systems:

1. with a long-term perspective (decades to a century);
2. for large-sized areas (ten of thousands to hundreds of thousands of square kilometers);

3. with a sound knowledge on the sizes and dynamics of the regional and global material stocks;

4. in the context of a development strategy for urban systems; and

5. considering the idiosyncrasies of every region (tailor-made design).

For P as an essential element that cannot be replaced, it is mandatory to adopt a management strategy that observes a long-term view even beyond a century. Based on the contemporary P flows and stocks, and considering the above five design principles, the following goals for P management are put forward:

1. *Decrease the amount of P required for producing food.* From a global perspective, this goal—which is equivalent to increasing agricultural yield regarding P—has two advantages: The less P is used to nourish people, the longer the depletion time of P resources. Also, if less P is used directly in METALAND and indirectly in its hinterland, more P is available for nourishment in less-affluent countries. Note that specific goals that apply to other substances (cf. the following case study on urban mining) do not apply to P: A decrease of P import only makes sense as long as the physiologic limit of 1–2 g P/person and day is not undercut. Because P is still comparatively inexpensive and only temporarily limited by high prices, a shift of a high-P-throughput agriculture to a more sparing and efficient agriculture might not be economic in the short run. However, in the long run, the early development of plant production systems that are based on low P input can be an important comparative advantage in a global economy. Also, it fosters independence of META-LAND from P imports, reducing vulnerability in unstable times.

2. *Decrease losses to the environment.* It is inherent to both biological systems *Plant production* as well as *Humans* that they waste nutrients. In densely populated areas such as METALAND, this may—according to the hydrologic dilution capacity—result in severe eutrophication. The goal regarding P management is here to reduce P losses to water and soil to a level ensuring oligotrophic or mesotrophic conditions in surface waters. To find an appropriate P reduction strategy is not just an economic exercise of optimizing cost–benefit curves of P reduction and P concentrations in surface waters. Because of the high costs, it also requires a commitment to a certain water quality level. This commitment reflects the status of mind of METALAND society. In addition to the necessary economic development, this commitment takes usually large efforts of public policy and several decades to evolve (see also chapter 2, "Learning by Adapting the Anthroposphere").

3. *Long-term accumulation and concentration of remaining P.* Technical as well as biological processes yield not only products but also wastes. In view of P availability and losses to the environment, it is of prime interest to keep such losses small. However, due to thermodynamic and economic reasons, there will always be P in wastes such as sewage sludge, MSW, and wastes from industry, commerce, and trade.

If P-containing wastes are mixed with other wastes, P is diluted. Yet, there are logistic and technical measures available to keep the concentration of P in residues high and to store and accumulate such wastes under controlled conditions for long-term reuse.

Based on today's P metabolism and the goals stated above, the following three topics are key elements of a future design of P management:

"New Agriculture"

SFA of P has pointed out the key role of agriculture for P management. Figure 5.5 proves that if P approaches scarcity, recycling of industrial and consumer wastes is not enough: The first focus of any strategy must be to change the production of food in order to necessitate less P. The net efficiency (output per input) of agriculture is 23%, which is much too low. New economic ways of growing plants and animals without losing P must be invented. Hydroponics is an example where plants are grown without soil in a closed system supported by a mineral substrate and fed by nutrient solutions not losing any P to the environment. Aquaponics expands the production of plants by adding aquatic organisms to the symbiotic system, thus producing both vegetal and animal protein in a closed-cycle environment without losing any P. Both of these systems are limited to the production of certain produce and are not yet suited for the mass production of major crops. Total recycling will not be possible; very small losses will be inevitable if energy demand and costs are to be kept within reasonable boundaries. Even if hydroponics and aquaponics are further developed into vertical farming (Despommier, 2010)— either embedded in façades of urban buildings or installed in separate constructions—it remains to be seen if these scenarios are appropriate solutions to global food production with high P efficiency. To a large extent, such utopian ideas depend on economic boundary conditions: If P will become scarce, and the price of P will rise, there will be more financial resources available for alternative crop production with less P losses.

Other approaches focus on soils and on plants: Measures are needed to increase mobility and availability of P in soils. A deeper understanding of the plant–root–soil system allows enhancement of the uptake of P by plants. More research is needed in both areas. A key issue is also the production of meat: intensive livestock farming results in a large surplus of manure. In METALAND, 42% of all P supplied to soils originates from manure of animal production. It corresponds with the amount of P that is annually accumulating in the soil. It is necessary to develop and implement regional concepts to establish closed P cycles without accumulation of surplus P in soils. Nutrient accounting as practiced by farmers is an important first step for identifying depletion or accumulation of P in soils. It can be immediately realized and has the potential to improve P efficiency substantially. If systematically applied, it can evolve into a monitoring and control system allowing more effective and economic application of fertilizer and manure.

In summary, the topic "new agriculture" is at this moment more an issue for a new comprehensive research agenda than for an immediate and massive change in P management: From a P scarcity point of view, there is enough time to look into new ways of feeding the world, to develop and implement the necessary new agro-technologies, to perform regional experiments, and to set up new production, monitoring, and control systems. However, because research and in particular implementation of entirely new production and management systems require decades rather than years, it is timely to start the development of a high-P-efficiency agriculture now.

Efficient Environmental Protection
According to figure 5.4, the main flows of P to the environment in METALAND concern the soil and the hydrosphere. Based on the commitment of the METALAND decision makers summarized at the beginning of this chapter, the following quantitative long-term goals can be set: (1) P anthropogenic inputs and outputs of the soil must be in balance (no accumulation or depletion), with only marginal anthropogenic flows to the surface waters that comply with the second goal. (2) Total P inputs from agriculture, industry, and private households into surface waters must yield mesotrophic—or better—water quality conditions (<0.02 mg P/L). For the soil, this means either an input reduction of 3.1 kg P/c·y or an output increase of the same amount. Because there are no feasible means available yet to achieve the latter, today's main focus must be on reducing inputs by technical (e.g., demand-oriented fertilizing, nutrient balancing by farmers), economic (e.g., tax on fertilizer fraction that is lost during crop production), or legislative (e.g., limitation for the application of fertilizer and manure on land) means.

For the hydrosphere, the two main sources are soil runoff and effluents from sewage treatment plants. Soil runoff is reduced by decreasing P accumulation in soil as discussed earlier and by changing agricultural practice toward prevention of runoff and erosion. Although these two measures are the most powerful means to reduce P inputs into surface waters, they are not further discussed here. For a detailed quantification of measures, data about specific agricultural practices as well as concrete METALAND data (soil properties and slope, hydrologic regime including precipitation pattern, etc.) are required. It is noteworthy that in some countries such as Switzerland, subsidies for agriculture are based on P balances of agricultural fields.

The total liquid P-load from animal production, industry, and private households to waste and wastewater management amounts to 1.4 kg P/c·y. As stated earlier, for an overall P-removal efficiency for sewage treatment of around 80%, the

METALAND lake will become eutrophic if all purified sewage is directed to the lake. If only one tenth reaches the lake, it will stay mesotrophic. For the eutrophic case, there are three design options: to decrease the input of P into the sewer, an option that is discussed later; to increase P-removal efficiency to 95%; or to discharge purified sewage into rivers instead of lakes. The choice of options is determined by costs and by downstream requirements: Lakes or estuaries downstream of METALAND may require that surface water exports of P are limited. For multiregion P management of large catchment areas, an interesting question arises: How much P may each region contribute in order to observe the limits of a common basin (e.g., a final river delta)? Which is the appropriate unit: mass of P per capita, or mass of P per area, or mass of P per net precipitation (precipitation minus evapotranspiration)? All three units will result in different permissible loadings of P. Because concentration of P is decisive for eutrophication, the third unit seems attractive: A region with a high precipitation rate has a large dilution potential for P and thus may emit more P than another, more arid region.

To assess the risk posed by the loss of P to the hydrosphere, the hydrologic regime is the most important boundary condition, determining the dilution potential for nutrients as discussed earlier. Thus, for protecting surface water quality in regions with a high population density such as METALAND and with lake water with residence times of several years, the efficiency of P removal must be high (>95%). In summary, for designing an appropriate wastewater collection and treatment system, the objectives for water quality must be given by societal consent, the source terms of P must be quantified, and the hydrologic regime must be known.

Solid industrial and municipal wastes are of minor importance regarding P and surface water quality and the environment. They are either incinerated and landfilled as bottom and filter ash or they are directly landfilled. In both cases, the mobility of P is small, and the landfill leachates contribute only marginally to surface water concentrations.

Increasing the Availability of P by Resource-Oriented Waste Management
Sewage sludge as well as MSW and industrial wastes are sometimes seen as important sources of P. Together they account for less than one fourth of the total amount of P required for food production. It becomes clear that even complete recycling of sewage and wastes will not solve the P scarcity problem in the long run unless a "new agriculture" is introduced. Regarding wastes, SFA shows the priorities for measures: The amount of P from *Industrial food processing and marketing* is similar to the total amount of P from consumer waste (sewage and MSW). Because there are ~1000 times more consumer sources than processing industries, a focus on the latter seems attractive. The same is true for animal wastes (cadavers, slaughterhouse residues, and the like) that contribute about 20% to total waste flows.

The management objective for these wastes is to facilitate economic recycling while reducing losses to the environment. In principle, there are two design options available: to collect highly concentrated individual wastes and upgrade them to a product useful for agricultural application or to use a single (or a few) collection system(s) and to extract P in a form that is useful in agriculture. The classical approach is the mixed collection, with sewage and municipal solid wastes as the two main conveyor belts. This concept has the advantage that the main problem, to dispose of (liquid or solid) wastes, is solved. To continue along this line means that existing large investments into collection and disposal infrastructure can still be used. The disadvantage is that new requirements emerge such as conservation of valuable resources or the elimination of specific hazardous constituents and that the traditional system must be adapted to those new requirements. One of the necessary changes is that new waste management systems must focus on the collection of valuable substances, such as nutrients or cellulose or individual metals, and not on functions of materials such as the packaging function (cf. the case study on waste management later in this chapter).

Thus, during the past decade, new concepts have been developed to overcome the disadvantages of earlier systems. One promising concept for enhanced P reuse is the concentration of P in sewage sludge, with subsequent incineration of the sludge and recovery of P from incineration ash (Schick et al., 2008; Adam et al., 2009; Lederer & Rechberger, 2010). This concept—which is in contrast to incinerating sludge in cement kilns where P is lost as a resource—enables that 80% to 90% of P contained in sewage is finally recycled back to agriculture. It also eliminates hazardous substances that are the main reason that sludge is not being applied in agriculture today: Organic constituents are completely mineralized at 800°C, and heavy metals are selectively reduced to insignificant levels. If done properly, P in the resulting fertilizer promises to be more soluble than in the original sewage sludge and in conventional fertilizers. To date, the process is expensive and has not yet proved to be economically feasible. Thus, facing high uncertainty about P scarcity, an intermediate solution would be to accumulate ash in monodeposits. The full concept could be designed as follows (note the decreasing number of treatment plants and the increasing accumulation of P): As many P-containing wastes as possible are connected to existing sewer systems. In hundreds of sewage treatment plants, 95% of P is removed and transferred to dewatered sewage sludge. Sludge is reduced to ashes in several centralized sludge incinerators that accept also other "clean" biogenic wastes such as cadavers or agricultural waste biomass. The ash, containing all P of the incinerator input, is stored—without P-diluting binders such as cement—in a few central ash deposits. These deposits are maintained until P becomes scarce and are entered into the market when economic conditions (economy of scale of recovering accumulated P; price of scarce P) allow making a profit.

Another approach that makes use of the existing collection system but in contrast to the ash example collects P separately is the Novaquatis approach (Larsen, Rauch & Gujer, 2001). It is noteworthy that the concept is based on SFA of P and nitrogen from private households showing that the main contribution of nutrients stems from urine and not from feces. The idea is to separate urine at the source, the toilet. Urine stored in-house is discharged into the conventional sewer system at predetermined times when other loads of sewage are small. At the sewage treatment plant, urine of all households is separately collected and converted into a fertilizer for reuse in agriculture. The sewage treatment plant is relieved from nutrients and nutrient peaks. It remains to be seen if the challenges associated with this concept such as economic constraints, consumer acceptance, or pharmaceutical and endocrine substances in urine can be overcome.

Other concepts that were developed in the 1950s focus on composting of the biogenic fraction of MSW. It is important to take into account the results of SFA: The load of nutrients in MSW is about 10 times smaller than that in sewage or sewage sludge. Thus, priorities have to be set clearly in the recovery of P from sewage. A viable design concept is to combine the various nutrient sources on a regional scale (Lampert & Brunner, 2000). Biomass residues from farming, local industry and trade, and consumers (gardening and biogenic kitchen waste) are regionally collected and treated. For reasons of greenhouse gas emissions, anaerobic treatment resulting in useful biogas is to be preferred over composting. Such a concept permits farmers to generate new income and to recycle nutrients efficiently and regionally. A precondition is that the biogenic wastes are comparatively clean, so that they can be applied in agriculture. Hence, input materials such as sewage sludge or MSW that contain hazardous constituents are not suited for such concepts.

Other Measures
In addition to the actions proposed above, input- and consumer-oriented strategies should also be taken into account. New methods for prospecting, exploitation, and production will be developed when P becomes less available and thus more expensive. The extraction of P from residues of iron production has been mentioned above (Matsubae et al., 2010; Yamasue et al., 2010). These methods will on one hand increase the reserve base. On the other hand, they are likely to be more efficient than today's methods, thus reducing energy requirements and maybe even costs. Consumer preference for food might also change: The trend toward more meat on the dining table has come to an end in the most affluent societies. It remains to be seen if this trend continues, resulting in a decrease in animal production and thus a significant decrease in P utilization in agriculture.

Urban Mining

Moving Geogenic Ores to Anthropogenic Resource Stocks

Mining is one of the basic technological skills in the cultural evolution. Archeological findings from prehistoric times indicate that *Homo sapiens* developed very early the capacities to exploit special minerals and to transform them mechanically and chemically to tools that served their needs. With the first urban cultures 5000 years before the present, mining of various metal ores as a special technological and economic branch became established (see also chapter 1). Since then, the control of sites of metal ores, carbon, and hydrocarbons and their exploitation and physical–chemical transformation to goods became an essential strategic element to gain political power. During the past 150 years, research in the emerging geosciences has produced an impressive set of facts and models on the dynamics of the earth's crust formation and of its chemical composition. Today, geological survey is an established national and global enterprise. Any material that cannot be produced in agricultural and forestry processes (or from its substances extracted in industrial fabrications) is mined. In a wider sense, the notion *mining* comprises the extraction of nonrenewable resources from the earth's crust. Mining is usually a local engineering event of regional economic importance, coupled with relatively high risks of environmental impacts (see also the "cadmium case" in chapter 2). In most developed countries, mining operations must follow strict guidelines with regard to protected areas, waste management, and land restoration. For global players in politics, the economic control of mining sites is an indispensable tool in their strategic arsenal.

At the end of the twentieth century, the term "urban mining" popped up and has become, within a decade, an attractive slogan in the field of waste management. In the context of metabolic studies, urban mining describes "the exploration and exploitation of material stocks in urban systems for anthropogenic activities." The buildup of anthropogenic infrastructure (see chapters 2 and 4) is a transfer and transformation of material from the earth's crust to the urban system. The material is "not lost, but often out of account" (see chapter 4). Urban mining is "urban geology combined with urban engineering" (Lichtensteiger & Baccini, 2008). At present, this view is not predominant. In contemporary practice, the term "urban mining" is mainly a surrogate for waste management with a touch of label cheating; that is, a new envelope for well-known contents (see also the case study on waste management later in this chapter). Thereby the focus is laid, as usual, on the flows at the end of the pipes leading into waste management but not on the dynamics of the stocks. It is the rediscovered alchemistic fascination to make gold out of junk (e.g., to regain the gold stored in old mobile phones); it is the dream to dig out the treasures hidden in old landfills.

Copper Stocks and Flows: A Brief Overview

The anthropogenic material stocks in developed urban systems amount to 400 Mg/c (see chapter 4). Roughly 80% to 90% of this stock is made of rock materials, silicates, and carbonates. Their origin is mostly from regional or even local sources. Metals, such as iron, aluminum, zinc, and copper, form a group of widely applied construction materials. Their mean concentration in the total anthropogenic stock is <2%. The corresponding ores in the earth's crust are found in spots of special geological formations and are exploited to be traded on a global scale. Copper metal, the most expensive among these four (between US$5/kg and US$10/kg), can be considered as a trace metal with approximately 1 g per Kg of the anthropogenic stock. However, it is evident that within this stock there must be "clusters" of high copper concentration. On the basis of first rough estimations (Zeltner et al., 1999; Wittmer, Lichtensteiger & Baccini, 2003; Rauch, 2009), the anthropogenic copper stock in developed countries (200–300 kg/c) is of the same order of magnitude as the explored and mineable geogenic copper stock (as copper ores in the earth's crust) per capita of the world population (see also table 2.6). Therefore, copper is an excellent case with which to:

1. Evaluate the characteristics of the anthropogenic stocks in comparison with the geogenic stocks. It is one of the most valuable trace elements.
2. Illustrate the design of strategies for a material management on a regional scale. It covers the whole building sector due to its diverse applications.

Copper is probably the first metal used by man and has been processed since 7000 BP. Until the beginning of modern times, copper was primarily used as copper metal (Cu^0), namely as pure copper, and as an alloy with tin (bronze) and later with zinc (brass), due to its mechanical and aesthetic properties. Since the end of the nineteenth century, its use as an electrical conductor started an exponentially growing copper flow. Within three to four human generations, the move of copper from the geosphere to the anthroposphere shows geological dimensions. The flows at the end of the twentieth century are between 10 and 20 kg/c.y in developed countries (Zeltner et al., 1999; Rechberger & Graedel, 2002) and roughly 2 to 3 kg/c.y worldwide (a recent summary of the copper household studies is given in Bader et al., 2010). In other words, the copper consumption of developed countries (20% to 30% of the global population) is at present an order of magnitude higher than that of developing countries.

To grasp the dynamic character of the copper stocks, a result of the first regional metabolic study, based on a dynamic MF model (see chapter 3), is given in figure 5.6.

The MFA system shows three processes for the production of copper metal (*Mining, Crushing, Smelting/Refining*), two processes for applying the metal

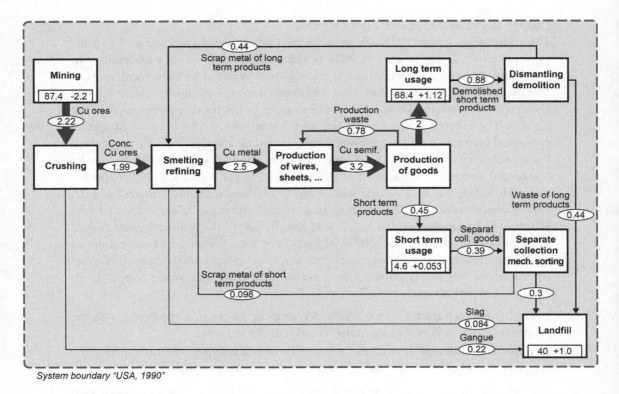

Figure 5.6
Real-time inventory of copper flows and stocks in the United States for the year 1990 (after Zeltner et al., 1999). Flows in Tg/y and stocks in Tg.

(*Production of Wires and Sheets, Production of Goods*), two processes for consumption (*Long-term Usage, Short-term Usage*), and three processes in waste management (*Dismantling/Demolition, Separate Collection/Mechanical Sorting, Landfill*), a total of 10 processes. It was assumed, based on copper import–export statistics, that the input and output flows are approximately in balance. Therefore, in the observed time period, the U.S. copper net household is based on the domestic stocks of the copper mines. The flow characteristics (in 1990) are as follows:

• The main streams are headed toward the Long-term Usage.

• The recoil flows of copper metals (0.5 Tg/y, mainly from Long-term Usage) contribute about 20% of the total demand of 2.5 Tg/y.

• The transfer of copper-containing materials and goods (from metal production and from waste management) to landfills amounts to 1 Tg/y. This is almost half of the excavated copper (2.2 Tg/y). The main fraction (70%) stems from waste management and is in metallic form (Cu^0).

• For a first approximation, the flows to the environmental compartments (hydrosphere and pedosphere) are negligible (<0.01 Tg/y) regarding the quantities. However, their eco-toxicology can be significant (see later).

The dynamics of stocks show the following pattern:

• In the copper system chosen, only 4 of the 10 processes show significant stocks.
• The accumulated quantities in the usage processes (68.4 Tg for long-term and 4.6 Tg for short-term, a total of 73 Tg) are almost as large as the mineable quantities (87 Tg) still available in domestic Mining stocks.
• The stock in Long-term Usage (approximately 15 times larger than the Short-term Usage stock) has a mean residence time of 50 years and that in Short-term Usage of 10 years.
• The second largest copper stock (40 Tg) has been accumulated in the Landfill.
• The annual growth of the stocks (around 1990) varies between 1% (Short-term Usage) and 2% (Long-term Usage and Landfill).

In conclusion, the regional situation for copper management illustrates not only the impressive transfer of copper from the geosphere to the anthroposphere. The results in figure 5.6, a flash in a momentary metabolic situation (1990) giving just a rough summary on stock depletion and stock formation in the twentieth century, stimulate the following questions:

1. What are the physical and chemical characteristics of the growing anthropogenic copper stocks and what is their potential to serve as new copper mines for future generations?
2. Is the current copper management, for a developed country and for developing countries with a growing demand, a sustainable way of handling this substance?
3. Which are the essential criteria and tools to design a "sustainable copper management"?

Exploration and Evaluation of Anthropogenic Material Stocks
Contrary to that on some ecosystems (e.g., forests), the data on material stocks of the anthroposphere are relatively poor. There are sufficient data on areas (known since antiquity), on money flows and stocks (necessary to control a national economy), and on a wide variety of statistics concerned with importing and exporting goods, supported by production and sales statistics of goods. The latter are very rarely combined with information on the chemical compositions of the goods. Therefore, any large-scale evaluation of anthropogenic material stocks is confronted

with poor data sets, with regard to quantity, composition, and quality. The MFA method and its mathematical models can, to a certain extent, overcome the shortcomings of data (see chapter 3). For a rough overview regarding the order of magnitudes of substance stocks on a global scale, statistical analysis is applied (e.g., Rauch, 2009). Based on national surveys, metal stocks (Al, Fe, Cu, and Zn) are put in relation to the economic indicator GDP. It supports earlier findings, as expected from regional metabolic studies (see also chapter 4, summarized in Wittmer, Lichtensteiger & Baccini, 2003), that the higher the state of economic wealth, coupled with a higher state of physical infrastructure, the higher the anthropogenic stocks of these metals. In other words, all developed countries operate with similar technical equipment having similar material compositions. The global mapping of the metal stocks visualizes the fact that the highly concentrated geogenic stocks of the four metals are globally dispersed (distributed in in-ground deposits of developed and developing countries). The anthropogenic stocks are more diffused, located primarily in developed countries.

The aspect of diffusing and concentrating materials was evaluated by applying statistical entropy analysis (SEA; Rechberger & Graedel, 2002). An illustration is given for copper in figure 5.7. The method operates with the relative statistical entropy (RSE), a dimensionless value between 0 (the substance copper is only in one

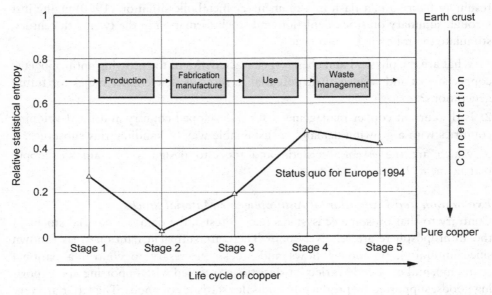

Figure 5.7
Development of the RSE along the life cycle of copper for the status quo in Europe 1994 (open system) (after Rechberger & Graedel, 2002).

of all possible flows/stocks within a given MFA system) and 1 (the substance copper is evenly distributed among all possible flows/stocks within a given MFA system). The MFA system applied for RSE is simplified from 10 processes to 4 processes in comparison with the U.S. copper system in figure 5.6. The process *Production* in the RSE system comprises the first three in the U.S. system, and the process *Fabrication & Manufacture* combines the two production processes of wires/sheets and goods. The process *Use* does not distinguish between long-term and short-term usage. Finally, *Waste Management* embodies the three processes Dismantling/Demolition, Separate Collection/Mechanical Sorting, and Landfill.

Throughout the life cycle of copper, the statistical entropy varies considerably among the above-mentioned processes and covers about 50% of the possible range between total dissipation and maximal concentration. However, the copper system as a whole neither dissipates nor concentrates copper significantly with regard to the original ore. This finding supports the hypothesis of a high copper resource potential of the anthropogenic copper stocks, similar to those in the earth's crust. In other words, copper is not diluted because of human use. It is just in another place with another chemical environment. This finding supports also the priority list of the three questions given above. First, we have to know more about the physical and chemical qualities of copper in anthropogenic stocks, their functions, quantities, and time of use.

Copper Exploration

Analogous to geological exploration, a method for surveying anthropogenic stocks was developed (Lichtensteiger et al., 2006; Wittmer & Lichtensteiger, 2007a; see also chapter 4). Generally and specifically for copper, the following key parameters were chosen:

1. Installation density of goods (functional units) applying copper (but not exclusively).
2. Frequencies: "frequency of material" (i.e., market share of copper in a certain good).
3. Mass: mass quantity per functional unit/good and substance quantity (Cu) per mass unit.

The data sources are as follows:

• National trade statistics.
• Statistics of various economic branches such as electrical power suppliers, transportation enterprises, telecommunications companies, associations of craftsmen (plumbers, electricians, heating technicians, etc.).
• Scientific literature.

• Field studies of selected buildings.
• Interviews with experts of the production companies and of the construction branches.

In a special effort, the copper stocks were investigated more in detail (Wittmer, Lichtensteiger & Baccini, 2003). This investigation was supported with a mathematical MFA model (described in detail by Bader et al., 2010) and produced the following main results in Switzerland in the year 2000 (figure 5.8 and figure 5.9):

1. The historical development of the copper metabolism in the twentieth century shows more than a doubling of the net input flow (from 1.5 to 3.8 kg/c.y) in the first 50 years and a fivefold increase of the total stock per capita (from 19 to 95 kg/c). Noteworthy is the fact that the infrastructure stock takes the lead (from 2 to 50 kg/c), due to the expansion of power grids during this time period. The inner recoiling flow (secondary copper) is increased by a factor of 5 (from 0.3 to 1.5 kg/c.y) and contributes almost 40% to the inner demand of 3.8 kg/c.y. The annual transfer to landfills is about 20% of the outputs from the in-use stocks (0.5 of 2.0 kg/c.y).

2. The second half of the century shows a further four- to fivefold increase of the import and export flows because of an expanding *Trade & Production* activity with copper-containing goods. The inner net flows to the in-use stocks are roughly doubled (from 3.8 to 8 kg/c.y). The contribution from recoil flows is 50% (4.1 of 8 kg/c.y). The movables (or short-term goods) show the strongest growth (from 0.7 to 2.7 kg/c.y). This is mainly due to the strong growth of transport vehicles (mainly automobiles, see also "TO TRANSPORT&COMMUNICATE" in chapter 4) during this time period. The relatively smallest growth shows the infrastructure that had its largest expansion (the power grids) in the first half of the century. The size of the three in-use stocks has doubled (from 95 to 220 kg/c). The "winners" are the *Buildings*, with a copper stock having almost reached the same size as that in the *Infrastructure* (80 ± 10 vs. 105 ± 25 kg/c). During this period, the building growth per capita was the highest. The distribution of copper in buildings (year 2000) is given in figure 5.9. This pattern of applications did and does not stay constant (Wittmer, 2006), because technological changes and lifestyles replace copper by other materials (e.g., by organic polymers in water pipes), and copper is newly introduced (e.g., substituting tiles in roofs).

Comparing these findings with some other surveys available (Wittmer & Lichtensteiger, 2007b), it becomes clear that the dynamics of copper flows and stocks can only be understood in the context of the whole regional metabolism. There are strong differences in the copper management due to the specific development characteristics of the regional anthroposphere. The copper household of the United States in 1990 (see figure 5.6) shows similar properties with respect to the total in-use stocks (approximately 240 kg/c) and the total flow into the processes of Usage (approximately 8 kg/c.y). The main differences of the two systems are twofold: (1) The United States is, in a net balance, transferring its own earth crust copper to the anthropogenic stocks, whereas the geogenically "copper-poor Switzerland" imports the substance in various goods and exports again around 80% of it in refined goods. The difference of 20% is used to build up the domestic stocks. Thereby it gets richer in copper every year. (2) The fourth stocks in both systems, the *Landfills,* differ significantly in size. The United States in 1990 had a quantity of approximately 130 kg/c, Switzerland only 50 kg/c (year 2000). This is probably due to the following two reasons (see figure 5.6 and figure 5.8): (1) Mining, not existing in Switzerland, contributes significantly to the landfill input (30%). (2) The recycling rate from waste management is higher in Switzerland (51% for the year 2000) than in the United States (21% for the year 1990). The separation efficiencies E (defined later), a parameter to evaluate the performance of waste management (mining wastes excluded), are also different (Switzerland 0.7, United States 0.4).

The first answers to the first question posed earlier ("What are the physical and chemical characteristics of the growing anthropogenic copper stocks and what is their potential to serve as new copper mines for future generations?") are summarized as follows: All copper (>99%) is installed as copper metal (Cu^0) in a broad variety of goods and functions, primarily in buildings and infrastructure, secondarily in commodities of shorter residence times. In developed countries, the anthropogenic copper stocks have reached relatively high values and serve already as important copper sources. However, the physical growth process of the anthroposphere still asks for more copper from primary sources (*Mining*). The data situation on the anthropogenic copper stocks is relatively poor (i.e., up to now, the exploration means a great expenditure).

The second question posed earlier is oriented toward the sustainability of the contemporary copper handling. Copper, due to its chemical properties (in metallic form relatively inert under normal atmospheric conditions), is not diluted and therefore "still available in anthropogenic stocks." From a physical and chemical point of view, there is not a "growing scarcity" of copper. On the contrary, the chemical transformation of Cu^{2+} components in ores to Cu metal by an endothermic reduction makes anthropogenic copper stocks, from an energetic and economic point of view, more valuable. In the anthropogenic copper stocks lies also a theoretical energy stock. Predominately fossil energy was invested to make copper sheets and wires. However, these stocks, at present mainly in the built anthroposphere of the developed countries, are not readily available. Copper availability depends on the lifetimes of these stocks. The growing scarcity is mainly a problem of the developing countries demanding higher copper flows for the buildup of their anthroposphere, depending on primary sources. The small losses of copper to the environmental compartments are, quantitatively seen in a regional copper system, negligible.

a)

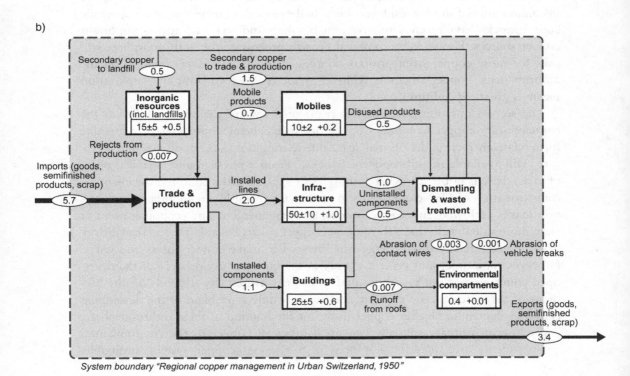

Secondary copper to landfill 0.1

Secondary copper to trade & production 0.3

Inorganic resources (incl. landfills) 2±1 +0.1

Mobile products 0.3

Mobiles 5±1 +0.1

Disused products 0.2

Rejects from production 0.001

Trade & production

Imports (goods, semifinished products, scrap) 1.8

Installed lines 0.8

Infra-structure 2±0.5 +0.8

0.01

Uninstalled components 0.2

Dismantling & waste treatment

Abrasion of contact wires 0.0001   0.0003 Abrasion of vehicle breaks

Installed components 0.4

Buildings 10±1 +0.2

0.003 Runoff from roofs

Environmental compartments 0.07 +0.003

Exports (goods, semifinished products, scrap) 0.6

System boundary "Regional copper management in Urban Switzerland, 1900"

b)

Secondary copper to landfill 0.5

Secondary copper to trade & production 1.5

Inorganic resources (incl. landfills) 15±5 +0.5

Mobile products 0.7

Mobiles 10±2 +0.2

Disused products 0.5

Rejects from production 0.007

Trade & production

Imports (goods, semifinished products, scrap) 5.7

Installed lines 2.0

Infra-structure 50±10 +1.0

1.0

Uninstalled components 0.5

Dismantling & waste treatment

Abrasion of contact wires 0.003   0.001 Abrasion of vehicle breaks

Installed components 1.1

Buildings 25±5 +0.6

0.007 Runoff from roofs

Environmental compartments 0.4 +0.01

Exports (goods, semifinished products, scrap) 3.4

System boundary "Regional copper management in Urban Switzerland, 1950"

c)

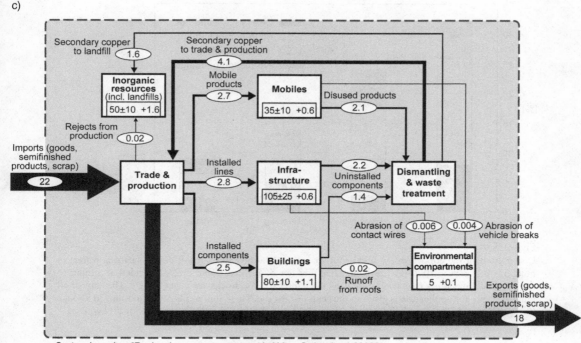

System boundary "Regional copper management in Urban Switzerland, 2000"

Figure 5.8
Development of the copper material system of Switzerland in the twentieth century, with snapshots for the years 1900 (a), 1950 (b), and 2000 (c). The stock and flow values are based on aggregated results from a dynamic copper model (Bader et al., 2010). Stocks in kg/c, flows in kg/c.y. Stock increases (kg/c.y) are indicated by positive digits in the boxes. The widths of flow arrows are proportional to the flow values (reprinted with permission from Wittmer & Lichtensteiger, 2007b).

However copper, although an essential element, has a high toxicity for unicellular organisms like algae and bacteria. Therefore, copper is also applied as a pesticide. According to environmental research findings, the most sensitive ecosystems with respect to higher copper impacts are aquatic ecosystems (Sigel, 1984). Environmental protection laws have set threshold values for copper concentration in sewage to prevent pollution. In all countries that have an effective sewage flow control, copper pollution of aquatic systems can be strongly reduced or even omitted. Nevertheless, certain accumulations of copper due to long-term emissions from infrastructures such as railway installations represent an environmental problem (Burkhardt, Rossi & Boller, 2008; Kral & Brunner, 2010).

It follows that the answer to the third question ("Which are the essential criteria and tools to design a 'sustainable copper management'?") is mainly hidden in the handling of the anthropogenic copper stocks in the developed countries.

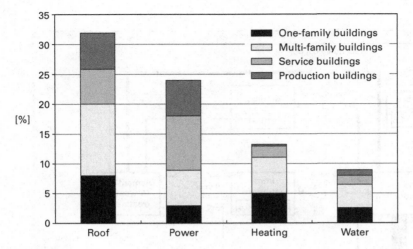

Figure 5.9
Distribution of the copper stocks within the major application ranges (roof, power, heating, water) in buildings in the year 2000 (Switzerland; after Wittmer & Lichtensteiger, 2007a), divided according to the four buildings types (see also figure 4.17 in chapter 4). The stocks are given in kg/c. The sum of all building types for application is given in the center of the pie. These four applications amount to 78 kg/c (97% of all applications in buildings, see figure 5.8, year 2000, with a building stock of 80 kg/c).

Designing a Copper Management Strategy

The design of a material management strategy follows the same political guidelines already introduced in the phosphorus case study. In the case of copper, based on the results of the exploration and evaluation presented earlier, the main targets for the twenty-first century (exemplified with Switzerland) are as follows:

1. *Reduce the net input of primary copper.* This goal is based on the long-term perspective on a global scale, where the growing demand of developing countries has to be satisfied primarily from a limited stock of copper ores. It is neither a pure altruistic nor a pure economic motivation. It is a sober strategic element to maintain a sound national sovereignty regarding the functioning of the metabolic system of a country, neither disposing of primary copper sources nor of enough political power to enforce the access to any resource stock. It is a politically and economically motivated goal, with a bonus for global solidarity in sharing the limited primary copper stocks.

2. *Minimize waste deposits diluting copper to higher states of RSE values.* Landfills for all sorts of solid wastes (see the case study on waste management later in this chapter) reduce strongly the resource potential for its trace components. If copper, leaving the in-use stocks, cannot be recycled for further applications, it should be deposited in stocks having low RSE values (see figure 5.7). It is a

long-term oriented goal, based on the sustainable development commitment in a national constitution. In principle, the political leaders and institutions are obliged to "conserve copper in concentrated forms" for coming generations.

3. *Reduce the copper flows to the environmental compartments.*   A growing copper stock exposed to abrasion and corrosion leads to more copper impacts in the hydrosphere, the atmosphere, and the pedosphere. Because of the eco-toxic effects of copper, measures have to be taken to stay below the threshold values. It is an ecologically motivated measure and is mandatory according to all environmental protection laws.

Based on the exploration presented above, we know the copper past of the Swiss region and the status quo at the beginning of the twenty-first century. The past helps to calibrate a regional dynamic copper system (Bader et al., 2010). This model allows studying the dynamics of a regional copper household, applying the scenario technique, various measures with regard to their effects on attaining the above three targets. They serve as a metabolic basis for the political design, as the implementation of these measures has to become part of the whole regional anthropogenic system.

Metabolic Scenarios for a Regional Copper Management (after Bader et al., 2010)
The dynamic copper model was formed according to the mathematical procedures presented in chapter 3. The regional copper system chosen is given in figure 5.10. It consists of eight processes (the system shown in figure 5.8 is an aggregated form). The process *Buildings* is internally differentiated by two types of copper applications; namely, the *inner* and the *outer* (see also figure 5.9). The process *Mobiles* is subdivided by *Cars* and *Other mobiles*. At present, the cars are still the the major copper carriers within the mobile goods.

Figure 5.11 shows calibrated logistic growth curves (for the inner processes, not the landfill) based on data from the twentieth century and the assumption that the per capita stocks reach a steady-state level around 2050. The growth of the consumption stocks in the twentieth century is different for buildings, mobiles, and infrastructures. The data records showed that buildings and mobiles grew linearly up to 1920 and then changed to strong logistic growth. The reason is the electrification of households and the improvements in building technologies (roofs, gutters, installations) around 1920–1940. Infrastructures show a two-step logistic growth: the first step reflects the electrification age and the second step, with a growth rate peaking at about 1960/1990, a large growth of infrastructure (power stations, new grids, and telecommunications systems).

Landfills grew exponentially up to 2000 and changed to a linear growth pattern after 2000 for the "business as usual 2000" scenario; that is, not changing the

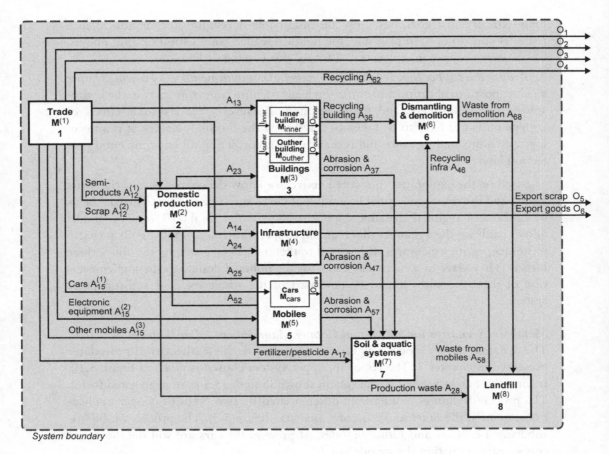

Figure 5.10
Regional copper system for Switzerland (reprinted with permission from Bader et al., 2010).

metabolic pattern shown in figure 5.8 for the year 2000. Clearly, for other scenarios, the landfills would either continue to grow exponentially (for continuing growth in stocks) or more slowly (with substitution of copper).

Induced by the growth of stocks, the input and output flows of buildings, infrastructures, and mobiles and the consumption loss also show large growth in the twentieth century (figure 5.12). (The two peaks in *Infrastructures* are due to the double logistic growth; see above.) Because of the large residence time in buildings and infrastructures, the time delay between input and output flows is quite large. As a consequence, a large increase of the output, especially from buildings, is expected in the coming decades. This is important for planning dismantling capacities.

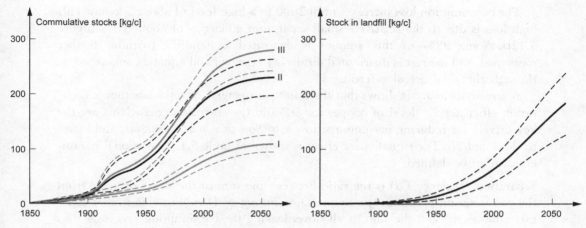

Figure 5.11
Cumulative copper stocks (kg/c): I, buildings $M^{(3)}$; II, buildings plus infrastructures $M^{(3)} + M^{(4)}$; III, buildings plus infrastructures plus mobiles $M^{(3)} + M^{(4)} + M^{(5)}$. Stock of landfill $M^{(8)}$ (kg/c). The dashed lines show the standard deviations (reprinted with permission from Bader et al., 2010). For the definitions of the abbreviations $M^{(i)}$ see figure 5.10.

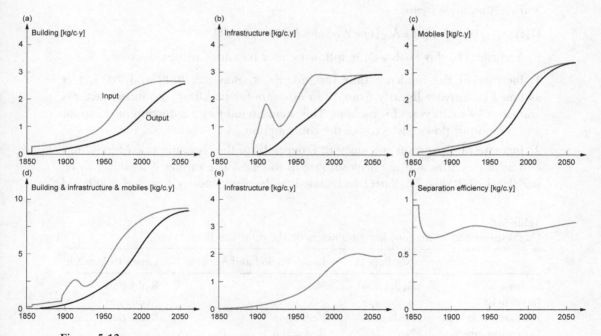

Figure 5.12
Copper input and output flows (kg/c.y) for (a) buildings, (b) infrastructures, (c) mobiles, (d) buildings and infrastructures and mobiles, (e) consumption loss, and (f) separation efficiency E. The upper curves show the input flows; the lower curves show the output flows (reprinted with permission from Bader et al., 2010).

The consumption loss increases until 2000 to a high level of about 2 kg/c.y. This high loss is due to the relatively small separation efficiency of about 70% (figure 5.12f). Some 98% of this amount is deposited in landfills (without further measures), and the rest is distributed diffusely to the soil and aquatic compartments through abrasion/corrosion/fertilizers.

A sensitivity analysis shows that the four "parameters" of residence time τ, separation efficiency E, "level of copper usage," and the corrosion coefficients are the key drivers for reducing net imports, losses to *Soil & Aquatic Systems*, and losses to the *Landfills*. Their qualitative effect is shown in table 5.1. Two special parameters are to be defined:

Separation efficiency E(t) is the ratio between the sum of the flows recycled from the consumption processes (partly via *Dismantling & Demolition* and other separate collections) and the sum of all flows leaving the consumption processes into waste management (see figure 5.10 for flow and process indices):

$$E(t) = \frac{A_{62}(t) + A_{52}(t)}{A_{62}(t) + A_{68}(t) + A_{52}(t) + A_{58}(t)}.$$

Consumption loss U is the sum of all internal copper flows to the *Landfill* and the *Soil & Aquatic Systems*:

$$U(t) = A_{28}(t) + A_{58}(t) + A_{68}(t) + A_{57}(t) + A_{47}(t) + A_{37}(t).$$

According to this analysis, the following four measures will be discussed.

1. Increase of the residence time (or lifetime) τ: Between 2000 and 2025, τ is assumed to increase linearly from 40 to 60 years for buildings and infrastructures and from 14 to 20 years for mobiles. This scenario induces a reduction of the input and the output flows (delayed) of the consumption.

2. Increase of separation efficiency E: E depends on the recycling rate of buildings and infrastructures and the disposal rate of mobiles. The former is assumed to be 0.9 (level of 2000). The latter is assumed to decrease linearly between 2000 and

Table 5.1
Qualitative effect of the four key parameters on the three key flows

	Net Import	Losses to Soil and Aquatic	Losses to Landfills
Increase of τ	Reduction	No	Reduction
Increase of E	Reduction	No	Reduction
Decrease of stocks	Reduction	Reduction	Reduction
Corrosion rate	No	Reduction	No

2025 from 0.6 to 0.1. There are already technologies available to realize this increase (Rem et al., 2004; Böni, 2010).

3. This measure affects only the distribution pattern of the outputs of the consumption for recycling and disposal. There is no effect on input, output of consumption, and recycling rate.

4. Decrease of stocks: The two measures "copper ban" and "forced dismantling" are designed to decrease the "level of copper usage" (stock) by substituting copper for other materials. A ban stops the input into consumption. As a consequence, the corresponding stock and output flows decrease slowly according to the lifetime. Forced dismantling alone (i.e., without an input ban) is not considered as a reasonable scenario. The effect would be similar to a decrease in lifetime. For the "ban" scenario, a copper ban for the envelope of buildings after 2010 was assumed. For the "dismantling" scenario (forced dismantling) a specific dismantling rate for the envelope of the building was assumed. Because the average lifetime of the envelope is 40 years, this dismantling rate means roughly a doubling of the rate induced by a "copper ban" alone.

5. Decrease of corrosion rate: Such a measure would clearly reduce the losses to aquatic systems and soils. However, this measure is difficult to realize. It is known that copper–zinc alloys have smaller corrosion rates. At present, it seems not possible to cover all copper sheets, gutters, and so forth with corrosion-reducing films. Therefore, this measure is not discussed further. It is partly a problem delegated to waste management with regard to sewage channeling and sewage treatment (see chapter 4).

The first three measures are integrated in four scenarios (table 5.2) and compared with the reference scenario, that is, no measures (see also figure 5.11 and figure 5.12).

The effect of the scenarios are shown on four system properties (figure 5.13); namely, on the (a) landfill stocks; (b) consumption loss; (c) import flows to the

Table 5.2
Synopsis of the four copper management scenarios

Scenario	Measures
Reference	No
Longer residence time of consumption stocks	Increase of τ
Improved recycling	Increase of E (separation efficiency)
Combination I	Increase of τ and E and a copper ban in buildings
Combination II	Combination I plus forced dismantling

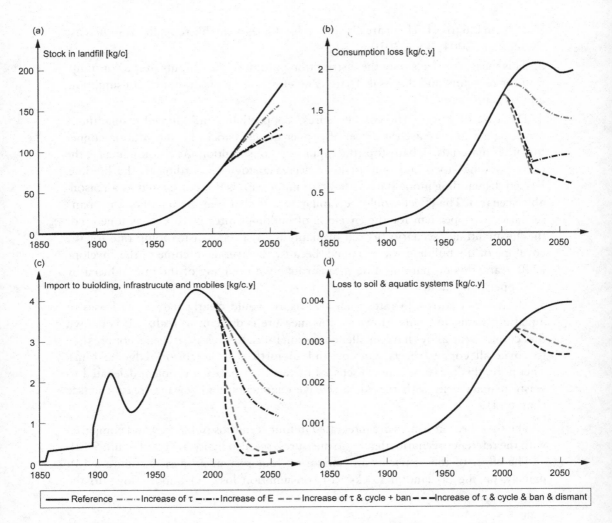

Figure 5.13
Results of the first four scenarios in a regional copper system: (a) the landfill stock (kg/c), (b) consumption loss (kg/c.y), (c) import to *Buildings, Infrastructures, Mobiles* (kg/c.y), and (d) loss to *Soil & Aquatic Systems* (kg/c.y).

three consumption processes (*Buildings*, *Infrastructure*, and *Mobiles*); and (d) losses to *Soil & Aquatic Systems*.

Three of them (a, b, and c) describe the residual flows to the environmental compartments. All three illustrate indirectly the system behavior with regard to its capacity and performance to keep copper within the internal consumption at a low RSE level. The fourth indicator (d) shows the changing behavior in copper use.

All four scenarios can reduce the growth of landfill stocks. It shows that the increase of E (separation efficiency) and not τ (lifetime) has the strongest effect. Consequently, the stock of landfill is directly related to the consumption loss.

The effects on consumption loss U are also diverse. An increase in separation efficiency immediately reduces this flow. This is because about 98% of the consumption loss is disposal to landfill, which is controlled by the recycling rate. Note that the increase after 2025 for this scenario is due to the fact that after 2025, the separation efficiency is constant at 0.9, but the output flows from consumption still increase (figure 5.13). An increase in lifetime or a copper ban reduces the consumption loss only after a time delay. Indeed, the full effect is reached after a delay of the order a lifetime. Additional forced dismantling first increases the consumption loss, as the output flows of consumption increase before they decrease because of reduced stocks.

All scenarios reduce the import to *Buildings*, *Infrastructures*, and *Mobiles* without any delay in time. Only the reasons are different. For an increased lifetime or a copper ban, the input to consumption is reduced and zero, respectively. For increased separation efficiency or additional forced dismantling, the recycling flows are increased. Reduced inputs and increased recycling flows both induce a reduction of import.

The loss to *Soil & Aquatic Systems* is only reduced by the "copper ban" and the combination "copper ban" plus "forced dismantling." Again, the full effect is reached only after a time delay of the order a lifetime of the copper-containing goods in question.

It is evident that forming a landfill stock of this type, namely with a high RSE level, built up mainly in the twentieth century, is an unacceptable procedure with regard to the criteria of a sustainable development. In other words, the copper example illustrates also the weakest part in the anthropogenic system concerning the regional metabolism, the waste management policy (see also the case study on waste management in this chapter).

Concluding Remarks

This investigation of a dynamic material system, exemplified with copper, shows the following: It is easy to identify the key drivers that can reduce flows such as consumption loss, loss to environmental compartments, or the import flows to the consumption processes. However, a dynamic model is crucial for understanding, discussing, and comparing their quantitative effect as a function of time on these flows. This is due to the consumption dynamics, which relates the stocks and outputs to the whole history of the input flows. As a consequence, possible measures to reduce input flows affect emissions from the stocks and the output flows with possibly large time delays. This is in contrast to stationary systems where input changes cause an immediate change in emissions and output flows.

Designing Waste Management

At the back end of the metabolic system, waste management forms the main interface between the anthroposphere and the environment (Brunner & Bogucka, 2006). Materials entering the anthroposphere have four options: they stay in the man-made stock, they leave as emissions, or they are recycled or disposed of as wastes. Because emissions are kept small by legislation and technology, the amount of wastes reflects—after the lifetime of goods—more or less the full amount of materials entering the anthroposphere. Thus, waste management is of primordial importance for environmental protection especially in societies with a high material turnover. For long-lived investment goods—which form the bulk of anthropogenic material turnover—there is a long delay between production and disposal. The full extent of future waste production does not become discernible until the day when these materials actually turn into waste. Hence, in societies with large growth rates, current waste production is not an appropriate indicator for future waste amounts, and the importance of waste management as a key filter and control valve for hazardous as well as valuable substances is often underestimated (cf. figure 2.19).

Historically, the main objective of waste management was to protect human health from detrimental impacts of wastes (cf. activity TO CLEAN in chapter 4). Hence, public cleansing focused on transporting wastes out of human settlements as quickly as possible, thus reducing by sanitation (collection and landfilling) the main risks, such as enteric diseases, emanating from wastes. This practice relocated the hygienic problem from urban areas to rural areas where wastes were applied to fields. Microorganisms closed the cycle when they were incorporated into produce that was consumed in cities. To ameliorate this intolerable hygienic condition, sanitary landfilling as well as waste incineration was introduced in the second half of the nineteenth century. Together with public drinking water supply and modern combinations of sewer systems and wastewater treatment plants, this resulted in complete elimination of typhoid fever in large parts of the modern world.

Because of the phenomena of modern affluent metabolism summarized in chapter 2, in particular growth of material turnover, change of materials, and material complexity, a new problem arose in the 1950s: emissions of heavy metals and synthetic organic substances from waste treatment polluted the environment. In addition, it became obvious that the large amount of wastes was not only a nuisance but also represented valuable secondary resources. Thus, recycling, which was already an important part of the economies of the First and Second World Wars, became popular again in the 1960s and 1970s.

Today, waste management in affluent countries is organized in various ways: whereas in the United States, landfilling is still the main disposal means, many countries in Europe and Japan have chosen to incinerate their wastes and to landfill incineration residues. Driven by volatile resource prices, discussions about potential scarcity, and corresponding subsequent legislation, recycling rates are on the rise in all countries. New logistic methods and technologies are introduced into the market. Institutions responsible for waste management have now numerous options to choose from. The challenge of designing an appropriate waste management system is substantial, as it considers not only technological and economic questions but also social and behavioral issues.

In the following, the design of a waste management system based on a metabolic MFA/SFA approach is exemplified for the case of METALAND. The focus is on metabolic processes, but economic boundary conditions as well as social topics such as acceptance of treatment plants and the like are discussed.

Key Issues for Designing Waste Management Systems

The following three basic requirements must be fulfilled for designing a waste management system:

1. Definition of objectives.
2. Definition of economic boundary conditions.
3. Establishment of a comprehensive physical knowledge base on the premise of MFA/SFA.

It is important to note that the three requirements are a condition sine qua non, and that they are linked together: The definition of objectives is affected by the economic boundary condition, the economic boundary conditions determine the establishment of the knowledge base, and the knowledge base influences the definition of the objectives.

Objectives of Waste Management

For METALAND, it is assumed that the following goals have been adopted for waste management:

1. Protection of mankind and the environment.
2. Conservation of resources such as materials, energy, land, and biodiversity.
3. Aftercare-free waste treatment systems (e.g., landfills).

These goals were developed in the late 1980s by countries such as Switzerland and Austria and have been applied and followed for two decades by now. Thus, it is safe to say that these goals can find a societal consensus and that they are well suited as an operational base for practical waste management. Whereas the first two are straightforward goals developed from past experience, the third objective requires some explanation. The background of this goal is the precautionary principle based on ethical arguments: Waste management should not postpone problems caused by today's generation until future times, thus preventing that upcoming generations will have to pay for remedial action resulting from today's practice. This goal, which became later part of the guidelines for a sustainable development, is of particular importance for landfilling, as Belevi and Baccini have shown that—even best practice—sanitary landfills require aftercare for several centuries (Belevi & Baccini, 1989). But it applies also to recycling: If hazardous substances are kept in cycles, they will have to be disposed of in some distant time, requiring "aftercare."

Thus, the practical implications of the goals stated above are that only three types of materials may be produced by waste management: (1) "clean" materials suitable for recycling, (2) immobile materials safe for landfilling with no negative impact on the environment for long time periods, and (3) emissions that are compatible with environmental standards not changing geogenic concentrations and loads. Note that a "zero emission" strategy is not required to reach the goals of waste management. Neither is a "waste hierarchy" approach needed such as "prevention before recycling and disposal." These are three means to reach the goals, and they should be applied taking economic criteria into account. If prevention is the most economic way of reaching a particular waste management objective, then this method should be chosen. If the disposal option is more economic, the waste hierarchy should not be used to justify avoiding this choice. It should be noted that to provoke a significant change by prevention, such as reducing cadmium input into waste management by 50%, massive interventions into markets and products are required, whereas filters are able to reduce the load of atmospheric emissions far more efficiently; for example, in the case of cadmium by more than 99.9% (see "TO CLEAN in Waste Management" in chapter 4). The waste hierarchy is particularly inappropriate for emerging economies that are often characterized by significant waste-derived health problems. Under such circumstances, the disposal alternative is by far superior to any prevention strategy (Brunner & Fellner, 2007).

The types of materials to be produced by waste management have been classified in a different way, too: In addition to clean recycling materials and environmentally compatible emissions, waste management must generate Earth crust–like materials and ore-like materials only. Crust-like materials can either be safely landfilled or used as a construction material, and ore-like materials are useful as a new secondary resource. Thus, the main purpose of waste management is separation and transformation or accumulation of substances, either by logistics (separate collection) or by chemical and mechanical technologies (thermal treatment, separation and size reduction). In summary and independent of the classification of products, an appropriate waste management strategy for reaching the objectives is to establish "clean cycles" and direct remaining materials to safe "final sinks."

The term *final sink* applies to the following: sinks are defined as water, air, and soil receiving anthropogenic material flows. They are necessary for accommodating emissions and residues from anthropogenic activities such as primary production (e.g., tailings), processing and manufacturing, and consumption. The permissible capacity of a sink for holding a particular substance is determined by the geogenic concentration of the material in the sink: For example, the atmosphere is an appropriate sink for molecular nitrogen, the oceans for water and chlorides, and the soil for carbonates. For trace metals such as cadmium or lead, the capacity of air, water, and soil as a sink is very limited. Overloading these spheres results in damaged ecosystems. Substances in sinks can be mobile (chloride in soil) or immobile (carbonates in sediments), thus they may move from one environmental compartment to another. If a substance reaches a sink where the substance has a residence time of >10,000 years, the sink is called a final sink (Döberl & Brunner, 2004).

The fate of materials in sinks is determined by the chemical speciation of the substance as well as by changes in the biogeochemical environment such as erosion, weathering, and changes in redox and pH conditions that constantly relocate surface material. Landfills are also sinks. Beginning in the 1970s, technology changed from so-called open dumping to sanitary landfilling with controlled inputs and outputs. Mass balances and SFA of sanitary landfills showed that because of biogeochemical reactions taking place in landfill bodies, such constructions require monitoring and emission control for centuries (Belevi & Baccini, 1989). Thus, the so-called final storage concept was developed (Baccini, 1989). This concept was based on the fact that the main emission source was reactions of the landfilled waste and that it was not possible to confine landfilled wastes "forever." Thus, the key measure to prevent long-term emissions was to transform wastes into "stone-like materials" that are more or less immobile and do not leave the landfill body during leaching.

To become final storage landfills (i.e., final sinks), landfills have to meet two criteria: First, before landfilling, wastes must be mineralized and immobilized in order to become inert and sedentary for long time periods under changing geochemical conditions. And second, the landfill must be situated in an area with low erosion rates ensuring that the deposited material remains in place for long time periods. Final-sink landfills (also called final-storage landfills) are per definition aftercare-free landfills.

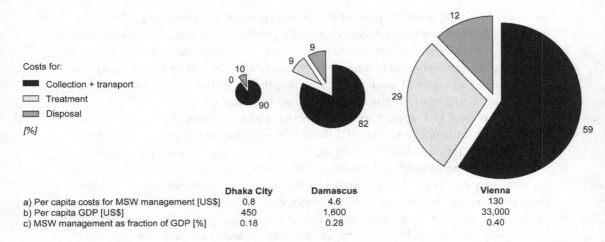

	Dhaka City	Damascus	Vienna
a) Per capita costs for MSW management [US$]	0.8	4.6	130
b) Per capita GDP [US$]	450	1,600	33,000
c) MSW management as fraction of GDP [%]	0.18	0.28	0.40

Figure 5.14
Per capita costs of MSW management, GDP, and fraction of GDP for MSW management in Dhaka City, Damascus, and Vienna for the year 2002 (after Brunner & Fellner, 2007).

Definition of Economic Boundary Conditions

In figure 5.14, the expenditures for municipal solid waste management in three cities of various economic power are summarized (Brunner & Fellner, 2007). The absolute amount per capita and year ranges between US$ 0.8 for Bangladesh (data for capital Dhaka not available) and US$130 for Vienna. The relative percentage with regard to GDP reaches from 0.18% in Dhaka City to 0.4% in Vienna. Three observations are noteworthy: First, the spread of expenditures for waste management over the whole range from poor to affluent countries is large and covers two orders of magnitude from about US$1 to US$10 to US$100 US$ per capita and year. Second, when compared with the total GDP, the costs of waste management are small. Third, the fraction of income that people are willing to spend for waste management depends on their wealth. The lower the income, the less is spent for waste disposal, which is—considering other needs of the people—conceivable.

Based on these figures, the economic boundary conditions can be selected for the design of a waste management system. The GDP in affluent METALAND amounts to US$4 × 10^{11}, or US$40,000 per capita. Thus, it can be assessed that the population of METALAND is willing to spend about US$160 per person and year for proper waste management. Considering that about 600 kg of MSW are produced per capita, the resulting per Mg figure amounts to US$270/Mg. Thus, the overall economic limit for the design of the METALAND waste management system is US$270 per Mg of waste.

Table 5.3
Average MSW generation assumed for METALAND (calculated from Brunner & Fellner, 2007)

Waste Fraction	kg/Capita and Year	Percentage
Biogenic wastes	180	30
Paper and cardboard	150	25
Plastics	90	15
Remainder	54	9.0
Metals	48	8.0
Glass	36	6.0
Bulky waste	30	5.0
Wood	12	2.0
Total MSW	600	100

In addition to the 600 kg of MSW presented in table 5.3, other waste categories such as construction wastes, industrial wastes, and hazardous wastes have to be cared for, too. Their management induces different costs for collection and treatment that are not included here. In the following case study of waste management in METALAND, MSW consisting of household waste and similar waste from trade, business, and commerce is taken into account only. The procedure for designing a waste management system for the other wastes not discussed here is alike.

Establishment of a Comprehensive Physical Knowledge Base on the Premise of MFA/SFA
The main question for the design of waste management systems is as follows: Which information is necessary for a rational, cost-effective design? There is no "handbook" or other rule to start an informed and goal-oriented design process in waste management. The information requirements of waste management are determined by the objectives of waste management, including issues important for protection of mankind and the environment, resource conservation, and aftercare-free waste management. In addition, information from other fields besides waste management such as the anthroposphere and the environment are required. In the following, it is shown that an MFA approach is instrumental to set up a knowledge base for designing waste management systems.

There are three kinds of information needed:

1. Data about *wastes* such as waste generation rate, waste composition with respect to waste goods, substance concentrations in wastes, variations of corresponding flows and concentrations with time and space, and the like. MFA-based

methodologies are available that expand the traditional approaches of direct waste sampling and analysis by indirect methods (Brunner & Ernst, 1986): Mass balances of waste treatment processes are established to determine transfer coefficients, and these transfer coefficients are then used to calculate waste composition by analyzing only one product of the treatment process (Morf & Brunner, 1998). And waste mass flows and compositions can be calculated based on the analysis of production, consumption, and lifetime of products.

2. Data about *waste processes* like the performance of waste collection, treatment, and disposal systems. MFA/SFA-derived mass balances and transfer coefficients are the most appropriate information that characterizes processes in a reproducible and rigid way. Data about emissions or recycling products, without complete substance balances, are of less use. They do not allow cross-checking whether the input into the treatment at the time of the emission measurement corresponds with the actual input of a new plant in another setting. As presented in the section on MFA methodology of chapter 3, transfer coefficients may be input or time dependent. It is crucial that the information that is used for establishing a waste management system consists of robust and up-to-date mass balance data from MFA.

3. *General metabolic data* about METALAND, such as import flows and stocks of substances. This information allows early recognition of future waste amounts deriving from imports and stocks and enables setting of priorities by putting waste management in relation to other economic sectors. It will go beyond traditional waste databases because it will include information that is beyond the boundaries of waste management. To give an example: In a traditional waste database, information about the concentration of lead in a particular waste is included. This information per se does not yet permit assessing the significance of the waste in view of environmental protection or resource conservation. Thus, the information about the concentration of lead in a particular waste must be combined with the mass flow of the waste, with other anthropogenic and geogenic flows and stocks of lead, and with costs for lead production from ores and from recycled wastes. Such a knowledge base:

- allows setting priorities (is lead in a particular waste of overall relevance as a resource or a pollutant?);
- is instrumental for early recognition of management constraints (when will lead concentrations, e.g., in the soil due to atmospheric deposition of airborne emissions, come close to limits set for soil protection?); and
- is indispensable for designing waste management and treatment schemes (which combinations of logistic systems and treatment processes allow cost-effective reuse and will direct lead to environmentally safe final sinks?)

To determine the data that must be included in the knowledge base, the global waste management goals have to be broken down to operational criteria that can be used for design purposes. Table 5.4 presents an example for such a procedure. Note that the selection and weighting of specific and operational goals may vary according to the stakeholders that engage in the design of a waste management system. In addition to this table, information that allows judging the relevance of individual substances for waste management are needed, too. The result of an SFA of national imports and waste flows is given in figure 5.15. It shows the relevance of selected substances and energy for waste management in Switzerland in the 1990s: the amount of Cd and Hg in MSW amounts to close to 50% of the national import, documenting the high relevance of these two heavy metals for waste management. The design of processes such as recycling or incineration and flue gas cleaning has to be focused on these important substances. In contrast, MSW is not an important carrier of carbon or energy. Thus, waste to energy will not be able to play a decisive role in supplying energy, except on a regional basis when large amounts of MSW are concentrated in a few large incinerators that supply regional energy.

Designing Waste Management Systems

Based on the legal framework defining the goals, the economic conditions setting the boundaries for costs, and the actual knowledge base about the regional material and waste flows, a waste management system can be designed. The main elements of such a system are collection, treatment, and final disposal.

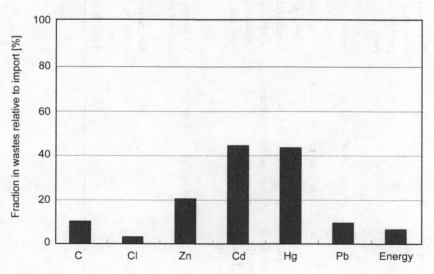

Figure 5.15
Fraction of selected substances and energy in combustible wastes relative to import (= 100%).

Table 5.4
Breaking down waste management goals to operational goals and corresponding indicators (Döberl et al, 2002)

Global Goal	Specific Goal	Operational Goal	Indicators
Protection of mankind and environment	Protection of air quality	Protection of regional air quality	Air quality standards
		Reduction of anthropogenic greenhouse effect	Greenhouse potential, CO_2, CH_4, SF_6
		Protection of ozone layer	Ozone depletion potential, chlorofluorocarbons (CFCs)
	Protection of water quality	Surface water quality standards Groundwater quality standards	Water quality standards, critical water volumes
	Protection of soil quality	No accumulation of pollutants in pedosphere	Soil quality standards
Conservation of resources such as materials, energy, land, and biodiversity	Conservation of land	Minimization of landfill space	Area
	Conservation of natural resources	Minimization of primary resource consumption	Fraction of indicator substances discarded by waste management
		Production of secondary resources	Fraction of total substance turnover replaced by recycling; substance concentrating efficiency (SCE)
	Conservation of energy	Substitution of traditional energy carriers by waste-derived fuel	Fraction of energy content of waste converted to useful energy forms by waste treatment (incineration, landfill gas, etc.)
		Minimization of energy demand for waste management and treatment	Total energy requirement for waste management
		Minimization of energy demand through recycling	Overall reduction of energy demand by recycling

Table 5.4
(continued)

Global Goal	Specific Goal	Operational Goal	Indicators
Aftercare-free waste treatment systems	Clean cycles	Elimination of hazardous constituents before recycling	Fraction of hazardous substances recycled
	Long-term sustainable substance flows to the environment	Minimization of the long-term reactivity of landfilled residues and of the mobility of pollutants within the landfill	Stone-like qualities
		Minimization of pollutants in landfilled residues	
	Final-sink landfills	Allocation of residues to appropriate biogeochemically stable landfill types	Percentage of wastes allocated to appropriate final sinks (e.g., salt mines)
		Mineralized and immobilized residues	Mobility of indicator substances under varying biogeochemical conditions

Collection

The original and first purpose of collection is to ensure public health. With increasing resource consumption, both the value of precious materials as well as the risks from hazardous materials in wastes have been recognized. Thus, collection must aim at (1) collecting all wastes and (2) accumulating specific substances by separately capturing relatively "pure" waste materials. In practice, waste collection systems focus on mixed wastes and on specific waste materials that are identified as crucial for (i) environmental protection or (ii) recycling. An example for (i) is hazardous wastes from private households. This example shows that the design of collection systems needs to be linked to treatment processes that follow collection: Whereas the separate collection of hazardous wastes from households (consisting mainly of waste oil, old paints and solvents, pharmaceuticals, and the like) is ecologically reasonable if wastes are landfilled, this makes no sense for waste management systems based on incineration. Thermal processes make good use of the energy contained in these hazardous materials and convert them highly effectively to harmless CO_2, H_2O, and a few other mineralized species.

Examples for (ii) represent polyethylene terephthalate (PET) bottles, newsprint, glass, and individual metals such as iron and aluminum. For a goal-oriented waste management, the focus on specific substances is more important than to center on functions of materials such as packaging. This is exemplified by an MFA of countrywide plastic flows and stocks in Austria (figure 5.16; the following per capita numbers have been calculated based on the actual number of ~8 million inhabitants): About 330 kg/c·y of plastic materials and precursors is imported, and about 200 kg/c·y is exported; 130 kg/c·y is added to the stock in use (40%) and in landfills (60%); 95% of the 95 kg/c·y of plastic wastes is consumed by end users in private households, the service sector, and for infrastructure. Plastic wastes are landfilled, incinerated, and recycled together with MSW as well as separately collected.

The first conclusion drawn from MFA is that future amounts of plastic wastes will be larger than today. Plastic materials are used as short-lived consumer goods with residence times in the anthroposphere <1 year [e.g., polyethene (PE; also called polyethylene) for packaging] and as long-lived investment goods such as polyvinylchloride (PVC) for construction purposes. In 1994, net plastic addition to stock amounted to 50 kg/c·y. Depending on the service life of the plastic, this amount will turn into waste, too, and will increase the future mass of plastic waste by 50% from 95 to 145 kg/c·y. Taking into account future growth of plastic consumption, the increase in plastic waste is likely to be even larger. This situation is not unique for plastic wastes but can be generalized for all goods with a long residence time and some growth rate of consumption.

The different behavior of plastic materials with short and long residence times is exemplified by table 5.5. Only 20% of PE that is often used for short-lived consumer

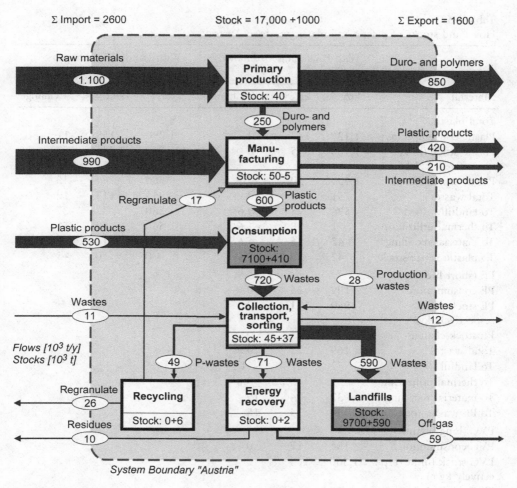

Σ Import = 2600   Stock = 17,000 +1000   Σ Export = 1600

Figure 5.16
Plastic flows (kt/y) and stocks (kt) in Austria in 1994 after introduction of the packaging ordinance (after Fehringer & Brunner, 1997).

goods is added to stock, whereas for the longer-lasting PVC, close to 60% is added to stock. Thus, the amount of PE waste is not only larger in absolute terms than that of PVC (169 vs. 77 kg/c·y) but also in relative-to-consumption terms (85% vs. 51%). It is safe to say that in the future, when the lifetime of PVC in stocks comes to an end, the amount of PVC wastes will grow faster than the amount of PE wastes. This has consequences: Because of the chlorine content of PVC, acid gases are produced during MSW incineration requiring neutralization by alkaline scrubbers. More PVC in MSW input signals higher costs for incineration due to increased

Table 5.5
Flows and stocks of plastic and plastic waste in Austria[a]

Material	With Packaging Ordinance (1994)			With Landfill Ordinance (2004)		
	kt/y	kg/c·y	% Change[b]	kt/y	kg/c·y	% Change[c]
Total plastic						
Plastic consumption	1,127	142	0	1,290	158	11
Plastic stock (in kt respectively kg/c)	7,100	895	0	11,000	1,346	50
Plastic stock change	400	50	0	360	44	−13
Total wastes	751	95	0	953	117	19
To landfill	596	75	−9.6	260	32	−37
To thermal utilization	71	8.9	−1.6	560	69	51
To material recycling	42	5	5.6	130	16	9
To plastic waste stock	42	5	5.6	0	0	−5
PE (short lifetime)						
PE consumption	199	25	0			
PE stock (in kt respectively kg/c)	960	121	0			
PE stock change	40	5	0			
Total wastes	169	21	0			
To landfill	82	10	−41			
To thermal utilization	25	3.2	3.8			
To material recycling	32	4	19			
To PE waste stock	30	4	18			
PVC (long lifetime)						
PVC consumption	134	17	0			
PVC stock (in kt respectively kg/c)	1,300	164	0			
PVC stock change	77	10	0			
Total wastes	69	9	0			
To landfill	63	8	0			
To thermal utilization	4	0.5	−2.8			
To material recycling	1	0	1.4			
To PVC waste stock	0	0	0.0			

[a]Data from Fehringer and Brunner (1997); Bogucka, Kosinska, and Brunner (2008). No data available for PE and PVC for 2004.
[b]Percent change with respect to status without packaging ordinance.
[c]Percent change with respect to status with packaging ordinance.

demand for neutralization agents and to dispose of larger amounts of incineration residues.

The second conclusion resulting from MFA concerns the effectiveness of regulation: The case of plastic wastes presented in table 5.5 exemplifies that the introduction of a stringent packaging ordinance for plastic materials contributes only little to the goals of waste management. Less than 10% of plastic wastes are diverted from landfills, and only 5% are recycled; the remaining 5% is stocked with unknown future toward incineration, cement kilns, or recycling plants. The more important legislative measure is the landfill ordinance (BMUJF, 1996) that—in order to come closer to the goal of aftercare-free landfilling—prohibits the disposal of organic materials in landfills. This measure has the effect of directing large amounts of plastic materials from landfills to incineration. Table 5.5 shows also that the packaging ordinance affects PE and PVC differently: Because of its economic recyclability, about 40% of PE is diverted from landfills and is directed to recycling and thermal treatment in incinerators and cement kilns (with intermediate storage for economic reasons), whereas PVC wastes are not affected.

Figures 5.16 and 5.17exemplify the power of MFA for waste management design: The classical approach focusing on packaging plastics misses the bulk of plastic wastes and leads to inefficient decisions. The MFA view discerns the large stock and hence the future growth of plastic wastes and allows setting of the right priorities regarding plastic waste management. The two stocks of consumption and in landfills are recognized at once, and the potential for plastic

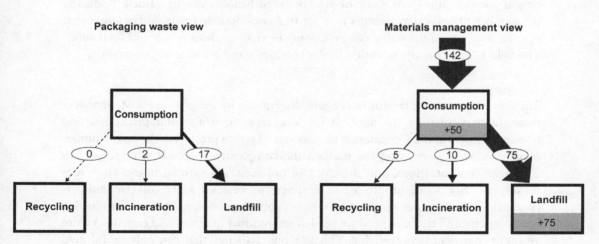

Figure 5.17
Two different views of plastic wastes (in kg/c·y). The MFA view of total plastic consumption and wastes yields information that is instrumental for early recognition of waste amounts and that enables setting of priorities regarding measures to reach waste management goals.

recycling and incineration and for diverting plastic wastes from landfills can better be assessed.

This case study exemplifies the main principle of separate collection; namely, to focus on substances and not on material functions. Goods that are composed of certain substances such as paper consisting of cellulose, glass consisting of silica (silicon dioxide), sheet metal consisting of iron or aluminum, beverage containers made of PET, and so forth are all well suited to be collected as more or less uniform substances. The term "well suited" means that their collection contributes to the goals of waste management, that it is generally profitable (with some variations due to volatile markets), and that it is feasible insofar as the consumer is able to recognize and is willing to collect such items. However, in the future it may be possible that new technologies such as sophisticated mechanical, thermal, or other separation techniques permit discrimination between substances with the same effectiveness at even lower costs. To compare the performance of the competing systems mixed versus separate collection, MFA/SFA represents an excellent backbone for evaluation ensuring the same boundary conditions and mass balance methodology for all new technologies to come. If such evaluations show the economic and ecologic superiority of a combination of mixed collection with sophisticated treatment, consumers will enjoy more convenience at lower costs.

It is important to note that collection and transport accounts for more than half of the costs of waste management (with transportation usually of minor importance) (cf. figure 5.14). Thus, separate collections are major cost factors that can be justified only if corresponding economic or goal-oriented benefits can be gained. Reducing the number of collection systems may set free considerable financial resources that may be used for additional waste processing in view of clean cycles and final sinks. The field is open for optimization under economic and waste goal constraints.

Treatment
The objective of waste treatment is again determined by the goals of waste management, in particular by the demand for long-term environmental protection and resource conservation. Considering the amount of goods produced, the large number of new substances entering the market, the complexity of goods with respect to substance concentrations, and the stock of old materials turning into wastes, it is no surprise that the input into waste treatment processes is highly diverse and heterogeneous. Waste processing must consist of robust technology that can handle this heterogeneity and that can produce predefined product qualities and emissions from an input that is not controllable. This is a big challenge that can only be fulfilled by sophisticated technology on a fairly large scale. MFA-based mass balances and transfer coefficients are instrumental to characterize treatment processes and to evaluate if they can fulfill the stringent requirements.

Table 5.6
End use and stocks of plastics and plastic additives in 1994[a]

Material	Total End Use kg/c.y	Packaging Plastic kg/c.y	Percentage[b]	Total Stock kg/c
Plastics	140	25	18	901
Softeners	1.8	0.024	1.4	17
Ba/Cd stabilizers	0.034	0.000024	0.07	0.3
Pb stabilizers	0.23	0.00024	0.11	2.2
Fire retardants	0.29	0	0	2.7

[a]MFA data from Fehringer and Brunner (1997).
[b]Fraction in packaging as percentage of total materials in all plastics.

In the following, the example of plastic waste is used to present design considerations for waste treatment. Plastic wastes are composed of a matrix substance, the polymer (PE, PVC, PET, etc.), and additives, such as softeners (e.g., phthalates), flame retardants (e.g., polybrominated diphenyl ethers), and stabilizers (e.g., heavy metals such as lead, zinc, and—in old stocks—cadmium). Average annual turnovers of additives in total plastics and in packaging plastics are presented in table 5.6. Because of constant progress in technology as well as legislation concerning hazardous materials, the materials used as polymers and additives are constantly changing. The basket of additives displayed in table 5.6 is representative for plastic manufacturing in 1994. Hence, concerning long-lived plastic stock with a lifetime of 10–20 years, a similar composition can be expected for today's plastic wastes. The table shows that there are only small amounts of hazardous additives in short-lived packaging materials and that the main risk of harmful materials is associated with the long-lived plastic stock.

In principle, plastic wastes can be recycled as plastic materials or other feedstock, can be used to generate energy, or can be landfilled. Taking into account plastic composition derived from table 5.6 and the goals of waste management, it appears that two distinctly different problems arise for waste management: On one hand there exist "clean" and "uniform" plastic fractions such as PET bottles or agricultural PE sheets for recycling. On the other hand, hazardous fractions of plastic wastes such as highly stabilized PVC have to be disposed of safely. Thus, plastic waste management requires tailor-made methodologies.

To reach the goal "protection of mankind and the environment," hazardous organic materials in plastic wastes have to be mineralized. MSW incinerators, cement kilns, as well as coal-fired power plants are all well suited for such mineralization. However, if inorganic stabilizers are highly concentrated, the latter two processes might not be appropriate because the products cement and synthetic gypsum may be spoiled by heavy metals.

Concerning the goal of "resource conservation," direct recycling of separately collected, clean fractions as discussed before is often economic. In addition, incineration offers also a path to resource conservation. Besides recovering the energy content of the plastic wastes—20% if electricity is produced and 70% for combined power and heat generation—materials can be recycled, too. MFA studies of incinerators (Morf & Brunner, 1998) show that volatile heavy metals fired with plastics can be transferred to filter ash, hence offering the possibility of concentrating and reusing these metals (cf. figure 5.18). MSW contains about 40% of the regional import of Cd, and 85% of Cd entering an MSW incinerator can be recovered from fly ash. Thus, about one third of the regional Cd import could be supplied by waste management. To make such a recycling scheme economically attractive, long-term strategies are necessary, accumulating filter residues from many incinerators for long time periods until economy of scale allows competitive recovery from such intermediate deposits.

The example of Cd shows how waste incineration can be used for both environmental protection as well as recycling. To fulfill both functions, the nonreusable residues of incineration such as bottom ash or filter residues must be transformed into "final storage quality," a material that cannot be mobilized in a landfill environment (Baccini, 1989). The example offers also new opportunities for alternative waste management design (figure 5.18): If substances such as iron, copper, aluminum, cadmium, and so forth can be recovered from incineration residues, it may be beneficial to use the MSW collection system as a common path for valuable substances and to add waste electric and electronic equipment, shredder residues of end-of-life vehicles, and other such wastes to this path. As mentioned earlier, MF-based evaluation is instrumental in comparing whether such a waste management design is economically and goal-wise superior to other strategies.

With respect to the goal "aftercare-free waste management," plastic wastes pose a special challenge. On one hand, when landfilled they release endocrine substances such as phthalates, bisphenol A, and brominated flame retardants for long time periods. On the other hand, when recycled, the hazardous additives are not eliminated from the cycle, thus posing a risk for the next user and for subsequent waste management (figure 5.19). Thus, processes such as incineration are required to eliminate potential hazards inherent to stabilized plastic materials.

Final Disposal
Although plastic recycling can minimize resource consumption and hence the need for sinks, the loss of substances during recycling cannot be completely prevented because of thermodynamic reasons. Thus, also a waste management scheme that is

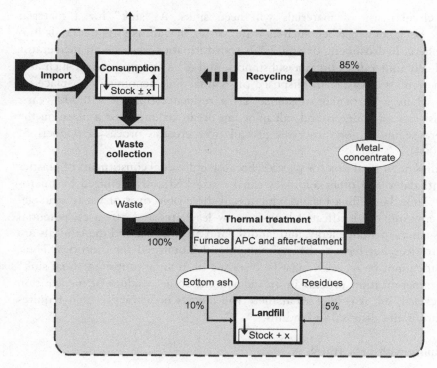

Figure 5.18
Cadmium recycling by MSW incineration. APC, air pollution control.

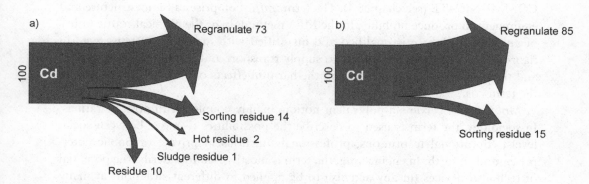

Figure 5.19
SFA of cadmium for recycling of (a) pure and (b) mixed plastic wastes: The goal of clean cycles is not achieved yet with 73% respectively 85% of Cd being transferred into recycling products (Fehringer & Brunner, 1997).

based on circular use of materials will need sinks. A "sink" for a material flow can be (1) a conveyor belt such as water or air that transports materials into the atmosphere, hydrosphere, or soil, (2) a transformation such as an incinerator that completely mineralizes organic substances, and (3) a storage process such as a landfill where plastic wastes are disposed of. A final sink denominates a place on the planet where a particular substance has a residence time of >10,000 years. It may consist of an underground salt mine, an ocean sediment, or a place on the globe where sedimentation processes prevail over erosion processes (Döberl & Brunner, 2004).

A landfill is not a final sink for plastics, because individual constituents of plastics such as phthalates and other additives can be attacked and mobilized by microorganisms. Thus, landfilling plastics violates all three objectives of waste management: the environment is affected long-term by leachates and off-gases; potential resources such as fuel oil, energy, and landfill land are wasted; and the landfills are not aftercare free but have to be maintained and monitored for centuries. Thus, plastics that cannot be recycled must be disposed of in an appropriate "final sink" such as a mineralization process in an incinerator. Some residues of incineration may be recycled, but it is likely that there will always be a fraction that requires immobilization and disposal in landfills.

More Mobility with Less Traffic

Mobility as a Benefit, Traffic as a Cost
Mobility describes all benefits elaborated by the activity TO TRANSPORT& COMMUNICATE (see chapter 4). The term *traffic* comprises all infrastructure and equipment to produce mobility. In the MFA methodology, the physical traffic utility of an anthroposphere is qualified and quantified with metabolic systems (e.g., in figure 4.26). The economic efforts to supply transport and communication services and the measures to mitigate or omit the harmful effects on life are summarized in the term *cost*.

Mobility has become a polyvalent notion, mainly coupled with positive ratings. In sociology, the term is used to describe the possibilities to move between social levels, educational institutions, professional functions in private or public enterprises, and so forth. In engineering, the term is mostly used to describe the potential of technical devices (in any activity) to be applied in different situations at many sites. Practically in all cases, an increase of individual and group mobility is considered as a positive development, a benefit for the individual and the society. A decrease of mobility (e.g., due to illness, handicaps, old age, poverty, lack of transport means, etc.) is considered as a loss of life quality. It is evident that mobility cannot be described solely with metabolic properties. It is a cultural property comprising a complex combination of sociological, psychological, and economic variables within a given system. On the contrary, traffic is less complex because most of its properties can be qualified and quantified within metabolic systems, economics included.

The Mobile Phone Story

The mobile or cellular phone, a relatively new device in communication, is an excellent illustration for the essence of mobility. Its predecessors were "handie-talkies" and "walkie-talkies," which stayed for decades in a marginal position to the dominance of the fixed net telephone. The rapid global spread of mobile phones started in the middle of the 1990s. Within less than 15 years it reached a quantity of approximately 6 billion phones (Reller, 2009), or roughly one phone per capita produced worldwide. The growth is not finished yet. The rapid technological innovations and extensions reduce the average lifetime of such a device to less than 5 years. At present, it is a relatively cheap device. It belongs to the few ubiquitous commodities (beside a famous soft drink) that are already available in the poorest and remotest quarters of any developing country. The internationally engaged providers of wireless communication systems invest in equipment around the globe. First comes, first cashes. The high benefit/cost ratio for the economic mobility of small entrepreneurs in developing countries is evident. Most of them could not operate as they do now, because there is no parallel or reasonable alternative on a wire grid. The huge extension of social communications by mobile phones can be observed in any public situation. The device is omnipresent. Possible negative health effects (from transmitters and phones) are still a controversial topic. The first metabolic studies estimate the consequences on the global management of certain metals such as indium and lithium (Reller, 2009). It is too early to evaluate seriously the cultural implications of this young development. However, mobile phones show the highly attractive role of a device that can improve human mobility, independent of culture. One who has a mobile phone is "in," meaning socially connected, accessible from anywhere anytime. One who does not have it is "out," isolated from the tribe of electronic nipples. In developed countries it is—like computers and the Internet— not yet a consequent replacement of other transport systems for information. First of all, it is an additive equipment, received as a welcome increase of mobility.

In urban systems, traffic has become, more and more, a negative connotation. It is the other side of the coin of the activity TO TRANSPORT&COMMUNICATE. During the past three decades, numerous studies on the implications of traffic have come to the following conclusions (e.g., SRU, 2005):

• Road transport is still, in spite of significant improvements, a heavy emitter of noise and air pollution. Air transports contribute more and more to the overall emission.

• The transport of persons and freight in automotive vehicles on the road and on airplanes is still less efficient, from a metabolic point of view, than the transport on rails and water.

• The corresponding health risks in areas of high traffic densities are too high. There are still too many accidents on roads causing injuries and death.

• Especially children and old people are mostly disadvantaged, not only due to higher health risks, but also due to their decrease in mobility within their local environment.

• The contribution of combustion engines in transport vehicles (primarily on roads and in the air) to the global climate change is at least 20% and is still growing worldwide. The increase in energy efficiency of vehicles has been compensated by higher transport distances per vehicle and year and by a growing stock of moving vehicles worldwide.

• Transport networks had negative ecological effects on habitats of various types, due to cutting ecosystems in too small parts, reducing thereby the biodiversity. They reduced also, within a few decades, the aesthetic qualities of cultural landscapes that evolved over centuries.

• It is possible to have more mobility with less traffic. One has to accept the strong interdependence of settlement structure, mobility, and traffic. A traffic policy that concentrates on solving acute traffic problems (e.g., traffic jams, etc.) but is separated from a long-term–oriented urban development policy is a failed undertaking.

The main conclusion of these studies, simplified in a graph (figure 5.20), is as follows: Urban systems (as defined in chapter 1 and applied in chapter 4 for META-LAND) have gone through a development (within the past 150 years) in which the benefit/cost curve for the traffic/mobility relation passed a maximum for a majority of inhabitants. Those who cannot afford to choose their residential and work sites in a traffic-poor environment (meaning low impacts at sites becoming more and more exclusive in any urban system; meaning also no time losses in commuting due to traffic jams), suffer from more traffic, losing mobility.

In this context, mobility of persons can be described as the sum of all possibilities to have access to the culturally defined services of an urban system.

The interdependence between mobility and traffic is a complex topic. The activity TO TRANSPORT&COMMUNICATE connects the other three activities with its networks and vehicles. The networks (roads, rails, water canals) change the landscape on a regional scale and give it a "new morphological pattern." New anthropogenic flows are introduced changing the "physiologic pattern" of large areas. For a grown urban system, two questions are to be answered:

1. With which morphological and metabolic properties must an urban system be equipped in order to have an optimal mobility/traffic ratio?

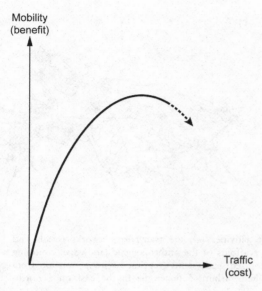

Figure 5.20
Schematic relation between mobility (giving a benefit) and traffic (creating costs) (after Frey, 1996). Many urban systems have already crossed the peak and lose mobility while increasing traffic.

2. How can an existing urban system with an unfavorable mobility/traffic ratio be reconstructed to reach more mobility with less traffic?

Analysis and Evaluation of the Mobility of a Region and Its Traffic Status

Choosing a Study Region (after Baccini & Oswald, 1998)
The region chosen lies within the Swiss Lowlands as part of the urban system Switzerland (figure 5.21). Its actual settlement structure is a combination of (1) late medieval founding of small cities (ca. 800–700 years BP) connected with a few terrestrial connections and one major waterway; (2) a traffic extension with railways established in the nineteenth century, connecting the region to the continent in all directions; (3) two thruway crossings, built in the twentieth century, of European importance. The region, having a total area of approximately 200 km², went through a landscape metamorphosis during the past 150 years. The mobility of the inhabitants grew continually, mainly due to the fact that larger national and international settlements chose their region as a transit land. The region's transport network became more and more sophisticated, and their access to national and international centers grew. Also, more and more enterprises, especially from the logistics branch, chose the region as a national or international trade center. At the end of the twentieth century, the inhabitants (with a

a)   b)

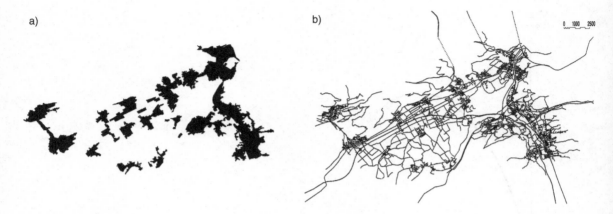

Figure 5.21
The study region Synoikos: the settled sites of activities (a); the transport networks (roads and rails) (b). The two when superimposed give the scaffolds of the anthropogenic landscape, confining and surrounding the terrestrial and aquatic (colonized) ecosystems (reprinted with permission from Baccini & Oswald, 1998). The three towns Olten-Aarburg-Zofingen lie in the east on a north-south strip.

population density of 500 c/km²) considered themselves as privileged urban people having the best mobility, although more and more inhabitants started to suffer from traffic. They were confronted with the special phenomenon of the so-called agglomerations in the urban development. More and more people moved to quieter areas with affordable real estate, although with the same craving for high mobility, increasing thereby the local and regional traffic on the roads again, reducing eventually the life quality and mobility of their own residential quarters. At the same time, the gentrification of the old densely populated centers led to traffic-free zones, with more shopping opportunities and less residential areas, which became very expensive. A typical *netzstadt* was formed (see chapter 1), experiencing a decline of mobility due to more traffic. The transport system got caught in a vicious circle (Oswald & Baccini, 2003). The morphological and metabolic properties of this urban system are similar to those of METALAND (see chapter 4).

Evaluation Method and Results (after Blaser & Redle, 1998)
The goal is to increase the mobility, or at least keep it on a high level, and to reduce the traffic.

The following working hypotheses were chosen:

• Find an optimal mixture of activities at one location.
• Shorten the connections between the sites of activities.

The following "traffic indicators" were chosen:

- Density of activity sites (d_{AS})
- Distribution of activities
- Diversity of connections
- Road areas
- Use of gravel
- Energy consumption.

Because of the relatively low quantitative contribution of other transport networks to the overall flows and stocks (e.g., for water, electricity, telephones; see also figure 4.26), these traffic systems are not considered here.

The density of activity sites (d_{AS}) can be quantified with the sum of inhabitants plus places of work per settlement areas (AS/km²). In figure 5.22, a graphical

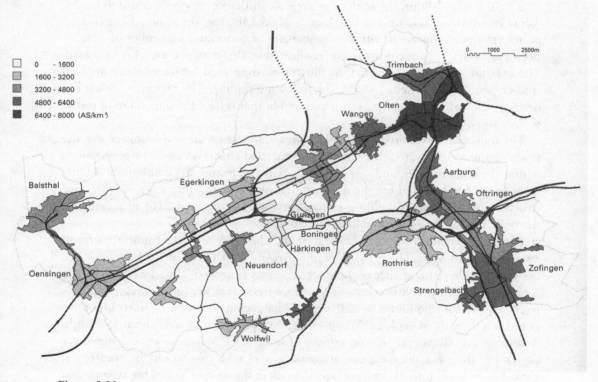

Figure 5.22
Density of activity sites (d_{AS}) of Synoikos. AS, inhabitants and workplaces (reprinted with permission from Baccini & Oswald, 1998).

illustration differentiates the character of the settlement areas given in figure 5.21. The settlements on the east (the strip Olten-Aarburg Zofingen) shows densities with a factor 2 and 3 times higher than those in the middle and the west of the region. If one studies the daily commuting by comparing the ratio of workplaces (WP) to inhabitants (Inh) in each community (w_i = WP/Inh), the traffic picture gets an additional facet, illustrating the spatial distribution of activities, together with d_{AS}. The lower the ratio w_i, the higher is the daily flow of commuters leaving the community. The higher this ratio, the more commuters enter daily the community. The daily regional "pulse of commuting" is in most cases the most important producer of traffic and not the transit traffic. The activity density is also a first measure of the regional freight traffic, excluding the freight transport in transit.

The diversity of connections is given in figure 5.23. The mean daily residence time of an inhabitant in traffic (sum of all moves) was, at the time period of the investigation, approximately 1.4 hours. Figure 5.23a shows the connections in the order of the admitted maximum velocities. The connections for automobiles dominate (≥ 50 km/h). Within the settlement areas, a sufficient supply of connections for "slow motions" is lacking. The graph of figure 5.23b gives the regional network of public transport, indicated with the frequencies of connections (number of connections per hour). Settlement areas not connected to this network are also indicated. The combination of the two graphs illustrates, supported by many more detailed studies (see Blaser & Redle, 1998), the following finding: The region Synoikos has developed a traffic system for person and freight that is based mainly on road traffic with automobiles.

The indicator "road area" is based on area statistics of the communities. For the study region Synoikos, roughly half of the total road area is occupied by connections within the communities, and one fourth is used connecting the communities. Only 10% of the total is used for the national and international thruways. It is the fine structure of distribution within the road network (see figure 5.21b) that demands the bulk area of the total road area (ca. 90 m^2/c).

For the dominant transport network "road," the stock of gravel (indicator gravel) amounts to 140 Mg/c. The network is still growing with a flow of 1.4 Mg/c.y, or 1% per year. For the maintenance of the stock, 1 Mg/c.y is needed. The sum of 2.4 Mg/c.y is provided by roughly 90% primary (or geogenic) gravel, because the recycling flow from dismantling buildings is still small. The region disposes of autochthonous gravel stocks with an exploitable capacity for road construction of about 150 Mg/c. Assuming a moderate population growth of less than 1% and the same demand per capita for the next decades, the geogenic gravel reservoir would be transferred almost completely into the anthropogenic stock in the second half of the twenty-first century. This is, from a metabolic point of view, not a major problem (see also the case study in this chapter on urban mining and see Redle & Baccini, 1998).

a)

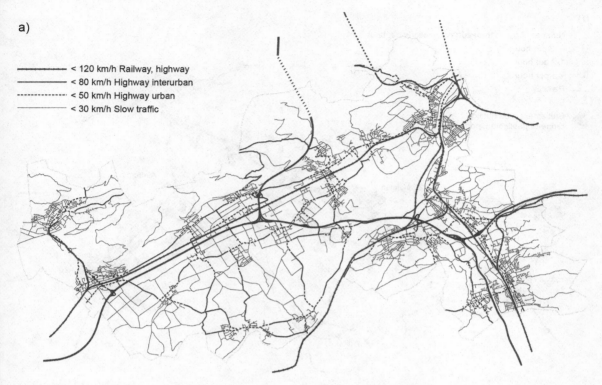

──────── < 120 km/h Railway, highway
──────── < 80 km/h Highway interurban
---------- < 50 km/h Highway urban
──────── < 30 km/h Slow traffic

Figure 5.23
The regional connections of Synoikos. (a) Connections given in the order of permitted velocities. (b) Connections of the public transport system given in the order of connection frequencies and of accessibility (within 350 m walking distance from residential and work sites) (reprinted with permission from Baccini & Oswald, 1998).

A major metabolic problem is the energy consumption of traffic. For the study region, five traffic categories were chosen to differentiate the energy consumption (based on statistical data); namely,

- Public transport of persons on rails
- Public transport of persons on roads
- Private transport of persons on roads
- Freight transport on rails
- Freight transport on roads

In figure 5.24, the results are given in km/c.y for persons and Mg of freight in the left pie chart and in energy consumption units (MJ/c.y) in the right pie chart. For a comparison, the "person kilometers" were transformed to "ton kilometers"

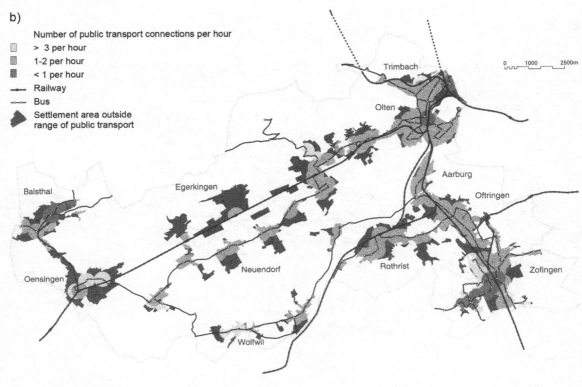

Figure 5.23
(continued)

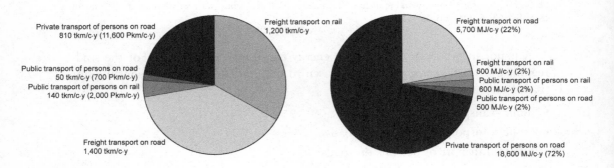

Figure 5.24
Traffic in the region Synoikos given as performance in t.km/c.y in the left pie chart and as energy consumption in MJ/c.y in the right pie chart. For the transformation of "person-km" into "ton-km," a mean weight of 70 kg/person was assumed (reprinted with permission from Baccini & Oswald, 1998).

assuming a medium weight per person of 70 kg. Thus, the traffic performance (left pie chart) is presented as a "mass-distance" unit per capita and year. This synopsis illustrates the following metabolic idiosyncrasy of an urban system: The traffic performance is dominated by the transport of freight (about three fourths of the total of 3600 t.km/c.y). On the contrary, the energy consumption (right pie chart) shows a dominance of the person transport (about three fourths of the total of 26 GJ/c.y). Ninety percent of the total person transport energy is used by private cars on the road. It must be underlined that these figures do not contain the through traffic, and the air and water traffic of the inhabitants leaving and the travelers passing the region. The comparison is restricted to the endogenous traffic activities of the region.

For the transport of persons and freight on the road, the energy source is practically 100% fossil fuel. On rails, the energy source is 100% electrical power (approximately 60% from hydroelectric power stations and 40% from nuclear power). A comparison of the transport efficiency, defined as MJ/t-km, MJ/person-km, is given in figure 5.25. It shows that public transport on the rails is more efficient by a factor 5 times that of private transport on the roads. For freight, the railway is roughly an order of magnitude more efficient than the road. An efficiency comparison between persons and freight makes no sense, because persons are not transported

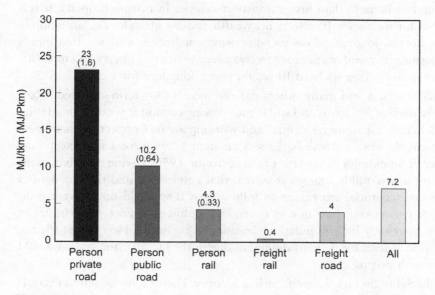

Figure 5.25
Comparison of the energy efficiency (in MJ/person-kilometer, respectively in MJ/ton-kilometer) for the traffic in the region Synoikos (1 ton = 1 Mg) (reprinted with permission from Baccini & Oswald, 1998).

densely packed as, for example, glass bottles (although there are some similarities in underground trains during rush hours).

The Status Quo: Summary and Conclusions
The Synoikos region, representing tens of thousands of similar urban systems in the developed world, has evolved into a traffic system for persons and freight that is mainly run by transport vehicles moving on roads driven by combustion engines with fossil fuels. After 50 years of experience with continually growing traffic, mobility starts to decrease for a majority of the inhabitants. Although there are more energy- and material-efficient alternatives available with less environmental impacts, this type of traffic system is still dominant. There are several reasons for this:

• Energy for transportation has become cheaper and cheaper during the twentieth century in relation to the available average net income of private households (less than 5%). There is no economic pressure yet to change or adapt the system.

• Private individual transport with automobiles became one of the relevant pillars of the settlement policy of most nations in the twentieth century. During the same period, automobile production and maintenance developed to a powerful economic branch. All regions that depend strongly on this branch are reluctant to promote alternative transport systems.

• The automobile is more than just a transport vehicle. In private property, it is a status symbol for its owner. It reflects his wealth and his lifestyle. The automobile replaced the horse and gives the owner (supposing one lives in a society that builds roads and permits its members to move freely) a feeling of great liberty and mobility. Today, Shakespeare's King Richard III would give a kingdom for a car.

• The Synoikos region and many others did not have a long-term–oriented urban planning. The driving force was and still is the growing economic wealth stimulating territorial demands for more residential and working areas. Connecting them with transport networks was and still is the second step. There were a few exceptions (e.g., the city of Stockholm in the late 1930s up to the 1950s having an urban planning based on a new public transport system that comes first and thereby defines the locations of territorial expansion to follow), but they could not prevent individual private person transport in cars overtaking public transport. Only in urban centers with extremely high population densities (e.g., Tokyo, Hong Kong) is the public transport system for persons dominant due to the fact that the system would break down with private cars.

• Regions like Synoikos lack a specific urban identity. Their radius of action exceeds largely the competence radius of their political institutions (communities, counties, departments). These institutions still operate as if their inhabitants live in separated villages. It is the "agglomeration syndrome."

Designing a Regional Reconstruction for More Mobility with Less Traffic

Premises and Guidelines

To reach and keep optimal mobility with less traffic, the following urban quality is chosen:

All relevant urban services, such as markets, schools, hospitals, cultural centers, recreational facilities, and so forth, are accessible, within half an hour, starting from residential sites, with public transport means or by foot or bicycle.

For the reconstruction of the given regional settlement pattern, the following morphological guidelines are chosen (see figure 5.26):

1. The smallest settlement units are modules (M). A module comprises facilities for the basic daily needs (e.g., supplies of groceries, health care, primary schools). All these facilities are able to be reached by foot within 10 minutes. Thus, a module has a mean diameter of roughly 350 m and an area of 0.4 km². A module is connected with public transport means to at least one center (C). A module has a population of at least 2000 inhabitants, offering 0.3 workplaces per capita. The

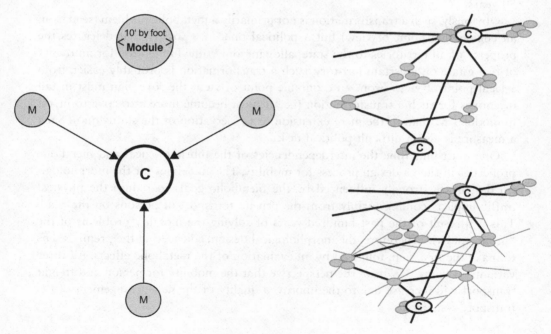

Figure 5.26
Settlement diagrams for the structure of a reconstructed region, a network with modules (M) and centers (C) (reprinted with permission from Baccini & Oswald, 1998).

corresponding density of activity sites (d_{AS}) (see figure 5.22) amounts to at least 7000. For a population between 100,000 and 130,000 in the region Synoikos in the year 2050, at least 50 modules are needed.

2. Centers (C) comprise facilities with larger urban services, not needed daily, of regional importance. They form also traffic nodes with connections to fast transit lines.

3. Between modules M there are several connections for slow-motion traffic (pedestrians, bicyclers) with a high recreational value, but only one connection for motorized vehicles.

The Morphological Design

Three variants to the status quo, each following the above-sketched guidelines, are presented in figure 5.27. Each variant chooses centers (C) as "urban developing areas." There are modules within this area and outside of them. Variant A chooses three existing centers of the status quo, variant B operates with two new centers, and variant C concentrates its center along the north–south axis in the region's east. Such a reconstruction should take approximately two generations, or at least 50 years.

Obviously, such a transformation is not primarily a metabolic problem (seen from an engineering point of view) but a political one. In a society that delegates the property of all territories to the state, allowing only time-limited use for an owner of real estate on a certain territory, such a transformation is probably easier, from a legal point of view. From an economic point of view, the sovereign must install incentives for such a transformation (i.e., it must become more attractive to invest in modules M than in the mere extension or conservation of the status quo). Such a measure is also a difficult political task.

One must underline the interdependencies of the morphological and metabolic properties in such a design process for mobility. The guidelines for the morphological design are strongly influenced by the metabolic goals to reduce the physical traffic impacts coming mainly from the private transport of persons on the road. This is a result of the past hundred years of solving the mobility problems of the "urban sprawl." Therefore, the morphological design, considering the premises, has to make the first step, followed by an evaluation of the metabolic effects. All three variants in figure 5.27 give the perspective that the mobility for person and freight transport will increase due to the improved quality of the new arrangement of the transport systems.

Metabolic Effects

The total road area of 7.4 km^2 (or approximately 90 m^2/c) gets reduced by 25%. The concept presented in figure 5.26 cuts the intermodular road areas by

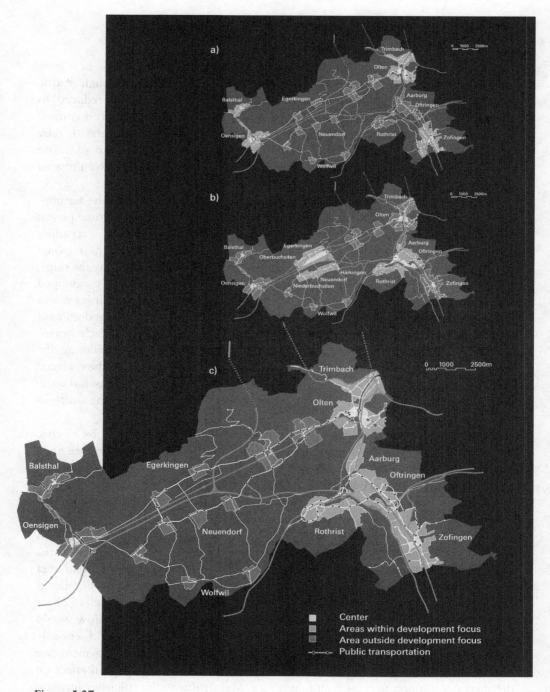

Figure 5.27
Morphological design variants for the Synoikos region: (a) with three existing centers (see figure 5.21); (b) with two new centers; (c) with one center on one axis in the east (reprinted with permission from Baccini & Oswald, 1998).

a factor of 2 and the streets within the modules by a factor of 1.5. Less area is needed for parking lots. Only the thruway areas for the through traffic keep their sizes. As a consequence, the geogenic gravel flows are reduced to almost zero because the demand can be satisfied from the recycled gravel coming from the out-of-function roads and serving now as gravel stocks. The thereby formed pits can be refilled with inert inorganic materials, coming out of waste management, finally covered with soil materials, to be integrated as recultivated territory.

The most important change will take place in the energy household. The assumptions chosen for the transport vehicles are given in table 5.7. Within the time period of 50 years, the vehicles on the road get equipped with electric motors. According to this scenario, the overall mobility is not reduced. In other words, the regional traffic performance (given in t.km/c.y, see also figure 5.24) stays practically the same, partly with reduced velocities. The traffic is rearranged and technically equipped with energy-saving vehicles that cause much less environmental impact. It is evident that such a reconstruction is coupled with a change of behavior, as the dominant status role of the combustion engine is discharged (a fate that the "vehicle horse" already experienced 100 years ago). One has to keep in mind, considering the "inertia of the human condition," that such an important device has to be replaced by other status symbols coupled with mobility. The automobile with classical combustion engine could become the vehicle of motor sport, similar to the horse in equestrian sport.

Based on today's technology, a substitution of combustion engines by electric motors would increase the demand for electricity and for the material copper. Such a scenario challenges the energy policy with regard to the production of electrical power. For the region Synoikos, the scenario of table 5.7 increases the electrical energy consumption from 1 GJ/c.y to 8 GJ/c.y. An illustration for an additional supply by photovoltaic cells is given: If we assume a net production within the region of 370 MJ/m^2.y (10% of solar input of 3700 MJ/m^2.y), the region would have to install 20 m^2/c photovoltaic cells, or other solar e-producers (be it in its own area or as investors in other regions). This is an area corresponding with the road area to be reduced. With regard to copper, the additional demand would be significant because the average copper content of current combustion engine vehicles (25 kg Cu) is doubled or tripled (ISI, 2010). The effect on the overall copper flow would result in an increase of 20% to 30% within the coming decades and would seriously challenge the regional and, if copied worldwide, the global copper management (see also the earlier case study on urban mining in this chapter). The overall effect on the energy consumption is given in figure 5.28. Mainly due to the efficiency increase of the transport of persons on the road, the energy consumption is reduced by a factor of 3.

Table 5.7
Assumptions regarding the energy consumption of transport vehicles[a]

Transport	Performance	Specific Energy Consumption
Freight on roads	Half of the current freight flow is moved from roads to rails	The large lorries get substituted by small lorries, driven by electric motors. Therefore, no net increase of energy efficiency, however a decrease of traffic noise and air pollution.
Freight on rails	Increase of transport capacity	no increase of efficiency
Persons, private on roads	30% is moved to public transport systems, 20% is realized by foot and bicycle	Because of electric vehicles, the efficiency is increased by a factor of 3
Persons, public transport, roads	15% of the current private transport is taken over	A better capacity utilization increases the efficiency by a factor of 2
Persons, public transport, rails	15% of the current private transport is taken over	A better capacity utilization increases the efficiency by a factor of 1.5

[a]After Blaser and Redle (1998).

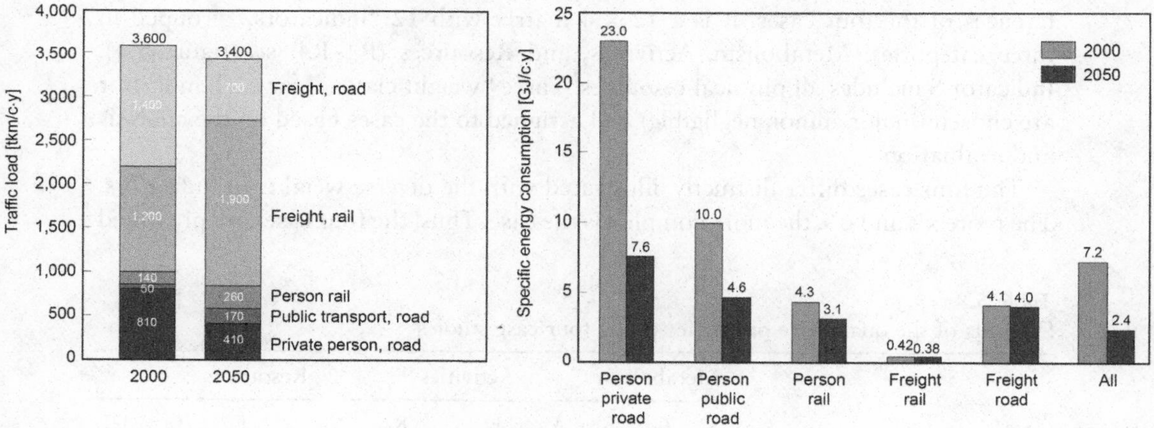

Figure 5.28
The effects on energy consumption due to the traffic-reduction scenario 2050 (reprinted with permission from Baccini & Oswald, 1998).

Concluding Remarks

The mobility study tackles the relatively most complex problem with regard to the four design cases presented. The consequences of a significant traffic reduction challenge the whole anthropogenic system. There is a fundamental difference between the design of new urban systems and the design for a reconstruction of an existing urban system. The merit of such design studies is a new view into urbanism gathering more intellects with transdisciplinary competencies to develop large-scale–oriented urban projects.

Summary and Conclusions

Conclusions Based on a Synoptic Analysis

All four cases operate with a set of normative guidelines for the design of metabolic scenarios presented in the first case on phosphorus; namely, to operate

1. on a long-term perspective (decades to a century);
2. for large-sized areas (ten of thousands to hundreds of thousands of square kilometers);
3. with a sound knowledge about the sizes and dynamics of the regional and global material stocks;
4. in the context of a development strategy for urban systems; and
5. by considering the idiosyncrasies of every region (tailor-made design).

A synopsis in table 5.8 gives a qualitative pattern regarding similarities and differences of the four cases. It is a 12 × 4 matrix with 12 "indicators," grouped in three categories: Metabolism, Activities, and Resources (R1–R4; see figure 5.1). Indicator S includes all physical resources. Three "weight classes" for each indicator are chosen (major, minor, negligible) and assigned to the cases based on the analysis and evaluation.

The four cases differ distinctly, illustrated with the diverse weights of indicators. The more x's and o's, the more complex is the case. Thus, the four cases are presented

Table 5.8
Synopsis of the qualitative properties of the four case studies

Case	Metabolism				Activities				Resources			
	S	G	E	A	N	C	R	T	Le	Ec	Te	UL
Phosphorus management	x	o	n	x	x	o	n	n	o	o	x	o
Urban mining (copper)	x	o	o	o	n	n	x	x	o	o	x	o
Waste management (plastics)	x	x	o	n	o	x	o	o	x	o	x	n
Mobility and traffic	o	x	x	x	o	n	x	x	o	x	x	x

S, substance; G, good; E, energy; A, area (territory); N, TO NOURISH: C, TO CLEAN; R, TO RESIDE&WORK; T, TO TRANSPORT&COMMUNICATE; Le, constitution and law; Ec, economy (business/trade); T, technology; UL, land use/landscape; x, indicator of major weight; o, indicator of minor weight; n, negligible weight.

in the order of increasing complexity. It is not surprising that the two indicators "substance" and "technology" are dominant, looking at their strong position in all four cases. In metabolic studies, based on MFA/SFA, they must be omnipresent. The diverse contribution of the activities to the individual cases follows from the detailed analysis and evaluations given in chapter 4. For each design case presented, the set of key parameters differs significantly.

The phosphorus metabolism is mainly driven by the activity TO NOURISH. The essential element phosphorus is bound chemically within the phosphate molecule, be it in geogenic or anthropogenic materials. Its main good as anthropogenic carrier is as fertilizer for agricultural production. For physiologic reasons, P cannot be substituted by other substances. Its main sink is the pedosphere. Worldwide, the actual phosphorus management has a relatively low efficiency, producing high RSE values (i.e., the phosphate, from concentrated geogenic apatite stocks, is continually dispersed and diluted on large soil areas). After one or two seasons, most of it is not available anymore for further agricultural use. The negative impact on the surface waters (eutrophication) is directly proportional to the growing P stocks in the pedosphere. In most countries, the only legal measures to control P flows stem from water protection laws. It becomes evident that the key process for an efficient and effective P management is agriculture. Any recycling efforts from the residuals of food consumption could not solve the problems of upcoming P scarcity. Therefore, the design recommendation focuses mainly on "new agricultural technologies." Because the P scarcity signals are very weak yet, the economic weight of P fertilizer costs in food production and the energy demand are very small; there is hardly any political pressure to design now a "sustainable P management" based on a new technology. There is not enough reliable data today to predict future P scarcity, thus P scarcity is highly speculative. Decisions must take into account this fact. There can be tremendous economic drawbacks if wrong decisions are taken. A political postponement of a strategic P management can be considered as a risky attitude because a new technology needs at least 50 years until it is installed and functioning on a global scale

Urban mining is urban geology combined with urban engineering. It is the renaissance of a procedure well known from nonaffluent societies; namely, to reuse the material in anthropogenic stocks. In developed countries, within the past 100 to 150 years, the high growth rate has led to a multiplication of these stocks per capita and a much more complex chemical composition. Exemplified with copper, it becomes clear that we have, during this growth period, lost track of the quantity and quality of these stocks. Urban geology explores these stocks. In the future, supported with dynamic MFA models, the generation and management of these stocks could be handled in a much more efficient way. Almost all copper is installed as copper metal (Cu^0) in a broad variety of goods and functions, primarily in buildings and infrastructure, secondarily in commodities of shorter residence times. In developed countries, the anthropogenic copper stocks have reached relatively high values and serve already as important copper sources to be exploited by urban engineering. From a physical and chemical point of view, there is not a "growing scarcity" of copper. On the contrary, the chemical transformation of Cu^{2+} components in ores to Cu metal by an endothermic reduction makes anthropogenic copper stocks, from an energetic and economic point of view, more valuable. In comparison with the phosphorus case, there are two main differences: (1) copper, although physiologically essential for life in trace amounts, is in its "technological role" replaceable by other materials, practically in all functions; (2) the design of urban mining (e.g., for copper) is not oriented toward a new technology but toward the interface consumption–waste management (see the following discussion on waste management). It is the politically based legal framing of the waste management quality. If waste management allows the dilution of copper in mixed wastes to be deposited in reactor landfills, the optimal handling of copper in anthropogenic stocks (e.g., in scenarios like "increasing separation efficiency") is not achievable. If copper is to be replaced by other materials, a "final sink" has to be chosen properly. Urban mining is, analogously to the P-management design, a long-term project. Within a strong growth period, its role in the overall metabolism is of minor importance (see also figure 4.29). However, it has to be developed over two to three generations to become eventually the resource leader of anthropogenic systems.

Waste management comprises, to a certain extent, the problems already discussed in urban mining. To run its processes in an optimal way, one has to dispose of data about wastes such as waste generation rate, waste composition with respect to waste goods, substance concentrations in wastes, variations of corresponding flows and concentrations with time and space, and the like. Furthermore, one needs data about waste processes like the performance of waste collection, treatment, and disposal systems. Last but not least, general metabolic data about regional import flows and stocks of substances are indispensable for basing waste management on an early recognition strategy. It follows that this case is more complex than urban mining. Exemplified with plastics, a design of waste management is oriented toward the sequence "collection–treatment–final disposal." In this metabolic "triptych," the main effort was and still is put on treatment. However, the weakest part of this sequence is found in the final disposal concept. An optimal waste management design has to operate with a consequent "final sink concept" for all outputs of an anthropogenic system. (For radioactive wastes, such a concept is well known but difficult to realize in practice, mainly due to strong political resistance of people either rejecting totally nuclear power plants or rejecting final deposits in their region.) The practical consequences for the products of waste management are as follows: only three types of materials may be produced; namely, (1) "clean" materials suited for recycling, (2) immobile materials safe for landfilling with no negative impact on the environment for long time periods, and (3) emissions that are compatible with environmental standards not changing geogenic concentrations and loads. Such a design strategy makes a number of popular "industrial ecology concepts" superfluous. A "zero emission" strategy is not required to reach the goals of waste management. Neither is a "waste hierarchy" approach needed such as "prevention before recycling and disposal." If prevention is the most economic way of reaching a particular waste management objective, then this method should be chosen. If the disposal option is more economic, a "waste management hierarchy" should not be used to justify avoiding this choice. On one hand, the new design is a waste management freed from ideologically shaped goals, solely based on societal and economic guidelines that lack a sufficient understanding of the metabolic system properties. On the other hand, it gives an open access to tailor-made solutions on a regional scale. New waste management shows an emancipation away from just "cleaning the urban system" into the role of "managing and controlling the flows to the final sinks."

The most complex case is on mobility. From a metabolic point of view, the topic is situated between the two big anthropogenic stocks (TO RESIDE&WORK and TO TRANSPORT&COMMUNICATE). Looking at its position in an anthropogenic system, it includes crucial economic and political aspects of any region; namely, settlement and real estate policy, private and public transport policies, and energy and environmental policy. In other words, a metabolic contribution to the design of mobility seems rather unimportant. Its main contribution lies in the elucidation of the relation between mobility and traffic. A study region (Synoikos, similar to METALAND, not in scale but in metabolic properties) was chosen to study this relation. The design goal is to get more mobility with less traffic. In the status quo, a majority of the inhabitants suffer under a decreasing mobility (the benefit) and an increasing traffic (cost). Mainly responsible for this situation is the transport of persons on the road. A major metabolic problem of traffic is the energy consumption with regard to nonrenewable gasoline, casualties, noise, and air pollution. It is obvious that the energy policy for such a system with the goal to maintain and even to extend it has to push and secure (1) fossil energy availability, (2) enough affordable cars, and (3) enough roads. This combination exerts worldwide the strongest political power and still profits from its mighty inertia. An alternative is only to be found by reconstructing the whole settlement system. Therefore, a morphological design has priority in which the concept of mobility is newly defined on the basis of scaled supply structures. In developed countries, most people live already in built urban systems. Therefore, a reconstruction to gain more mobility with less traffic is a three-generation or a century project, comparable with the growth project in the twentieth century. It is like rebuilding a big ship while crossing the ocean. Therefore, it is evident that such a project is a matter of the political design of the whole anthropogenic system, not yet found in practice, contrary to the three former cases that show already first applications in the status nascendi. What happened to the scenarios in the real study region? Observable are some political processes that bring a new view into urbanism, stimulating alternative projects to improve mobility. There is more opposition against projects favoring more roads. New or reconstructed residential areas reduce traffic and promote public transport means. However, the region has not yet reached a critical mass of cooperating intellects to design the new mobility for the whole region.

Summarizing the Essentials of Metabolic Designing

The strength of the metabolic design is its potential to grasp the key processes and to choose a system-oriented set of priorities. However, metabolic studies, based on the MFA/SFA methodology, contain uncertainties regarding the long-term development of the corresponding anthropogenic system. There is neither enough reliable data to make precise flow and stock prognosis nor is there, by definition, a guarantee regarding technological innovation, not to speak of socioeconomic and political imponderability. Decision makers have to be aware of the limits of the scenario technique. Referring to figure 5.1, one can postulate that, in the past, anthropogenic systems have suffered from the lack of sound metabolic understanding and consequently from lacking any metabolic design on a long-term and large-scale perspective. Therefore, many shortcomings in resource management are still in practice, and the strategy of late recognition in the metabolism of the anthroposphere is still the state of the art. One can postulate (referring to two examples in chapter 2) that the "lead story" (introducing leaded gasoline in the 1940s, polluting air and soil for decades) and the "cadmium story" (polluting rice fields since the 1950s) could have been prevented by applying the instruments of metabolic analysis and evaluation in a political strategy of early recognition. Already at that time, the scientific knowledge was sufficient to do it. There was and still is a cultural resistance to do it, namely to operate simultaneously in small and large scales (in time and space), considering dynamic processes of growth and shrinking.

The case studies presented underline the fact that anthropogenic systems are much more complex than their metabolic subsystems. In many cases of disturbed anthropospheres, the metabolism shows the first symptoms, but not necessarily the causes. There is a tendency to fight the symptoms by installing new technical devices, by installing new ordinances, and/or offering new economic incentives without studying their effects on the causes. The suitability of a technical device or the effect of a normative rule can be tested with models of metabolic systems. Sustainability asks for robust anthropogenic systems that are flexible and resilient, flexible to adapt in time to changing environmental conditions and to anthropogenic innovations, resilient to maintain the functioning of large systems in spite of the processes of adaptation. It is evident that adaptation on the metabolic level is easier when certain societal boundary conditions are favorable and extremely difficult when crucial interests conflict.

Metabolic design operates with scientific facts based on the methods of natural science, which permit distinguishing between "correct" and "incorrect" with respect to predefined goals and hypotheses by verification and falsification. Designing the anthroposphere is not within this tight frame. It oscillates not between "right" and "wrong"; it swings between "making sense" and "making nonsense." Metabolic design serves the design of the anthroposphere, the design of futures, an essential part of cultural evolution.

References

Adam, C., Peplinski, B., Michaelis, M., Kley, G., & Simon, F. G. (2009). Thermochemical treatment of sewage sludge ashes for phosphorus recovery. *Waste Management*, 29(3), 1122–1128.

Adriaanse, A., Bringezu, S., Hammond, A., Moriguchi, Y., Rodenburg, E., Rogich, D., et al. (1997). *Resource Flows: The Material Basis of Industrial Economies. World Resources Institute Report*. Washington, DC: World Resources Institute, Wuppertal Institute, Netherlands Ministry of Housing, Spatial Planning, and Environment, Japan's National Institute for Environmental Studies.

Ayres, R. U., & Ayres, L. W. (1999). *Accounting for Resources, 2: The Life Cycle of Materials*. Cheltenham, UK: Edward Elgar Publishing Limited.

Ayres, R. U., Norberg-Bohn, V., Prince, J., Stigliani, W. M., & Yanowitz, J. (1989). *Industrial Metabolism, the Environment, and Application of Materials—Balance Principles for Selected Materials*. Laxenburg: International Institute for Applied Systems.

Azar, C., Holmberg, J., & Lindgren, K. (1996). Socio-ecological indicators for sustainability. *Ecological Economics*, 18(2), 89–112.

Baccini, P. (1985). Der Phosphorhaushalt der Schweiz. Möglichkeiten und Grenzen aktueller Gewässerschutzmassnahmen. *Gewässerschutz-Wasser-Abwasser*, 69, 133–146.

Baccini, P. (Ed.). (1989). *The Landfill—Reactor and Final Storage*. Lecture Notes in Earth Sciences. Berlin, Heidelberg: Springer.

Baccini, P. (1996). Understanding regional metabolism for a sustainable development of urban systems. *Environmental Science and Pollution Research*, 3(2), 108–111.

Baccini, P. (1997). A city's metabolism: Towards the sustainable development of urban systems. *Journal of Urban Technology*, 4(2), 27–39.

Baccini, P. (2002). Der Stoffwechsel urbaner Systeme. In A. Klotz, O. Frey, & W. Rosinak (Eds.), *Stadt und Nachhaltigkeit*. Vienna: Springer-Verlag.

Baccini, P. (2006). Überleben mit Umweltforschung? *Gaia*, 15(1).

Baccini, P. (2008). Zukünfte urbanen Lebens mit Altlasten, Bergwerken und Erfindungen. In A. Gleich & St. Gössling-Reimann (Eds.), *Industrial Ecology*. Wiesbaden: Vieweg & Teubner.

Baccini, P., & Bader, H.-P. (1996). *Regionaler Stoffhaushalt: Erfassung, Bewertung und Steuerung*. Heidelberg: Spektrum Akademischer Verlag.

Baccini, P., & Brunner, P. H. (1985). Behandlung und Endlagerung von Reststoffen aus Kehrichtverbrennungsanlagen. *Gas-Wasser-Abwasser, 65*(7), 403–409.

Baccini, P., & Brunner, P. H. (1991). *Metabolism of the Anthroposphere.* Berlin, Heidelberg, New York: Springer.

Baccini, P., & Oswald, F. (1998). *Netzstadt: Transdisziplinäre Methoden zum Umbau urbaner Systeme.* Zurich: vdf Hochschulverlag an der ETH Zürich.

Baccini, P., & Oswald, F. (1999). *Netzstadt: transdisziplinäre Methoden zum Umbau urbaner Systeme: Ergebnisse aus dem ETH-Forschungsprojekt SYNOIKOS "Nachhaltigkeit und urbane Gestaltung im Raum Kreuzung Schweizer Mittelland."* Zurich: vdf Hochschulverlag AG an der ETH.

Baccini, P., & Oswald, F. (2007). Designing the urban: Linking physiology and morphology. In G. Hirsch-Hadorn, H. Hoffmann-Riem, S. Biber-Klemm, W. Grossenbacher-Mansuy, D. Joye, C. Pohl, U. Wiesmann, U. & E. Zemp (Eds.), *Handbook of Transdisciplinary Research.* Springer.

Baccini, P., & Pedraza, A. (2006). Die Bestimmung von Materialgehalten in Gebäuden (Measuring Material Contents in Buildings). In Th. Lichtensteiger (Ed.), *Bauwerke als Ressourcennutzer und Ressourcenspender in der langfristen Entwicklung urbaner Systeme.* Zurich: vdf Hochschulverlag an der ETH Zürich.

Baccini, P., & von Steiger, B. (1993). Die Stoffbilanzierung landwirtschaftlicher Böden—Eine Methode zur Früherkennung von Bodenveränderungen, Z. Pflanzenernähr. *Bodenkunde, 156,* 45–54.

Baccini, P., & Zimmerli, R. (1985). *Bestimmung und Beurteilung der physikalish-chemischen Eigenschaften von Abfällen der Auto-Verwertung Ostschweiz AG. EAWAG Project No. 30–323.* Dubendorf: EAWAG.

Baccini, P., Kytzia, S., & Oswald, F. (2002). Restructuring urban systems. In F. Moavendzadeh, K. Hanaki, & P. Baccini (Eds.), *Future Cities: Dynamics and Sustainability.* Dordrecht/Boston: Kluwer Academic Publishers.

Baccini, P., Henseler, G., Figi, R., & Belevi, H. (1987). Water and element balances of municipal solid waste landfills. *Waste Management & Research, 5*(1), 483–499.

Baccini, P., Daxbeck, H., Glenck, E., & Henseler, G. (1993). *METAPOLIS, Güterumsatz und Stoffwechselprozesse in den Privathaushalten einer Stadt, Nationales Forschungsprogramm "Stadt und Verkehr"* (Vols. 34A and 34B). Zürich: NFP 25 Stadt und Verkehr.

Bader, H.-P. (2010). SIMBOX. Retrieved August 14, 2010, from <http://www.eawag.ch/organisation/abteilungen/siam/software/simbox/index_EN>.

Bader, H.-P., & Baccini, P. (1993). Monitoring and control of regional material fluxes. In R. Schulin, A. Desaules, R. Webster, & B. Von Steiger (Eds.), *Soil Monitoring.* Basel, Boston, Berlin: Birkhäuser Verlag.

Bader, B., & Scheidegger, R. (1995). *Benutzeranleitung zum Programm SIMBOX (User Manual for the Program SIMBOX).* Dubendorf: EAWAG.

Bader., H.-P., Scheidegger, R., Wittmer, D., & Lichtensteiger, Th. (2006). A dynamic model to describe the copper household in Switzerland. In Th. Lichtensteiger (Ed.), *Bauwerke als Ressourcennutzer und Ressourcenspender in der langfristigen Entwicklung urbaner Systeme.* Zurich: vdf Hochschulverlag an der ETH Zürich.

Bader, H.-P., Scheidegger, R., Wittmer, D., & Lichtensteiger, Th. (2010). Copper flows in buildings, infrastructure and mobiles: A dynamic model and its applications to Switzerland. *Clean Technology Environmental Policy, 13*(1), 87–101.

Baehr, H. D. (1989). *Thermodynamik (7. Auflage)*. Berlin, Heidelberg, New York: Springer.

BAFU. (2010). *Luftschadstoff-Emissionen des Strassenverkehrs 1990–2035, Aktualisierung 2010. Umwelt-Wissen Nr. 1021*. Bern: Bundesamt für Umwelt BAFU.

Bahn-Walkowiak, B., Bleischwitz, R., Bringezu, S. W., Kuhndt, M., Lemken, T. (2008). Resource Efficiency: Japan and Europe at the Forefront. Synopsis of the Project and Conference Results and Outlook on a Japanese-German Cooperation, Federal Environment Agency, Wuppertal Institute for Climate, Environment, Energy, UNEP/ Wuppertal Institute Collaborating Centre on Sustainable Consumption and Production (CSCP), Wuppertal.

BAL. (1983). *Landwirtschaftliches Jahrbuch der Schweiz*. Bern: Bundesamt für Landwirtschaft (BAL).

BAS. (1987). *Statistisches Jahrbuch der Schweiz 1978/88*. Basel: Bundesamt für Statistik (BAS).

Becker-Boost, E., & Fiala, E. (2000). *Wachstum ohne Grenzen*. Vienna: Springer.

Belevi, H. (1998). *Environmental Engineering of Municipal Solid Waste Incineration*. Zurich: vdf Hochschulverlag an der ETH Zürich.

Belevi, H., & Baccini, P. (1989). Longterm behaviour of municipal solid waste landfills. *Waste Management & Research, 7*(1), 483–499.

Bergbäck, B., Anderberg, S. T., & Lohm, U. (1994). Accumulated environmental impact: the case of cadmium in Sweden. *Science of the Total Environment, 145*, 13–28.

Bergbäck, B., Johansson, K., & Mohlander, U. (2001). Urban metal flows—A case study of Stockholm. Review and conclusions. *Water, Air, and Soil Pollution, 1*(3–4), 3–24.

Bernstein, W. J. (2008). *A Splendid Exchange, How Trade Shaped the World*. London: Atlantic Books.

Binder, C., de Baan, L., & Wittmer, D. (2009). *Phosphorflüsse in der Schweiz. Stand, Risiken und Handlungsoptionen. Abschlussbericht. Umwelt-Wissen Nr. 0928*. Berne: Bundesamt für Umwelt.

Blaser, C., & Redle, M. (1998). More mobility with less traffic—restructuring scenarios for the "transportation activity." In P. Baccini & F. Oswald (Eds.), *Netzstadt: Transdisziplinäre Methoden zum Umbau urbane Systeme (Summaries in English)*. Zurich: vdf Hochschulverlag an der ETH Zürich.

BMUJF. (1996). *VERPACKVO 1996, BGBL. NR. 648/1996 IDF BGBL. II NR. 364/2006*. Vienna: BMUJF.

Bogucka, R., Kosinska, I., & Brunner, P. H. (2008). Setting priorities in plastic waste management—Lessons learned from material flow analysis in Austria and Poland. *Polimery, 53*(1), 55–59.

Böni, D. (2010). Ausbeute von Inhaltsstoffen der Trockenschlacke bei ZAR Schweiz. In R. Warnecke (Ed), *VDI Wissensforum Fachkonferenz Feuerung und Kessel—Beläge und Korrosion—in Grossfeuerungsanlagen, June 22/23*. Frankfurt am Main: VDI. Retrieved June 6, 2011, from http://www.zar-ch.ch/images/stories/Praesentationen_Eroeffnung/vdi%20 wissensforum_23-06-2010.pdf.

Boyden, S., Millar, S., Newcombe, K., & O'Neill, B. (1981). *The Ecology of a City and Its People: The Case of Hong Kong.* Canberra: Australian National University Press.

Braudel, F. (1979). *Civilisation matérielle, economie et capitalsime, XVe—XVIIe siècle. Les structures du quotidie: le possible et l'impossible.* Paris: Libraisrie Armand Colin.

Bringezu, S. (1997). Accounting for the physical basis of national economies: Material flow indicators. In B. Moldan & S. Billharz (Eds.), *SCOPE 58—Sustainability Indicators: Report of the Project on Indicators for Sustainable Development.* Chichester, UK: John Wiley & Sons.

Bringezu, S., Schütz, H., Steger, S., & Baudisch, J. (2004). International comparison of resource use and its relation to economic growth: The development of total material requirement, direct material inputs and hidden flows and the structure of TMR. *Ecological Economics, 51*(1–2), 97–124.

Brown, I. R., & Jacobson, J. L. (1987). *The future of Urbanization: Facing the Ecological and Economic Constraints. Worldwatch Paper, 77.* Washington, DC: Worldwatch.

Brunner, P. H., & Baccini, P. (1981). *Schwermetalle, Sorgenkinder der Entsorgung. Neue Zürcher Zeitung Nr.70, Beilage Forschung und Technik 65.* Zurich: Neue Zürcher Zeitung.

Brunner, P. H., & Baccini, P. (1992). Regional material management and environmental *protection. Waste Management & Research, 10(2),* 203–212.

Brunner, P. H., & Bogucka, R. (2006). Waste Management, a Key Element for Environmental Protection and Resources Conservation. In W. Höflinger (Ed.): EMChIE 2006, 5th European Meeting on Chemical Industry and Environment, 28–36. Vienna: TU Vienna Institute of Chemical Engineering.

Brunner, P. H., & Ernst, W. R. (1986). Alternative methods for the analysis of municipal solid waste. *Waste Management & Research, 4*(2), 147–160.

Brunner, P. H., & Fellner, J. (2007). Setting priorities for waste management strategies in developing countries. *Waste Management & Research, 25*(3), 234–240.

Brunner, P. H., & Mönch, H. (1986). The flux of metals through a municipal solid waste incinerator. *Waste Management & Research, 4,* 105–119.

Brunner, P. H., & Rechberger, H. (2004). *Practical Handbook of Material Flow Analysis.* Boca Raton: CRC Press/Lewis Publishers.

Brunner, P. H., & Zimmerli, R. (1986). *Beitrag zur Methodik der Zuordnung von Abfallstoffen zu Deponietypen. EAWAG Project No. 30–4721.* Dubendorf: EAWAG.

Brunner, P. H., Daxbeck, H., & Baccini, P. (1994). Industrial metabolism at the regional and local level. In R. U. Ayres & U. E. Simonis (Eds.), *Industrial Metabolism Restructuring for Sustainable Development.* Tokyo: UN University Tokyo.

Brunner, P. H., Ernst, W., & Sigel, O. (1983). *Vorstudie 'Regionale Abfallbewirtschaftung.* Dubendorf: EAWAG.

Brunner, P. H., Müller, B., & Rebernig, G. (2006). Cities and their surfaces as sources of pollution. In P.H. Brunner, Th. Jakl, & E. Ober, (Eds.), *Conference Proceedings "Proceedings of Towards the City Surface of Tomorrow."* Vienna: TU Vienna and Lebensministerium.

Brunner, P. H., Capri, S., Marcomini, A., & Giger, W. (1988). Occurrence and behaviour of linear alkylbenzenesulphonates, nonylphenol, nonylphenol mono- and nonylphenol diethoxylates in sewage and sewage sludge treatment. *Water Research, 22*(12), 1465–1472.

Brunner, P. H., Beer, B., Daxbeck, H., Henseler, G., Piepke, G., & von Steiger, B. (1990). *RESUB: Der Regionale Stoffhaushalt im Unteren Bünztal.* Dubendorf: EAWAG.

Buitenkamp, M., Venner, H., & Wams, T. (Eds.). (1992). *Action Plan Sustainable Netherlands.* Amsterdam: Friends of the Earth Netherlands.

Burckhardt-Holm, P., Giger, W., Güttinger, H., Ochsenbein, U., Peter, A., Scheurer, K., et al. (2005). Where have all the fish gone? The reasons why the fish catches in Swiss rivers are declining. *Environmental Science & Technology, 39,* 441A–447A.

Burg, C., & Benzinger, R. V. (1984). Kunststoffe im Automobilbau—eine Herausforderung für das Recycling. In K. J. Thome-Kozmiensky (Ed.), *Recycling von Kunststoffen 1.* Berlin: EF-Verlag für Energie- und Umwelttechnik.

Burkhardt, M., Rossi, L., & Boller, M. (2008). Diffuse release of environmental hazards by railways. *Desalination, 226,* 106–113.

BUWAL. (2003). *Erhebung der Kehrichtzusammensetzung 2001/02. Schriftenreihe Umwelt Nr. 356.* Bern: Bundesamt für Umwelt, Wald und Landschaft.

Cemburau (2010). *World Statistical Review 1999–2009.* Retrieved November 28, 2010, from <www.cembureau.eu>.

Cencic, O. (2004). Software for MFA. In P.H. Brunner & H. Rechberger (Eds.). *Practical Handbook of Material Flow Analysis.* Boca Raton: CRC Press/Lewis Publishers.

Cencic, O., & Rechberger, H. (2008). Material flow analysis with software STAN. *Journal of Environmental Economics and Management, 18*(1), 3–7.

Chassot, G. M. (1995). *Recensement et evaluation du metabolism anthropique sur la base des boues d'épuration.* Thesis ETH Zurich, No. 11155. Zurich: Swiss Federal Institute of Technology.

Cleveland, C. J., & Ruth, M. (1998). Indicators of dematerialization and the materials intensity of use. *Journal of Industrial Ecology, 2*(3), 15–50.

Cordell, D., Drangert, J., & White, S. (2009). The story of phosphorus: Global food security and food for thought. *Global Environmental Change, 19*(2), 292–305.

Costa, M. M., Schaeffer, R., & Worrell, E. (2001). Exergy accounting of energy and materials flows in steel production systems. *Energy, 26*(4), 363–384.

Crutzen, P. J., & Stoermer, E. F. (2000). The "Anthropocene." *Global Change Newsletter, 41,* 17–18.

Dalemo, M., Sonesson, U., Björklund, A., Mingarini, K., Frostell, B., Jonsson, H., et al. (1997). ORWARE—A simulation model for organic waste handling systems. Part 1: Model description. *Resources, Conservation and Recycling, 21*(1), 17–37.

Daxbeck, H., Kisliakova, A., & Obernosterer, R. (2001). *Der ökologische Fussabdruck der Stadt Wien. Vienna, Austria* (p. 22). Vienna: Stadt Wien MA 48.

Daxbeck, H., Lampert, Ch., Morf, L., Obernosterer, R., Rechberger, H., Reiner, I., et al. (1996). *Der anthropogene Stoffhaushalt der Stadt Wien—N, C und Pb (Projekt PILOT).* Vienna: Institute for Water Quality, Resource, and Waste Management, Vienna University of Technology.

De Duve, C. (2003). The facts of life. In *The Cultural Values of Science, Scripta Varia 105 (Plenary Session 8–11 November 2002).* Vatican City: The Pontifical Academy of Sciences.

de Haen, H., Murty, K.N., & Tangermann, S. (1982). *Künftiger Nahrungsmittelverbrauch in der Europäischen Gemeinschaft*. Munster-Hiltrup: Landwirtschaftsverlag.

Despommier, D. (2010). *The Vertical Farm: Feeding the World in the 21st Century*. New York: Thomas Dunne Books.

Diamond, J. (1997). *Guns, Germs, and Steel. The Fate of Human Societies*. New York: W.W.Norton & Company.

Diamond, J. (2005). *Collapse: How Societies Choose to Fail or Survive* (p. 112). New York: Viking Adult.

Döberl, G., & Brunner, P. H. (2004). Substances and their (final) sinks—A new indicator for monitoring sustainability. In *Proceedings Symposium "Indicators for Evaluating Sustainable Development—The Ecological Dimension."* Berlin: Freie Universität Berlin, Environmental Policy Research Centre.

Döberl, G., Huber, R., & Brunner, P. H. (2001). Final storage quality as a necessary goal for waste management. In *Proceedings ISWA World Congress 2001 "Waste in a Competitive World."* Stavanger: ISWA.

Döberl, G., Huber, R., Brunner, P. H., Eder, M., Pierrard, R., Hutterer, H., et al. (2002). Long-term assessment of waste management options—A new, integrated and goal-oriented approach. *Waste Management & Research, 20*(4), 311–327.

Du, S., Lu, B., Zhai, F., & Popkin, B. M. (2002). A new stage of the nutrition transition in China. *Public Health Nutrition, 5*(1A), 169–174.

Duvigneaud, P., & Denaeyer-De Smet, S. (1977). L'Ecosystème urbs, l'écosystème urbain Bruxellois. In P. Duvigneaud & P. Kestemeont (Eds.), *Productivité biologique en Belgique SCOPE. Travaux de la Section belge du Programme Biologique International*. Gembloux: Duculot.

ecoinvent Centre. (2010). Database ecoinvent data v2.2. Retrieved June 2, 2011, from <http://www.ecoinvent.org/de/>.

Ehrenfeld, J. R. (2008). Kann Industrial Ecology die "Wissenschaft der Nachhaltigkeit" werden? In A. von Gleich & St. Gössling-Reisemann (Eds.), *Industrial Ecology*. Wiesbaden: Vieweg & Teubner.

Ehrenfeld, J. R. (2009). Understanding of complexity expands the reach of industrial ecology. *Journal of Industrial Ecology, 13*(2), 165–167.

Elias, N. (1939). *Über den Prozess der Zivilisation. Soziogenetische und psychogenetische Untersuchungen*. 2 Bde., Basel: Verlag Haus zum Falken.

Elser, J., & White, S. (2010). Peak phosphorus. *Foreign Policy* April, 20.

Elshkaki, A. (2000). *Modelling Substance Flow Analysis in Simulink.—Assistance to Use Dynflow*. Master's thesis. Leiden: Institute of Environmental Sciences (CML), Leiden University.

Elshkaki, A. (2007). *Systems Analysis of Stock Buffering—Development of a Dynamic Substance Flow-Stock Model for the Identification and Estimation of Future Resources, Waste Streams and Emissions*. Doctoral Dissertation. Leiden: Institute of Environmental Sciences (CML), Leiden University.

Euro Chlor. (1997). *Risk Assessment for the Marine Environment OSPARCOM Region— North Sea—Trichloroethylene*. Brussels: Euro Chlor.

Euro Chlor. (1999). *Risk Assessment for the Marine Environment OSPARCOM Region—North Sea—Dichloromethane.* Brussels: Euro Chlor.

European Communities (2001). *Disposal and recycling Routes for Sewage Sludge, Part 3 —Scientific and Technical Sub-Component Report.* Luxemburg: Office for Official Publications of the European Communities.

European Union. (2010). *ILCD Handbook—International Reference Life Cycle Data System.* Ispra: European Commission—Join Research Center—Institute for Environment and Sustainability.

Eurostat. (2001). *Economy-wide Material Flow Accounts and Derived Indicators—A Methodological Guide.* Luxembourg: Office for Official Publications of the European Communities.

Faist, M. (2000). *Ressourceneffizienz in der Aktivität Ernähren (Resource Efficiency in the Activity TO NOURISH).* Thesis ETHZ, No. 13884. Zurich: ETH Zurich.

Faist, M., Kytzia, S., & Baccini, P. (2000). The impact of household food consumption on resource and energy management. *International Journal of Environment and Pollution, 15*(2), 183–199.

Falbe, J. (Ed.). (1987). *Surfactants in Consumer Products, Theory, Technology and Application.* Berlin: Springer-Verlag.

FAO. (2010a). AQUASTAT—Country Fact Sheet Switzerland. Retrieved November 22, 2010, from <http://www.fao.org/nr/water/aquastat/data/factsheets/aquastat_fact_sheet_che _en.pdf>.

FAO. (2010b). Product Quantity of Paper and Paperboards. Food and Agricultural Organization of the United Nations. Retrieved June 7, 2011, from <http://faostat.fao.org/site/626/ default.aspx#ancor>.

Fehringer, R., & Brunner, P. H. (1997). Flows of plastics and their possible reuse in Austria. In *Proceedings to the Con-Account Workshop from Paradigm to Practice of Sustainability.* Leiden: Wuppertal Institute and Science Centre.

Fiedler, H. J., & Rösler, H. J. (Eds.). (1993). *Spurenelemente in der Umwelt.* Jena, Stuttgart: Gustav Fischer Verlag.

Fischer, E. P., & Wiegandt, K. (Eds.). (2010). *Evolution und Kultur des Menschen.* Frankfurt: Fischer.

Fischer-Kowalski, M., Haberl, H., Hüttler, W., Payer, H., Schindl, H., Winiwarter, V., et al. (Eds.). (1997). *Gesellschaftlicher Stoffwechsel und Kolonisierung von Natur. Ein Versuch in Sozialer Ökologie.* Amsterdam: Overseas Publishers Association.

Forrester, J. W. (1971). *World Dynamics.* Cambridge, UK: Wright-Allen Press.

Förstner, U., & Müller, G. (1974). *Schwermetalle in Flüssen und Seen.* Berlin: Springer-Verlag.

Franke, M. (1987). Auto und Umwelt. *Entsorgungspraxis, 7*(8), 336–341.

Frey, R. L. (1996). *Stadt: Lebens- und Wirtschaftsraum.* Zurich: vdf Hochschulverlag an der ETH Zürich.

Gerst, M. D., & Graedel, T. E. (2008). In-use stocks of metals: Status and implications. *Environmental Science & Technology, 42,* 7038–7045.

Giger, W., Brunner, P. H., & Schaffner, Ch. (1984). 4-Nonylphenol in sewage sludge: Accumulation of toxic metabolites from nonionic surfactants. *Science, 225*(4662), 623–625.

Giger, W., Alder, A. C., Golet, E. M., Kohler, H.-P. E., McArdell, C. S., Molnar, E., et al. (2003). Occurrence and fate of antibiotics as trace contaminants in wastewaters, sewage sludges, and surface waters. *Chimia, 57,* 485–491.

Global 2000 & Barney, G.O. (Eds.). (1980). *The Global 2000 Report to the President.* Washington, DC: U.S. Government Printing Office.

Global Footprint Network. (2010). *Methodology and Sources.* Retrieved June 7, 2011, from <http://www.footprintnetwork.org/en/index.php/GFN/page/methodology/>.

Goedkoop, M., & Spriensma, R. (2001). *The Eco-indicator 99. A Damage Oriented Method for Life Cycle Impact Assessment (Methodology Report).* Amersfoort: Product Ecology Consultants.

Gordon, R. B., Bertram, M., & Graedel, T. E. (2006). Metal stocks and sustainability. *Proceedings of the National Academy of Sciences of the United States of America, 103,* 1209–1214.

Gordon, R. B., Lifset, R., Bertram, M., et al. (2004). Where is all the zinc going: The stocks and flows project, Part 2. *Journal of the Minerals, Metals and Materials Society, 56,* 24–29.

GPRI. (2010). GPRI Statement on Global Phosphorus Scarcity, statement by the Global Phosphorus Research Initiative on September 26, 2010 Retrieved June 2nd, 2011 from <http://phosphorusfutures.net/files/GPRI_Statement_responseIFDC_final.pdf>.

Graedel, T. (2010). *Metal Stocks in Society—Scientific Synthesis.* Paris: International Panel for Sustainable Resource Management, United Nations Environment Program.

Graedel, T., & van der Voet, E. (Eds.). (2010). *Linkages of Sustainability.* Cambridge, MA: MIT Press.

Groh, D. (1992). *Anthropologische Dimensionen der Geschichte.* Frankfurt am Main: Suhrkamp Taschenbuch Wissenschaft.

Guinée, J. B., Gorrée, M., Heijungs, R., et al. (2001). *Life Cycle Assessment. An Operational Guide to the ISO Standards.* Leiden: Institute of Environmental Sciences (CML), Leiden University.

Güttinger, H., & Stumm, W. (1990). Ökotoxikologie am Beispiel der Rheinverschmutzung durch den Chemie-Unfall bei Sandoz in Basel. *Naturwissenschaften, 77,* 253–261.

Hale, R. C. (2003). Endocrine disruptors in wastewater and sludge treatment processes. *Environmental Health Perspectives, 111*(10), 550.

Hardin, G. (1968). The tragedy of the commons. *Science, 162*(3859), 1243–1248.

Harrison, R. P. (2008). *Gardens. An Essay on the Human Condition.* Chicago: The University of Chicago Press.

Hendrickson, C. T., Lave, L. B., & Matthews, H. S. (2005). *Environmental Life Cycle Assessment of Goods and Services: An Input–Output Approach.* Washington, DC: Resources for the Future Press.

Henseler, G., Scheidegger, R., & Brunner, P. H. (1992). Die Bestimmung von Stoffflüssen im Wasserhaushalt einer Region ["Determination of material flux through the hydrosphere of a region"]. *Vom Wasser, 78,* 91–116.

Herman, R., Ardekanis, S. A., Govind, S., & Dona, E. (1988). The dynamic characterization of cities. In J. H. Ausubel & R. Herman (Eds.), *Cities and Their Vital Systems*. Washington, DC: National Academy.

Hinterberger, F., Luks, F., & Schmidt-Bleek, F. (1997). Material flows vs. 'natural capital': What makes an economy sustainable? *Ecological Economics*, 23(1), 1–14.

Hirsch Hadorn, G., Hoffmann-Riem, H., Biber-Klemm, S., Grossenbacher, W., Joye, D., Pohl, C., et al. (Eds.). (2008). *Handbook of Transdisciplinary Research*. Springer.

Holford, I. C. R. (1997). Soil phosphorus: its measurement, and its uptake by plants. *Australian Journal of Soil Research*, 35, 227–239.

Hölldobler, B., & Wilson, E. O. (2008). *The Superorganism: The Beauty, Elegance, and Strangeness of Insect Societies*. London: W.W. Norton.

Howard, E. (1902). *Garden Cities of Tomorrow*. London: S. Sonnenschein & Co. Ltd.

Huntington, S. (1996). *The Clash of Civilizations*. New York: Simon & Schuster.

IFS. (2006). Pro Kopf Wohnfläche weiter gestiegen. Retrieved November 22, 2010, from <http://www.ifs-staedtebauinstitut.de/hi/hi2006/hi02.pdf>.

IFU Hamburg GmbH. (2010). Umberto. Retrieved August 14, 2010, from <http://www.umberto.de/en/home/introduction/index.htm>.

International Society for Industrial Ecology. (2010). Industrial Ecology: Tools of the Trade. Retrieved August 13, 2010, from <http://www.is4ie.org/about>.

ISI. (2010). *Kupfer für Zukunftstechnologien: Nachfrage und Angebot unter besonderer Berücksichtigung der Elektromobilität*. Karlsruhe: Frauenhofer-Institut für System- und Innovationsforschung.

ISO. (2006). International Organization for Standardization ISO 14044:2006 "Environmental management— Life cycle assessment—Principles and framework" [Environmental management—Life cycle assessment—Requirements and guidelines]. Geneva: The International Standards Organization. *ISO, 14044, 2006*.

Jansen, U., & Hertle, H. (2007). 1. *Zwischenbericht zum Forschungs- und Entwicklungsvorhaben "Aktivierung deutsch-amerikanischer und deutsch-japanischer Städtepartnerschaften für den Klimaschutz."* Dessau: UBA.

Johnstone, I. M. (2001). Energy and mass flows of housing: estimating mortality. *Building and Environment*, 36(1), 43–51.

Kasprzyk, M. (2007). MoCuBa—New Mobility Culture; Concepts for Urban Areas of Europe and Especially the Baltic Sea Region. Retrieved November 28, 2010, from <http://www.bauumwelt.bremen.de/sixcms/media.php/13/MoCuBa-Broschuere.pdf>.

Kelen, B. (1983). *Confucius*. Singapore: Graham Brash.

Kennedy, C. A., Cuddihy, J., & Engel Yan, J. (2007). The changing metabolism of cities. *Journal of Industrial Ecology*, 11(2), 43–59.

Kjeldsen, P., Barlaz, M. A., Rooker, A. P., Baun, A., Ledin, A., & Christensen, T. H. (2002). Present and long-term composition of MSW landfill leachate: A review. *Critical Reviews in Environmental Science and Technology*, 32(4), 297–336.

Koehn, H. A. (1987). Wohin mit dem alten Autoöl. *Entsorgungspraxis*, 7(8), 342–347.

Kral, U., & Brunner, P. H. (2010). Die Stoffflussanalyse zur nachhaltigen Bewirtschaftung von Bahnschotter. *ETR Eisenbahntechnische Rundschau, 59*(6), 388–391.

Kraut, H. (Ed.). (1981). *Stoffwechsel, Ernährung und Nahrungsmittelbedarf, Energiebedarf, Proteinbedarf. Beiträge zur Ernährungswissenschaft.* Current Topics in Nutritional Sciences, Vol. 1 Der Nahrungsbedarf des Menschen. Darmstadt: Steinkopf.

Kroiss, H., Morf, L. S., Lampert, C., Zessner, M., & Spindler, A. (2008). *Optimiertes Stoffflussmonitoring für die Abwasserentsorgung Wiens (OSMA—Wien), Bericht Phase C.* Vienna: Institute for Water Quality, Resource, and Waste Management, Vienna University of Technology.

Krotscheck, C., & Narodoslawsky, M. (1996). The sustainable process index: a new dimension in ecological evaluation. *Ecological Engineering, 6*(4), 241–258.

Kruspan, P. (2000). *Natürliche und technische Petrogenese von Puzzolanen (Natural and Technical Petrogenesis of Puzzollans).* Thesis ETH Zurich, No. 13904. Zurich: ETH Zurich.

Kytzia, S. (1997). Wie kann man Stoffhaushaltsysteme mit ökonomischen Daten verknüpfen? Ein erster Ansatz am Beispiel der Wohngebäude. In Th. Lichtensteiger (Ed.), *Ressourcen im Bau.* Zurich: vdf Hochschulverlag an der ETH Zürich.

Kytzia, S., Faist, M., & Baccini, P. (2004). Economically extended—MFA: A material flow approach for a better understanding of food production chain. *Journal of Cleaner Production, 12*(8–10), 877–889.

Lampert, Ch., & Brunner, P. H. (2000). *Ressourcenschonende und umweltverträgliche regionale Nutzung biogener Materialien.* Graz: Amt der Steiermärkischen Landesregierung.

Landes, D. S. (1969). *Unbound Prometheus, Technological Change and Industrial Development in Western Europe from 1750 to the Present.* Cambridge, UK: Cambridge University Press.

Laner, D., Fellner, J., & Brunner, P.H. (2010). Die Umweltverträglichkeit von Deponieemissionen unter dem Aspekt der Nachsorgedauer. *Österreichische Wasser- und Abfallwirtschaft, 62* (7–8), 131–140.

Larsen, T., Rauch, W., & Gujer, W. (2001). Waste design paves the way for sustainable urban wastewater management. In *Proceedings International Unesco Symposium "Frontiers in. Urban Water Management: Deadlock or Hope?"* Paris: UNESCO.

Larsen, A., Peters, I., Alder, A., Eggen, R., Maurer, M., & Muncke, J. (2001). Re-engineering the toilet for sustainable wastewater management—Could one answer be a new toilet that separates urine from wastewater streams? *Environmental Science & Technology, 35*(9), 192A–197A.

Lavoisier, A. (1789). *Traité élémentaire de chimie, présenté dans un ordre nouveau et d'après les découvertes moderns.* Paris: Chez Cuchet [Reprinted Cultures et Civilisations, Bruxelles, 1965].

Le Corbusier. (1970). *Manière de penser l'Urbanisme.* Paris: Denoel/Editions Gonthier.

Lederer, J., & Rechberger, H. (2010). Comparative goal-oriented assessment of conventional and alternative sewage sludge treatment options. *Waste Management (New York, N.Y.), 30*(6), 1043–1056.

Lees, A. (1985). *Cities Perceived.* Manchester, UK: Urban Society in European and American Thought.

Lentner, C. (1981). *Geigy Scientific Tables (Vol 1): Units of Measurement, Bodyfluids, Composition of the Body, Nutrition.* Basel: Ciba-Geigy.

Leontief, W. (1986). Input-Output Economics. Oxford: Oxford University Press.

Lichtensteiger, Th., & Baccini, P. (2008). Exploration of urban stocks. *Journal of Environmental Economics and Management, 18*(1), 41–48.

Lichtensteiger, Th., & Baccini, P. Henseler Henseler, G., Pedraza, A., & Wittmer, D. (2006). Exploration regionaler Gebäudelager. In Th. Lichtensteiger (Ed.), *Bauwerke als Ressourcennutzer und Ressourcenspender in der langfristigen Entwicklung urbaner Systeme.* Zurich: vdf Hochschulverlag an der ETH Zürich.

Lifset, R., Gordon, R., Graedel, T., Spatari, S., & Bertram, M. (2002). Where has all the copper gone: The stocks and flows project, part 1. *Journal of the Minerals. Metals and Materials Society, 54*(10), 21–26.

Lohm, U., Anderberg, S., & Bergbäck, B. (1994). Industrial metabolism at the national level: A case-study on chromium and lead pollution in Sweden, 1880–1980. In R. U. Ayres & U. E. Simonis (Eds.), *Industrial Metabolism: Restructuring for Sustainable Development.* Tokyo: United Nations University Press.

MacElroy, R. D., Tibbitts, T. W., Thompson, B. G., & Volk, T. (Eds.). (1989). Life sciences and space research XXIII (3). *Advances in Space Research, 9* (8).

Marius, R. (1994). Introduction. In R. Marius (Ed.), More, T. Utopia. London: Everyman.

Matsubae, K., Kajiyama, J., Jeong, Y., & Nagasaka, T. (2010). International phosphorus flow in Asia with focus on the accompanying substances. In *Proceedings of ISIE Asia-Pacific Meeting & ISIE MFA-ConAccount Meeting, 7–9 November 2010.* Tokyo: International Society of Industrial Ecology and ConAcount.

Mayhew, C., & Simmon, R. (2010). "Earth at Night." Retrieved November 18, 2010, from <http://antwrp.gsfc.nasa.gov/apod/ap081005.html>.

Mc Manus, T. (2006). Intel Corporation, personal communication [cited by Johnson, J., Harper, E.M., Lifset, R., & Graedel, T (2007). Dining at the periodic table: Metals concentrations as they relate to recycling. *Environmental Science and Technology, 41,* 1759–1765].

Meadows, D. H., Meadows, D. L., Randers, J., & Behrens, W. W. (1972). *The Limits to Growth.* New York: Universe Books.

Michaelis, P., & Jackson, T. (2000). Material and energy flow through the UK iron and steel sector. Part 1: 1954–1994. *Resources, Conservation and Recycling, 29*(1), 131–156.

Moll, S., Bringezu, S., & Schütz, H. (2005). *Resource Use in European Countries.* Wuppertal: Wuppertal Institute for Climate Environment and Energy.

Montag, D., Gethke, D., & Pinnekamp, J. (2009). Different strategies for recovering phosphorus: Technologies and costs. In K. Ashley, D. Mavinic, & F. Koch (Eds.), *International Conference on Nutrient Recovery from Wastewater Streams, 10-13. Mai 2009, Vancouver, Canada.* London, New York: IWA-Publishing.

More, Th. (1994). *Utopia.* London: Everyman.

Morf, L. S., & Brunner, P. H. (1998). The MSW incinerator as a monitoring tool for waste management. *Environmental Science & Technology, 32*(12), 1825–1831.

Müller, D. (1998). *Modellierung, Simulation und Bewertung des regionalen Holzhaushaltes.* ETH Thesis, No. 12990. Zurich: ETH Zurich.

Müller, D., Bader, H.-P., & Baccini, P. (2004). Long-term coordination of timber production and consumption using a dynamic material and energy flow analysis. *Journal of Industrial Ecology, 8*(3), 65–87.

Müller, D., Oehler, D., & Baccini, P. (1995). *Regionale Bewirtschaftung von Biomasse.* Zurich: vdf Hochschulverlag an der ETH Zürich.

Nakamura, S., & Kondo, Y. (2009). *Waste Input-Output Analysis: Concepts and Application to Industrial Ecology.* Amsterdam: Springer.

Nakicenovic, N. (1986). The automobile road to technological change. *Technological Forecasting and Social Change, 29*(4), 309–340.

Narodoslawsky, M., & Krotscheck, Ch. (1995). The sustainable process index (SPI): evaluating processes according to environmental compatibility. *Journal of Hazardous Materials, 41*(2–3), 383–397.

Obernosterer, R., Brunner, P. H., Daxbeck, H., Gagan, T., Glenck, E., Hendriks, C., et al. (1998). *Materials Accounting as a Tool for Decision Making in Environmental Policy—Mac TEmPo Case Study Report. Urban Metabolism, the City of Vienna.* Vienna: Institute for Water Quality, Resource, and Waste Management, Vienna University of Technology.

Obrist, W. (1987). Stoffbilanz der Kompostierung. In P. Baccini & P. H. Brunner (Eds.), *Die Umweltverträglichkeitsprüfung von Entsorgungsanlagen, Einführung in die Methodik der Stoffflussanalyse, EAWAG Seminar September 1987.* Dübendorf: EAWAG.

Odum, E. P. (1989). *Ecology and Our Endangered Life-Support Systems.* Sunderland: Sinauer Associates Inc.

Odum, E. P., & Odum, T. O. (1953). *Fundamentals of Ecology.* Philadelphia: W.B. Saunders.

OECD (1981a). *Agrarproduktion und Nahrungsverbrauch in Polen: Entwicklung und Aussichten.* Münster-Hiltrup: Landwirtschaftsverlag.

OECD (1981b). *Agrarproduktion und Nahrungsverbrauch in Ungarn.* Münster-Hiltrup: Landwirtschaftsverlag.

OECD. (2002). Material flow accounting and its application: The experience of Japan. In *Proceedings of Special Session on Material Flow Accounting.* Paris: OECD.

OECD. (2007a). *Measuring Material Flows and Resource Productivity—The OECD Guide.* Paris: OECD.

OECD. (2007b). OECD Country Summary Data. Assembled by the Victoria Transport Policy Institute. Retrieved November 22, 2010, from <http://www.vtpi.org/OECD2006.xls>.

Ohlsson, L. (1999). *Environment, Scarcity, and Conflict: A Study of Malthusian Concerns.* Gothenburg: University of Gothenborg, Department of Peace and Development Research.

Olsthoorn, A. A., & Boelens, J. (1998). *Substance Flow Analysis with FLUX.* Report No. W98\15. Amsterdam: Institute for Environmental Studies, Free University Amsterdam.

Öster. Normungsinstitut. (2005a). *ÖNORM S 2096–1 (Stoffflussanalyse—Teil 1: Anwendung in der Abfallwirtschaft—Begriffe).* Vienna: Austrian Standards Institute.

Öster. Normungsinstitut. (2005b). *ÖNORM S 2096–2 (Stoffflussanalyse—Teil 2: Anwendung in der Abfallwirtschaft—Methodik).* Vienna: Österreichisches Normungsinstitut.

Osterhammel, J. (2009). *Die Verwandlung der Welt. Eine Geschichte des 19. Jahrhunderts.* Munich: C.H. Beck.

Oswald, F., & Baccini, P., (2003). *Netzstadt: Designing the Urban*. Basel: Birkhäuser.

Oswalt, P., & Rieniets, T. (Eds.). (2006). *Atlas of Shrinking Cities*. Otsfildern: Hatje Cantz Verlag.

Pacione, M. (2005). *Urban Geography: A Global Perspective*. London: Routledge.

Palfi, R., Poxhofer, R., Kriegl, M., & Alber, S. (2006). *SEES Sustainable Electrical and Electronic System for the Automotive Sector. D6: Car Shredder Manual, STREP, TST3-CT-2003–506075*. Brussels: European Commission 6th Framework Program.

PE International. (2010). GaBi Software. Retrieved August 14, 2010, from <http://www.gabi-software.com/index.php?id=85&L=7&redirect=1>.

Pfister, F. (2005). *Resource Potentials and Limitations of a Nicaraguan Agricultural Region*. Thesis ETHZ, No. 15169. Zurich: Swiss Federal Institute of Technology.

Pfister, F., & Baccini, P. (2005). Resource potentials and limitations of a Nicaraguan agricultural region. *Environment, Development and Sustainability*, 7(3), 337–361.

Pfister, F., Bader, H.-P., Scheidegger, R., & Baccini, P. (2005). Dynamic modelling of resource management for farming systems. *Agricultural Systems*, 86, 1–28.

PlasticsEurope. (2008). *The Compelling Facts about Plastics—An Analysis of Plastics Production, Demand and Recovery for 2006 in Europe*. Brussels: PlasticsEurope.

Pretty, J. (Ed.). (2008). *Sustainable Agriculture and Food, Vol. 1*. London: Earthscan.

Rauch, J. N. (2009). Global Mapping of Al, Cu, Fe, and Zn In-Use Stocks and In-Ground Resources. Retrieved November 29, 2010, from <http://www.pnas.org/content/106/45/18920.figures-only>.

Rebernig, G., Müller, B., & Brunner, P. H. (2006). Plaster and coatings on mineral surfaces: Protection of objects and environment. In P. H. Brunner, Th. Jakl, & E. Ober (Eds.), *Conference Proceedings "Proceedings of Towards the City Surface of Tomorrow."* Vienna: TU Vienna and Lebensministerium.

Rechberger, H. (1999). *Entwicklung einer Methode zur Bewertung von Stoffbilanzen in der Abfallwirtschaft*. Doctoral Dissertation. Institute for Water Quality, Resource, and Waste Management, Vienna University of Technology.

Rechberger, H., & Brunner, P. H. (2002). A new, entropy based method to support waste and resource management decisions. *Environmental Science & Technology*, 36(4), 809–816.

Rechberger, H., & Cencic, O. (2010). STAN—Software for Substance Flow Analysis. Retrieved August 14, 2010, from <http://www.iwa.tuwien.ac.at/iwa226_english/stan.html>.

Rechberger, H., & Graedel, T. E. (2002). The contemporary European copper cycle: statistical entropy analysis. *Ecological Economics*, 42(1–2), 59–72.

Redle, M. (1999). *Kies- und Energiehaushalt urbane Regionen in Abhängigkeit der Siedlungsentwicklung*. ETH Thesis, No. 13108. Zurich: ETH Zurich.

Redle, M., & Baccini, P. (1998). Stadt mit weniger Energie, viel Kies und neuer Identität. *Gaia*, 7(3), 184–195.

Reller, A. (2009). The mobile phone: Powerful communicator and potential metal dissipater. *Gaia*, 18(2), 127–135.

Rem, P. C., De Vries, C., van Kooy, L. A., Bevilacqua, P., & Reuter, M. A. (2004). The Amsterdam pilot on bottom ash. *Minerals Engineering*, 17(2), 363–365.

Richter, M., & Wieland, U. (Eds.). (2010). *Urban Ecology, a Global Framework*. Oxford: Blackwell.

Ridgewell, J. (1993). *A Taste of Japan*. New York: Thomson Learning.

Ritthoff, M., Rohn, H., Liedtke, C., et al. (2002). *Calculating MIPS: Resource Productivity of Products and Services*. Wuppertal: Wuppertal Institute for Climate Environment and Energy.

Roosevelt, F. D. (1938). 64. Message to Congress on Phosphates for Soil Fertility on May 20, 1938 [cited by: J.T. Woolley & G. Peters, The American Presidency Project (online)]. Retrieved September 12, 2010, from <http://www.presidency.ucsb.edu/ws/?pid=15643>.

Rosen, M. A., & Dincer, I. (2001). Exergy as the confluence of energy, environment and sustainable development. *Exergy. International Journal (Toronto, Ont.)*, *1*(1), 3–13.

Rust, D., & Wildes, S. (2008). *Surfactants—A Market Opportunity Study Update. Prepared for the United Soybean Board*. Chesterfield: OMNITECH International.

Salhofer, S., Obersteiner, G., Schneider, F., & Lebersorger, S. (2008). Potentials for the prevention of municipal solid waste. *Waste Management (New York, N.Y.)*, *28*(2), 245–259.

Salvatore, M., Pozzi, F., Ataman, E., Huddleston, B., & Bloise, M. (2005). *Mapping Global Urban and Rural Population Distributions. Environment and Natural Resources Series, 24*. Rome: FAO.

Satsuki, T. (1999). Technology Trends in Laundry Products: Far East/Asian Countries. In A. Cahn (Ed). Proceedings of the 4th World Conference on Detergents 1998 in Montreux: Strategies for the 21st Century. Champaign, Ill: AOCS Press.

Schick, J., Kratz, S., Adam, C., Hermann, L., Kley, G., & Schnug, E. (2008). Agronomic potential of P-fertilisers made from sewage sludge ashes—The EU-project SUSAN. In W.E.H. Blum, M.H. Gerzabek, & M. Vodrazka (Eds.), *Proceedings of Conference Eurosoil 2008 (S28.G.01)*. Vienna: University of Natural Resources and Applied Life Sciences.

Schmidt-Bleek, F. (1997). *Wieviel Umwelt braucht der Mensch? Faktor 10—das Mass für ökologisches Wirtschaften*. Munich: Deutscher Taschenbuch Verlag.

Settle, D. M., & Patterson, C. C. (1980). Lead in albacore: Guide to lead pollution in Americans. *Science*, *207*(4436), 1167–1176.

Siebel, W. (2002). Ist die Stadt ein zukunftsfähiges Modell? In A. Klotz, O. Frey, & W. Rosinak (Eds.), *Stadt und Nachhaltigkeit*. Vienna, New York: Springer.

Sieferle, R. P. (1993). Die Grenzen der Umweltgeschichte. *Gaia*, *2*(1), 8–21.

Siegrist, H. (1997). Personal communication. Dübendorf: EAWAG.

Sieverts, Th. (1997). *Zwischenstadt: zwischen Ort und Welt, Raum und Zeit, Stadt und Land*. Braunschweig: Vieweg.

Sigel, H. (Ed.). (1984). *Circulation of Metals in Biological Systems* (Vol. 18). Metal Ions In Biological Systems. New York: Marcel Dekker.

Skutan, S., Vanzetta, G.M., & Brunner, P.H. (2009). Cadmium im Schrott aus Müllverbrennungsanlagen. *Österreichische Wasser- und Abfallwirtschaft*, *61*, 77–80.

Sloterdijk, P. (2004). *Sphären III, Schäume*. Frankfurt am Main: Suhrkamp.

Soares, A., Guieysse, B., Jefferson, B., Cartmell, E., & Lester, J. N. (2008). Nonylphenol in the environment: A critical review on occurrence, fate, toxicity and treatment in wastewaters. *Environment International*, *34*(7), 1033–1049.

Somlyódy, L., Brunner, P. H., & Kroiß, H. (1999). Nutrient balances for Danube countries: A strategic analysis. *Water Science and Technology, 40*(10), 9–16.

Spiess, E. (2008). *Schweizer Weltatlas*. Bern: Schweizerische Konferenz der kantonalen Erziehungsdirektoren (EDK).

SRU. (2005). *Umwelt und Straßenverkehr: Hohe Mobilität—Umweltverträglicher Verkehr, (Environment and Road Transport High Mobility- Environmentally Sound Traffic) Special Report of the German Advisory Council on the Environment (SRU)*. Berlin: NOMOS.

Stache, H., & Grossmann, H. (1985). *Waschmittel-Aufgaben in Hygiene und Umwelt*. Berlin: Springer-Verlag.

Steinberg, S. (1991). The Flat Earth. In S. Steinberg (Ed.), *The Disovery of America 1992*. New York: Alfred A. Knopf.

Stiglitz, J. (2006). *Making Globalization Work*. London: Penguin Books.

Stumm, W., & Baccini, P. (1978). Man-made chemical perturbations of lakes. In A. Lerman (Ed.), *LAKES Chemistry Geology Physics*. New York: Springer Verlag.

Sukopp, H. (Ed.). (1990). *Stadtökologie. Das Beispiel Berlin*. Berlin: Dietrich Reimer Verlag.

SUSTAIN. (1994). *Forschungs- und Entwicklungsbedarf für den Übergang zu einer nach-haltigen Wirtschaftsweise in Österreich*. Graz: Institute of Chemical Engineering / University of Technology.

Szargut, J., Morris, D., & Steward, F. (1988). *Energy Analysis of Thermal, Chemical, and Metallurgical Processes*. New York: Hemisphere Publishing.

Thünen von, J.H. (1826). *Der isolierte Staat in Beziehung auf Landwirtschaft und Nation-alökonomie*. Stuttgart: G. Fischer Stuttgart [1966 reprint].

Toffler, A. (1980). *The Third Wave*. New York: Bantam Books.

U.S. Geological Survey. (2009). Cement statistics. In T. D. Kelly & G. R. Matos (Eds.), Historical Statistics for Mineral and Material Commodities in the United States: U.S. Geological Survey Data Series 140. Retrieved November 22, 2010, from <http://minerals.usgs.gov/ds/2005/140/cement.xls>.

UN. (1986). *Annual Bulletin of Transport Statistics for Europe*, Vol. 30 (1980, 1985, 1986). New York: United Nations.

UN. (2004). *World Urbanization Prospects, the 2003 Revision*. New York: Department for Economic and Social Affairs, United Nations.

UN. (2008). *Annual Bulletin of Transport Statistics for Europe and North America*. Geneva: Economic Commission for Europe, United Nations.

UN. (2009). *World Urbanization Prospects: The 2009 Revision*. United Nations, Department of Economic and Social Affairs, Population Division, File 2: Percentage of Population Residing in Urban Areas by Major Area, Region and Country 1950–2050, POP/DB/WUP/Rev.2009/1/F2. Retrieved November 28, 2010, from <http://esa.un.org/unpd/wup/Fig_7.htm>.

UNECE. (2008). *Annual Bulletin of Transport Statistics for Europe and North America*. Geneva: United Nations Economic Commission for Europe.

UN-HABITAT. (2004). *Report State of the World's Cities 2004/5 - Globalization and Urban Culture*. Nairobi: United Nations Human Settlements Program.

USGS. (2008). *Land Use Data*. Retrieved November 23, 2010, from <http://nj.usgs.gov/flowstatistics/index.html>.

USGS. (2009). Historical Statistics for Mineral and Material Commodities in the United States. Retrieved June 2, 2011, from <http://minerals.usgs.gov/ds/2005/140/cement.pdf.>.

USGS. (2010a). *Phosphate Rock*. U.S. Geological Survey, Mineral Commodity Summaries. Retrieved August 18, 2010, from <http://minerals.usgs.gov/minerals/pubs/commodity/phosphate_rock/mcs-2010-phosp.pdf>.

USGS. (2010b). *Mineral Commodity Profiles*. Retrieved September 19, 2010, from <http://minerals.usgs.gov/minerals/pubs/commodity/mcp.html>.

Van der Voet, E. (1996). *Substances from Cradle to Grave. Development of a Methodology for the Analysis of Substance Flows Through the Economy and the Environment of a Region with Case Studies on Cadmium and Nitrogen Compounds*. Doctoral Dissertation. Leiden: Institute of Environmental Sciences (CML), Leiden University.

Van der Voet, E., van Oers, L., Moll, S., Schütz, H., Bringezu, S., de Bruyn, S., et al. (2004). *Policy Review on Decoupling: Development of Indicators to Assess Decoupling of Economic Development and Environmental Pressure in the EU25 and AC-3 Countries*. CML report 166. Leiden: Institute of Environmental Sciences (CML), Leiden University.

Viazzo, P. P. (1990). An anthropological perspective of environment, population and social structure in the Alps. In P. Brimblecombe & Chr. Pfister (Eds.), *The Silent Countdown*. Berlin: Springer.

von Braun, J. (2005). *The World Food Situation, an Overview*. Washington, DC: International Food Policy Research Institute.

von Weizsäcker, E. U., Hargroves, K., & Smith, M. (2010). *Faktor 5—Die Formel für nachhaltiges Wachstum*. Munich: Verlagsgruppe Droemer Knaur Verlag.

von Weizsäcker, E. U., Lovins, A., & Lovins, L. (1995). *Faktor vier: doppelter Wohlstand-halbierter Naturverbrauch: der neue Bericht an den Club of Rome*. Munich: Droemer Knaur Verlag.

Wackernagel, M., & Rees, W. (1996). *Our Ecological Footprint: Reducing Human Impact on the Earth*. Gabriola Island, Canada: New Society Publishers.

Wall, G. (1977). *Exergy—A Useful Concept Within Resource Accounting (Report No. 77-42)*. Gothenburg: Institute of Theoretical Physics—Chalmers University of Technology and University of Gothenburg.

Wall, G. (1993). Exergy, ecology and democracy—Concepts of a vital society or a proposal for an exergy tax. In J. Szargut & G. Tsatsaronis (Eds.), *Proceedings of International Conference on Energy Systems and Ecology*. Krakow: EPFL.

Wall, G., Sciubba, E., & Naso, V. (1994). Exergy use in the Italian society. *Energy, 19*(12), 1267–1274.

Warren-Rhodes, K., & Koenig, A. (2001). Escalating trends in the urban metabolism of Hong Kong: 1971–1997. *Ambio, 30*(7), 429–438.

Warwick, O. (2005). *Defra: Risk Reduction Strategy and Analysis of the Advantages and Drawbacks for Tetrachloroethylene. Final Report Stage 4 for Daveshini Padayachee Chemicals & GM Policy*. London: Entec UK Limited.

Weisz, H., Macho, Th., Nicholini, M., & Sieferle, R.P. (Eds.). (1997). *Gesellschaftlicher Stoffwechsel und Kolonisierung von Natur. Ein Versuch in Sozialer Ökologie.* Amsterdam: Overseas Publishers Association.

White, S., & Cordell, D. (2008). *Peak Phosphorus: The Sequel to Peak Oil.* Global Phosphorus Research Initiative (GPRI). Retrieved June 7, 2011, from <http://phosphorusfutures.net/peak-phosphorus>.

Wilson, D. (1990). Recycling of demolition wastes. In F. Moavenzadeh (Ed.), *Concise Encyclopedia of Building and Construction Materials.* Oxford: Pergamon/Elsevier.

Wind, T. (2007). The Role of Detergents in the Phosphate-Balance of European Surface Waters. Retrieved September 29, 2010, from <http://www.ewaonline.de/journal/2007_03.pdf>.

Wittmer, D. (2006). *Kupfer im regionalen Ressourcenhaushalt—Ein methodischer Beitrag zur Exploration urbaner Lagerstätten.* Dissertation No. 16325. Zurich: ETH Zurich.

Wittmer, D., & Lichtensteiger, T. (2007a). Exploration of urban deposits: Long term prospects for resource and waste management. *Waste Management & Research, 25*(3), 1–7.

Wittmer, D., & Lichtensteiger, T. (2007b). Development of anthropogenic raw material stocks: a retrospective approach for prospective scenarios. *Minerals & Energy, 22*(1–2), 62–71.

Wittmer, D., Lichtensteiger, T., & Baccini, P. (2003). Copper exploration for urban mining. In G. E. Lagos, A. E. M. Warner, & M. Sanchez (Eds.), *Proceedings of Copper 2003. Health, Environment and Sustainable Development, 11.* Westmount: Canadian Institute of Mining, Metallurgy and Petroleum.

Wolman, A. (1965). Metabolism of cities. *Scientific American, 213*(3), 179–190.

Wutz, M. (1982). Entwicklungen im Automobilrecycling. In K. J. Thomé-Kozmiensky (Ed.), *Recycling Berlin '79.* Berlin: EF-Verlag.

Wydeven, T., & Golub, M. A. (1991). Waste streams in a crewed space habitat. *Waste Management & Research, 9*(1), 91–101.

Xu, M. (2010). Development of the physical input monetary output model for understanding material flows within ecological–economic systems. *Journal of Resources and Ecology, 1*(2), 123–134.

Yamasue, E., Matsubae, K., Nakajima, K., & Nagasaka, T. (2010). Total materials requirement of phosphoric acid reclaimed from steel making slag. In *Proceedings of ISIE Asia-Pacific Meeting & ISIE MFA-ConAccount Meeting, 7–9 November 2010.* Tokyo: International Society of Industrial Ecology and ConAcount.

Zeltner, C. (1998). *Petrologische Evaluation der thermischen Behandlung von Siedlungsabfällen über Schmelzprozesse (Petrological Evaluation of Thermal Treatment of Municipal Solid Waste by Melting Processes).* Thesis ETH Zurich, No. 12688. Zurich: ETH Zurich.

Zeltner, C., Bader, H.-P., Scheidegger, R., & Baccini, P. (1999). Sustainable metal management exemplified by copper in the USA. *Regional Environmental Change, 1*(1), 31–46.

Zhongjie, L. (2010). *Kenzo Tange and the Metabolist Movement: Urban Utopias of Modern Japan.* London and New York: Routledge, Chapman & Hall.

Zweite, A. (1986). *Beuys zu Ehren.* Munich: Lenbachhaus.

Glossary

Activity Defines a group of actions to satisfy the needs of human beings in a social context. Activities generate metabolic systems. For the description of metabolic systems, four basic activities are chosen: TO NOURISH, TO CLEAN, TO RESIDE&WORK, TO TRANSPORT&COMMUNICATE (see table 3.2).

Anthropogenic system Comprises a structured anthroposphere consisting of institutional and metabolic systems, the latter organized on the basis of material flow analysis (MFA).

Anthroposphere From Greek *anthropos* ("human being") and *sphaira* ("sphere"); describes the space in which biological and cultural activities of human beings take place.

Dissipation Flow of materials from the anthroposphere to the environment in a way that makes it impossible to recover the material by a finite amount of energy.

Educt Material entering a process or system. In analogy to "product" that leaves a process or system.

Evolution Biological evolution: comprises all processes that are involved in bringing forward new forms of life. Cultural evolution: comprises all processes that are involved in bringing forward new forms of anthropogenic systems.

Final sink A *sink* (see definition) where substances remain for residence times of >10,000 years. For organic substances, anthropogenic processes that are able to mineralize completely such compounds (e.g., incineration) act also as final sinks. This applies to the mineralized organic compound only and not to the resulting mineralization products water and carbon dioxide; for them, the mineralization process is a source.

Flow and flux Flow: amount of mass that flows per time through a conductor. Flux: amount of matter that flows per time unit through a unit area ("cross section"). Also, a flux is equal to the product of material density (mass per volume) and velocity of flow (distance per time).

Good Substances and substance mixtures (chemically defined) that have valued functions (economically defined) (see tables 3.3 and 3.4).

Hinterland An area or several areas that are outside of an MFA system and that are connected to and depend metabolically on the system by imports and exports.

Infrastructure The basic physical structures needed for the operation of a society or enterprise such as networks for transportation, energy, information, water, and wastes. The term is also applied to immaterial structures of a society, comprising organizational and institutional bodies such as education, health care, security, and so forth.

Material   Comprises all goods and substances within metabolic systems (MFA method, see table 3.3).

Material flow analysis (MFA)   A physical and mathematical method to qualify and quantify metabolisms of anthropogenic systems as well as of ecosystems.

Material system   Open system consisting of connected processes and goods through which substances are flowing (see table 3.3).

Metabolic system   Open system consisting of connected processes and goods through which substances and energy are flowing.

Metabolism   Comprises all physical flows and stocks of energy and matter within and between the entities of a living system. The original Greek meaning of the word is "conversion" or "transformation." The term is used mainly in biology and medicine for physiologic processes within individual organisms. In a metaphoric sense, it is used for larger entities, such as ecosystems and cultural systems.

Mobility   Literally the state of being in motion. Within metabolic studies of urban systems, it refers to the sum of benefits generated by the activity TO TRANSPORT&COMMUNICATE and is the antipode to the sum of traffic needed for mobility.

Morphology   Literally the knowledge of form. Within the netzstadt method, it refers to the spatial arrangement and characteristics of territories within urban systems due to the dynamic geogenic and anthropogenic influences.

Netzstadt   Netzstadt model: Uses the metaphor of a network (German *netz*) to describe an urban system (German *stadt*) as a structure of nodes and the connections between, which is spatially differentiated from a hinterland by a border and subdivided into scales with different organizational levels. An urban system is generated by four activities and spatially ordered by six territories.

Netzstadt method   A tool to analyze urban systems characterized in the terms of the netzstadt model and to provide a basis for shaping these systems (after Oswald & Baccini, 2003).

Process   Transport, transformation, and storage of goods and substances (MFA method, see table 3.3).

Resources   In the context of metabolic studies (see figure 5.1), four types of resources are defined. (R1) comprises materials and organisms that have evolved within the pre-human geosphere and biosphere and still further develop alongside the evolution of the anthroposphere. (R2) comprises the built physical stock within the anthroposphere. (R3) comprises all economic and technological know-how to produce, maintain, and trade goods (entrepreneurship). (R4) embodies all institutional equipment to guide, organize, and administer social entities (governments, nongovernmental organizations) and to cultivate knowledge (educational and cultural institutions).

Scenario   Representation of a hypothetical development that proceeds from existing conditions, demonstrating causal connections and effects on new conditions.

Sink   *Sink* is the antonym for the term *source*, which stands for the origin of an import of a substance into the anthroposphere. A sink can be one of the following: (1) A place on planet Earth where anthropogenic substances (emissions, products of corrosion and weathering, wastes) are disposed of in the environment such as a lake, agricultural soil, or a landfill. (2) A conveyor belt that transports anthropogenic materials from the anthroposphere to certain environmental compartments, such as a river, urban air, or soil erosion. (3) A transformation process such as incineration or microbial degradation that transforms a substance A into a different substance B. Sinks can be located within the anthroposphere (e.g., substance transformation in an incinerator), at the border between anthroposphere and environment (e.g., a river receiving treated wastewater), or in the environment (e.g., the stratosphere for halogenated hydrocarbons).

Stock   MFA term describing materials stored within a process.

Substance   Chemical element (atom) or chemical compound (molecule) (see table 3.3).

Substance flow analysis (SFA)   A physical and mathematical method to qualify and quantify flows and stocks of substances required for the metabolisms of anthropogenic systems as well as of ecosystems.

Sustainable development   Concept of societal development, which follows the principle of maintaining the reproductive capacity of human societies, according to which the needs of current generations may only be satisfied in a way that does not endanger the existential basis of future generations. The concept was formally recognized by the World Conservation Union in 1980. In 1987, the Brundtland Report for the United Nations adopted the concept. In 1992, a United Nations Earth Summit at Rio de Janeiro established the term as a label for a policy guideline challenging nations and private enterprises.

Synoikos   From Greek *syn* (together) and *oikos* (house), used by Aristotle to describe the formation of cities, taken as a project title to study urban systems leading to the netzstadt model (after Baccini & Oswald, 1998), giving also the label for participatory methods to develop urban systems.

System   A set of related constituents of processes and flows of materials forming an integrated entity, with system boundaries in time and space defining the elements within and outside the system.

Traffic   Comprises all metabolic processes and goods to generate mobility (see *Mobility*). It refers to the costs of mobility.

Transfer coefficient   Denominating the fraction of an *educt* (see definition) or of the sum of educts transferred to a particular output of a process. Transfer coefficients can be calculated for both levels goods and substances. The sum of all transfer coefficients for a particular material of a process equals 1.

Urbanization   The formation of anthropogenic settlements having high densities of population, infrastructure, flows of materials, energy, and information, controlling economically and politically a larger environment to secure their supplies and disposal.

Urban mining   The exploration and exploitation of material stocks in urban systems for anthropogenic activities; it is urban geology combined with urban design, engineering, and resource recovery.

Urban system   An urban system is composed of open geogenic and anthropogenic networks that are connected with each other. The nodes of these networks are places of high densities of people, goods (geogenic included), and information. These nodes are connected by flows of people, goods, energy, and information. The system boundary is given by political conventions in the case of anthropogenic subsystems and by climatic and environmental properties for geogenic subsystems.

Index